T0091906

Life Sculpted

LIFE SCULPTED

Tales of the Animals, Plants, and Fungi That
Drill, Break, and Scrape to Shape the Earth

ANTHONY J. MARTIN

The University of Chicago Press
Chicago and London

The University of Chicago Press, Chicago 60637
The University of Chicago Press, Ltd., London
© 2023 by Anthony Martin
All rights reserved. No part of this book may be used or reproduced in any manner whatsoever without written permission, except in the case of brief quotations in critical articles and reviews. For more information, contact the University of Chicago Press, 1427 E. 60th St., Chicago, IL 60637.
Published 2023
Printed in the United States of America

32 31 30 29 28 27 26 25 24 23   1 2 3 4 5

ISBN-13: 978-0-226-81047-8 (cloth)
ISBN-13: 978-0-226-81050-8 (e-book)
DOI: 10.7208/chicago/9780226810508.001.0001

Library of Congress Cataloging-in-Publication Data

Names: Martin, Anthony J., 1960–, author.
Title: Life sculpted : tales of the animals, plants, and fungi that drill, break, and scrape to shape the earth / Anthony J. Martin.
Description: Chicago : The University of Chicago Press, 2023. | Includes bibliographical references and index.
Identifiers: LCCN 2022038222 | ISBN 9780226810478 (cloth) | ISBN 9780226810508 (ebook)
Subjects: LCSH: Animals, Fossil. | Boring. | Paleontology.
Classification: LCC QE761 .M367 2023 | DDC 560—dc23/eng20221107
LC record available at https://lccn.loc.gov/2022038222

♾ This paper meets the requirements of ANSI/NISO Z39.48-1992 (Permanence of Paper).

To John K. Pope,
my first mentor in paleontology.
Thanks for all of the brachiopods.

How could a stone flourish in spring?
Crumble into soil to grow colorful flowers
You have been stony for so many years
Try something different, be soil for a while

Rumi, *Masnavi I*, verses 1911–1912

Contents

Preface

Not far from where I'm writing this, an enormous rock is fighting a battle against life, and life is winning. The rock's mass of igneous-born minerals crystallized deep underground more than 300 million years before it was uplifted and the land above worn down, exposing its silicate-bound bulk. Once under air, lichens colonized its surface and formed rudimentary soils and plant roots took hold in those soils. A little more than 10,000 years ago, humans arrived and likely treated this domed outcrop as a memorable landmark; indeed, it later became a gathering place for the Muscogee people. After colonizers forced the Muscogee off the surrounding land, they soon gave the outcrop the facile name "Stone Mountain," an appellation belying its impermanence. After all, this "mountain" continues to shed just a bit more of its mineral matter each day, a diminishing by weather but also accelerated by lichens, flowering plants, pine trees, and animals large and small traveling across it.

People rendered their own erosion of its surface, wearing trails into its sides by simply walking on it, their footwear carrying bits of minerals that abraded. Much of its surface was also altered radically in just a few decades by the humans' quarrying thick sheets of rock they used for buildings and curbstones, displaced chunks still apparent throughout the nearby city of Atlanta. In the twentieth century, a small group of people defaced it with faces, carving an enormous bas-relief sculpture on one side to honor a lost cause that will continue losing until it fades into the past and eventually vanishes. Much like the mountain.

The story of Stone Mountain in Georgia is but one example of how everyday actions of life—from the very slow to the sudden—change the hard parts of our world. In this instance the hard parts are rock, but others could be from life itself, represented by animals' shells or bones, or woody tissues in plants. A realization that life is breaking, scraping, drilling, or otherwise changing the solid to the not-so-solid—and has been doing so for more than a billion years—compelled me to write this book about that grand natural history. Moreover, I wanted to write about how life of the prehuman past wore down rocks, shells, bones, and wood in ways both familiar and alien, often leaving long-lasting clues of what happened and when. Such vestiges beckon us, their empty spaces in solid materials reflecting gaps in our understanding of life adapting to the hard parts of a world that is always transforming, whether from its own inner workings or from life itself.

The idea for this book started as a sequel of sorts, in that I wanted to follow up on the theme of a previous book of mine, *The Evolution Underground: Burrows, Bunkers, and the Marvelous Subterranean World Under Our Feet* (2017, Pegasus Books). That book explored how burrows and burrowing animals changed the earth, and how burrows helped many lineages of animals survive the worst extinctions in the history of life. In contrast, this book is broader, exploring bioerosion—the breakdown of solid substances by life—and the importance of living things as reducers and recyclers of those substances. In some instances, though, I highlight a few groups of organisms—parrotfishes, crabs, woodpeckers, clams, and deep-sea worms—to lend a greater appreciation of their evolutionary journeys and the roles they play in shaping modern ecosystems. At the same time, I wanted this book to balance an enthusiastic curiosity for natural history with increased concerns about the future of nature, especially as we reckon with climate change and its varied impacts on life in the rest of the twenty-first century.

Regardless of my intentions, I hope you learn much new from this book, including the perspective that although life may be boring, knowing how it is boring is anything but. Thank you for learning with me.

Chapter 1

A Boring History of Life

Rarely am I as comfortable as when my feet are firmly atop Cretaceous rocks, and these particular rocks were quite reassuring. The outcrop beneath me, composed of buff limestone beds on a Portuguese shore, was elevated high enough above ocean waves for me to revel in its support, while also providing a firm foundation that allowed for contemplating its antiquity. Yet in this instance of solace, the limestone surface felt rough, uneven, and worn. However solid this platform seemed, it was somehow incomplete, its irregularity sending a message of lost time.

Much like people asked to name their favorite child or pet, I am flummoxed whenever asked to choose my favorite geologic period, but the word "Cretaceous" pops out of my mouth the most often. Part of my fondness for all things Cretaceous is because it offers an uneasy blend of the familiar and the exotic, like an alternate reality that scrambles biological and geological elements from vastly different times. The Cretaceous was when life was soft, hard, and all textures in between. It also feels familiar in part because it was the last of the three geologic periods of the Mesozoic Era, spanning 145–66 million years ago and succeeding the more famous Jurassic Period, the latter lending its name to one (and only one) noteworthy movie. Granted, the Cretaceous is also best known for its dinosaurs, especially those made iconic through both science and pop culture, such as *Triceratops*, *Velociraptor*, and that perennially overexposed diva, *Tyrannosaurus*.

But these dinosaurs also shared their landscapes with flowering plants, insects, lizards, snakes, crocodilians, and birds that were both ordinary and bizarre by today's standards.[1]

Life in Cretaceous oceans was similarly odd, hosting its own set of gigantic reptilian denizens, such as Nessie-necked plesiosaurs and dragon-like mosasaurs, as well as gigantic, coiled nautilus-like ammonites. Still, the seas also held sea turtles, sharks, bony fishes, shrimp, crabs, sponges, corals, and other modern-appearing animals.[2] Moreover, these Cretaceous animals had ample room to swim, float, or otherwise move about because the oceans were extremely high and broad, the consequence of a greenhouse Earth that melted most continental ice while also enabling forests to grow near the poles.[3]

As dinosaurs stomped, bit, mated, and defecated on the land, a tiny portion of this vast Cretaceous seaway covered the southern portion of what is now Portugal. So, when a few dozen other paleontologists and I stood on the edge of the modern Mediterranean Sea in the summer of 2016, it was relatively easy for us to imagine azure Cretaceous skies and warm, tropical, aquamarine waters there about a hundred million years ago. We were also better enabled for different dimensions of imagination than most paleontologists, because we were ichnologists. Unlike most paleontologists who study bodily remains of past lives, ichnologists intuit trace fossils—the tracks, burrows, nests, feces, and other clues of past animal lives and behaviors. And in 2016, Portugal was the perfect place for us to assemble at a once-every-four-years conference called Ichnia, and to partake in its field trips. After all, Portuguese rocks hold an extraordinary array of trace fossils, ranging from the sublime, like minuscule 500-million-year-old worm burrows, to the stupendous, like enormous and exquisitely preserved 150-million-year-old sauropod dinosaur tracks.[4]

On the post-meeting field trip of this conference, the field-guidebook authors had helpfully labeled the physical strenuousness required of each visited site. Most were identified as "easy," a few as "moderate," and only one as "difficult," giving participants advance notice of what to expect. On this day the guidebook designated our first afternoon destination, the "Oura megasurface," as "easy." But before that, and true to Portuguese custom, our field-trip hosts ensured we were well fed, leading us to the open-air deck of a seaside restaurant for lunch. Lively conversations ensued, in which our merry band of ichnologists ate, drank, and bantered about trace fossils seen that morning, or any other time. However, such genteel revelry ended abruptly when the field-trip leaders pointed toward a coastal outcrop several hundred meters west of our deck and past a sandy beach, and informed us how we must hike up, onto, and along that distant rocky cliff to reach our next destination. After registering a brief moment of disbelief, we groaned, pushed ourselves away from the tables, properly adjusted our trousers, and began stumbling across the beach past reclining (and staring) tourists and toward the outcrop, all the while experiencing feelings of remorse and impending mortality.

As an ardent fan of *Star Trek* both old and new, I had whimsically envisaged the Oura megasurface as the name of a vast, planar energy field in a far-off corner of the galaxy, its discovery a major advance in our knowledge of the universe. Instead, it was boring, but I mean that in the best way possible. The Oura megasurface was one of the most spectacular examples in the world of a formerly soft ocean bottom that hardened into rock, then was much later drilled, rasped, scraped, or otherwise diminished by small animals, but by the millions, and for a long time. Hence this megasurface dealt less with astronomical planes and more with temporal anomalies, taking us back into a prehuman past that yes, told tales of primeval monsters, but also of much smaller and industrious underwater beings that tore apart solid stone.

The leader of our post-lunch slog was paleontologist Ana Santos, who had done her PhD dissertation on the Oura megasurface and a similar nearby surface at Foz da Fonte. For those of us who read the field-trip guidebook chapter beforehand, we recalled how she thoroughly documented trace fossils in the top surface of the Cretaceous limestone outcrop there.[5] We also remembered that these traces were made by a wide variety of marine animals during the Miocene Epoch about 20 million years ago. Among the borings Santos identified were those made by sponges, polychaete ("bristle") worms, another group of worms called sipunculids, barnacles, sea urchins, and clams, an under-sea menagerie of which little else remained but their traces.

After what felt like an hour of clambering over the craggy out-crops, we arrived on the megasurface. Once present, we listened to Santos and one of her colleagues explain how those marine animals carved the Cretaceous limestone and left distinctive bor-ings attributable to each. As she generously shared her expertise with us, we caught our collective breath and soon realized this was a very special place, and that having all of these trace fossils in the same location was extraordinary. We stooped, squatted, kneeled, and otherwise brought ourselves closer to these engrav-ings to learn more.

Reading such a complex surface is like interpreting a Jackson Pollock painting made on top of other paintings, in which thou-sands of overlapping splashes and strokes require separating them from those before to understand their order of emplace-ment. Fortunately, ichnologists are well trained at such mental unraveling, and after a few minutes of staring at the complex patterns, individual traces emerged so that we could discern who had done what. For instance, the sponge borings were akin to miniature shotgun blasts, consisting of equally spaced clusters of small holes in the rock. The polychaete worm borings were shallow U-shaped grooves on the top surface, whereas sipuncu-lid worm borings were evident as hollow vase-like depressions. Openings of barnacle borings looked like tiny eyes both open and nearly closed; those more pointed on one end outlined shapes of tears that would have flowed from such eyes. Vestiges of sea urchins were either roundish pits where these animals scraped down in one place, or meandering grooves with finely indented floors, hinting at how urchins moved laterally as they scraped these hard surfaces. The most striking of traces, though, were the clam borings. These traces were perfectly circular and cleanly defined holes that looked as if someone with a stout drill had carved them out that very day. A few even contained the origi-nal shells of their clam owners, showing how they died while in

The Oura megasurface of southern Portugal, preserving the trace fossils of Miocene sponges, clams, barnacles, sea urchins, and worms that bored into a Cretaceous limestone. A, Ichnologists strolling on the limestone megasurface, with the seaside town of Praia da Oura in the background. B, The bioeroded surface, with "gutters" made by sea urchins and holes of varying sizes carved by clams and other animals in Cretaceous limestone, but made about 80 million years later during the Miocene Epoch.

homes of their making. These and all other borings overlapped in various ways, their cross-cutting informing us which animal preceded the other and hinting at a long, complex biological history. Once exhumed from its submarine grave during the Miocene, this formerly inert Cretaceous limestone became alive once more, its sea-dwelling inhabitants heedless of the bodies and traces of marine environments that had passed more than 80 million years before them.

Our too-brief visit to the Oura megasurface that day left me with many lasting lessons and questions. The truth was, I knew pitifully little then about bioerosion—the biological eroding of hard substances—as well as borings both modern and ancient.[6] This is embarrassing for me to admit, as I have used "Ichnologist" as my Twitter handle for scientific outreach since 2010, and have also written weighty books about modern traces of the Georgia coast, dinosaur trace fossils, and the natural history of burrowing animals.[7] Although Santos had devoted a significant part of her life to understanding borings and bioerosion through the outcrops of Oura, I was shamefully ignorant of how the animals made these trace fossils, and what motivated them to break down rock. So these trace fossils lingered in my consciousness and nagged, imploring me to learn more.

In an attempt to address my ichnological inadequacies, in May 2019—almost exactly three years later—I returned to Santos's and her colleagues' research articles about the Oura megasurface. After studying their descriptions of the borings and the geologic history these trace fossils defined, I better understood not only each type of trace fossil and their creators, but also why they told a complex story more than a hundred million years in the making.

During the Cretaceous, this limestone was not rock, but a muddy-sandy seafloor that supplied refuge for an abundant array of animals, such as burrowing shrimp, lobsters, and sea urchins. Their distinctive burrows were preserved in the sediment, speaking of a warm-water community that thrived there while huge marine reptiles—plesiosaurs, mosasaurs, and sea turtles—as well as fishes and ammonites swam above. Then, a few million years later, these sediments were buried, cemented, or otherwise solidified into limestone, preserving both corporeal remains and burrows of the animals that lived there. Not all of this limestone endured, though, as it was eroded for more than 80 million years afterward. This meant the sedimentary record there of a mass extinction 66 million years ago, one that killed dinosaurs, plesiosaurs, mosasaurs, ammonites, and more, was deleted.[8] A mere 20 million years ago, when some apes were evolving into lineages that eventually led to our own, oceanic processes exposed these rocky remnants and bioeroding animals wore them down again. About 5 million years passed, and then another erasure occurred, as sponge, worm, barnacle, sea urchin, and clam descendants of the original stonecutters moved into the old neighborhood and remodeled it more to their liking.

Eventually the collision of tectonic plates near what we now call Portugal caused this geological evidence of sedimentation, lithification, and biological abrasion to lift above present-day sea level. This raising presented the limestone beds as craggy coastal ledges and cliffs along Praia da Oura ("Oura Beach"), the namesake of the Oura megasurface. Such uplift was fortunate, too, as it better allowed the rock- and fossil-worshipping relatives of Miocene apes (us) to visit the outcrops without having to use recently invented scuba gear. Best of all, sandy beaches adjacent to these outcrops encouraged restaurant owners to build outdoor decks capable of holding tables teeming with food and drink that went into ichnologists' bellies just before they became personally acquainted with this history.

Despite all of this paleontological postenlightenment, though, I was not satisfied, and annoyingly unanswered questions lingered. For example, how did something as soft as a sponge render shotgun-like patterns of holes? Likewise, worms are not among the first animals I would imagine drilling into limestone, and barnacles I had seen before then simply settled on hard surfaces and did not wear them away or go down into them. Clamshells are made of the same or similar minerals composing limestone, as are sea urchins, demonstrating feats that seemingly transcended the physical limits of these animals' skeletons. How did these Miocene invertebrates all manage to drill, scrape, or otherwise destroy solid rock?

The questions continued, reaching far beyond present-day Portugal. Just how far back into the geologic past did such rock-eroding behaviors happen? When did life begin attacking other hard stuff, such as shells, wood, and bones, and how did these biological superpowers evolve? Why and how did plants and animals incorporate minerals or other tough tissues into their bodies, and what is the evidence for evolutionary "arms races" between armoring defenses and penetrating offenses? What trace fossil evidence do we have for these innovations and their effects, whether in individual die-or-died scenarios or broad swaths of time? How does bioerosion as a process affect entire environments, or even global climate? These inquiries and others led me to dig (or rather, drill) deeper, seeking answers that also generated many new questions, impelling me to learn more about bioerosion as an essential facet in the history of life.

∴ ∴ ∴

Life is hard, but when it comes to evolution, life overcame by being boring. Rocks, shells, wood, and bones all presented barriers to life that once were impenetrable, but not for long, and never more. The list of living things that reduce rock-solid stuff into smaller bits or otherwise punch through solid substances is long and remarkable. Those who erode include bacteria, fungi, lichens, sponges, worms, clams, snails, octopi, barnacles, sea urchins, beetles, ants, termites, fishes, crocodilians, birds, monkeys, and even elephants. Moreover, the processes and evidence for borings and other diminution of densely compacted matter are not only all around us today—from the deep sea to mountaintops, and from pole to pole—but even within us. For example, bacteria cause enough tooth decay to keep dental workers happily employed. Predators past and present also had anatomical attributes strong enough to pierce seashells or break bones, demonstrating how boring behaviors can range from the devious to the dramatic. Wood was compromised almost as soon as it evolved, sometimes by seemingly unlikely allies of fungi and animals. The ways life has accomplished such feats are almost as varied as their perpetrators, involving acids, poisons, drill bits, files, gut bacteria, stout teeth, powerful jaws, or other bodily attributes and behaviors that softened the seemingly impassable.

The cast of characters introduced at the Oura megasurface, which included sponges, polychaete worms, sipunculid worms, clams, barnacles, and sea urchins, may have been from only 15–20 million years ago, but they embodied a longer heritage of boring activities. All of these animals' evolutionary lineages extend much further back, well beyond the Miocene and even surpassing the Cretaceous. For instance, the fossil records for bivalves, polychaetes, sipunculids, and barnacles show they originated in the Cambrian Period more than 500 million years ago as byproducts of a great diversification of animal life nicknamed the "Biological Big Bang."[9] Sea urchins were not far behind these other groups, as echinoids—the group of animals that includes sea urchins and sand dollars—had evolved before the end of the Ordovician Period, about 450 million years ago.[10] Marine animals produced by this and other diversifications, including vertebrates, later adapted to landward environments and wore down all hard substances there too, whether rock, shell, wood, or bone.

Hence the Miocene animals and their modern counterparts represent progenies that out-survived dinosaurs and giant marine reptiles, but also with ancestors that made it past four other mass extinctions before then. This means that life has been boring for a very long time, while also bestowing us with myriad successful bioeroders we are lucky enough to witness today. For much of this book, then, I hereby pledge to remark upon these bioeroding entities, explain what they do, and provide a glimpse of how they changed the world.

Chapter 2

Small but Diminishing

Few sounds in modern society are as terrifying as that of a dentist's drill at full throttle. Its high-pitched whine begins outside your mouth, a sonic commencement portending of awfulness, a warning that it soon will be inside your head, resonating and amplifying so it fills all spaces great and small with its malevolent intent. Then, it makes contact with the outer surface of your tooth. However dulled it and nearby soft tissues might be from localized narcotics, the tooth adopts a low-frequency basso profundo, absorbing and transmitting the drill's grinding vibrations of insidious doom.

Unfortunately, I have experienced these devices and other dental implements far too often. For one, I grew up as one of six children in a low-income family. Owing to this confluence of population density and economic deprivation, my parents could not afford for their children to regularly visit dentists. Combine this situation with an overabundance of cheap, sugar-infested breakfast cereals that (much like smoking) were completely normal during the 1960s and 1970s, followed by four years of college and eight years of graduate school with a subsistence income and no dental insurance. Little wonder that the rocks in my mouth became a festering personal experiment in bioerosion. Drillings, fillings, root canals, implants, crowns, bridges, and other fixes tried to confront the slow-motion war of attrition waged in my oral cavity. But the damage was done. Of my surviving teeth, I am lucky that incisors and bicuspids fronting authorial smiles were less affected, but the imbalance of premolars and molars affect and otherwise influence my chewing today, with the right side of my jaw doing most of the work. And because food is a great pleasure in my life, this reminder of past decay is constant.

Should I continue to feel guilty about lusting for a 75 percent cacao chocolate bar, its siren call hearkening to my Catholic upbringing, but one in which the internal onus was reserved for temptations far more impure, followed by weak-willed indulgences? Not necessarily, as past indiscretions cannot be undone. Yet I remained curious about why decay happened, as it seems far too easy to blame just poverty and sugar for this loss of toothsome integrity. The scientist in me knew there was more to the story of why my teeth acquired cracks and holes and otherwise became staging grounds for invasive degradation.

When looking at the details, you might be surprised and relieved to discover that at least one villain—sugar—is exonerated for its direct role in tooth decay. Sucrose and other simple carbohydrates are more like accomplices, acting like ready-to-burn woodpiles for the bacteria that actually break down teeth. The main culprit behind dental decline is *Streptococcus mutans* (*S. mutans*), a roundish bacterium living in all of our mouths and belonging to the same genus responsible for a common and painful throat infection nicknamed "strep throat," which I also experienced often as a child.[1] *S. mutans* lives in crevices, which is also why it functions as an anaerobe, not needing oxygen to function and reproduce. How tooth decay happens is that *S. mutans* consumes sucrose and produces wastes that form an organic film, plaque.[2] Plaque acts like a thick shag carpet glued onto your teeth, allowing more bacteria to hide and thrive there, especially if it is not whisked away by vigorously brushing and flossing tooth owners or scraped off by ever-vigilant dental hygienists. Still, if you give these bacteria more food as fermentable carbohydrates, such as glucose and fructose, they produce lactic acid. This acid is bad for teeth, corroding and generally weakening their normally tough, compact solidness. Given enough bacteria, sugary foods, plaque, and time, tooth surfaces weaken.

Attacks first manifest as white spots and then as cavities, often denoted by brown colors that the younger me will always associate with impending pain and oral disaster.

Teeth are mostly composed of apatite, which I learned from geology classes is a calcium phosphate mineral likewise making up bones.[3] Among those bones are vertebrae, which means an animal is not a vertebrate unless it produces apatite. Apatite is also more durable than the minerals making up shelled invertebrates, like clams and snails. Shells in most of these animals are made of calcium carbonate minerals, such as aragonite and calcite, which are the most common parts of limestone.[4] Although some limestones are hard enough that people use them for foundations and walls of buildings, the minerals composing limestone are even more susceptible to acid damage than apatite.

If you took a college-level introductory geology class, you probably dropped acid in it, and often. This is because you were doing a simple test to identify aragonite or calcite, which involved applying a drop or two of diluted hydrochloric acid onto a mystery mineral or rock and observing whether or not it fizzes. If bubbles burst out of the mineral, congratulations, your calcium-carbonate-reveal party was a success: your specimen is aragonite, calcite, or a limestone bearing either of those minerals. The bubbles formed by this reaction represent carbon dioxide liberated from its chemical bonds and saying goodbye to its calcium companions. In my experiences with students, though, they also become emotionally effervescent, and I soon must pry acid bottles from their hands to prevent their gleefully squirting acid on anything they suspect of harboring calcium carbonate.

But here's what's interesting about those same mineral- or rock-identification labs that is also pertinent to teeth. If my students applied hydrochloric acid to apatite or other phosphate minerals, or even fossil teeth, nothing happened. Another mineral identification test used in introductory labs—the Mohs hardness scale—also shows that apatite is relatively tougher than aragonite or calcite in another way. The Mohs hardness scale—named after nineteenth-century geologist Friedrich Mohs—ranks minerals from 1 (softest) to 10 (hardest). Minerals on this scale range from talc, which is so soft it is the main ingredient for baby powder, to diamond, which is so hard it is used to cut glass when heisting more diamonds.[5] As a 5, apatite is in the middle, whereas aragonite and calcite only rank as a 3. Hence apatite's greater hardness, the compact density of apatite in teeth, and its resistance to most acids collectively make for an impressive combination of durable properties. These qualities also help us better understand why vertebrates evolved these types of teeth, rather than ones made of aragonite, calcite, or other softer, fizzier minerals.

Nevertheless, *S. mutans* and similar bacteria found a way to overcome these mineral properties and naturally selected teeth traits, including those in our supposedly highly evolved oral cavities. Every day in billions of mouths, these simple, single-celled organisms attack, and attack relentlessly, wearing down mineral defenses with lactic acids, like unicellular versions of the acid-bleeding alien in the too-easily titled movie *Alien*.[6] Despite their diminutiveness, the cumulative impact of these microbes is massive, with dental ailments ranking third in all human health problems behind heart disease and cancer. Thus each time I brush my teeth or floss, I think about how I am fighting a microcosmic battle against organisms that multiply exponentially and will not relent if we let up our defenses, especially if we eat more chocolate.

: : :

It's easy being blue-green: just ask cyanobacteria. Although they won't reply, these one-celled photosynthetic survivors and their traces have a long history. For one, cyanobacteria persisted after all mass extinctions in Earth history. But what about trilobites, you say, those jointy-legged denizens of Paleozoic seas? Please. These animals only lived for 250 million years before going kaput. Ammonites? Those squid-like critters were around for more than 350 million years, so they did slightly better than trilobites, but also gone. Dinosaurs? Total losers. (Well, except for birds, which still rule.) In contrast, cyanobacteria have been living and thriving in their environments for more than 3.5 billion years.[7]

Near the beginning of their reign, cyanobacteria even formed huge, layered, dome-like colonies in shallow seas, called stromatolites. These wave-resistant structures even qualified as the first reefs, more than 3 billion years before corals.[8] Cyanobacteria are also credited for flipping a switch on the earth's atmosphere about 2.4 billion years ago, with their photosynthesis producing enough oxygen to kill many anaerobic microbes.[9] This cyanobacterial breath of fresh air eventually led to conditions amenable to the evolution of animals, including those who recognize cyanobacteria for their immense contributions to the evolution of the earth and its life.

And just what are cyanobacteria? While taking a biology class as a first-year undergraduate student, I dutifully learned and wrote on exams that these microorganisms were "blue-green algae." This simple definition later turned out to be as wrong as saying that manatees are sea urchins. Once the genetic relationships of cyanobacteria and algae were teased out, biologists realized that the former are from a far more ancient lineage of small, relatively simple, single-celled organisms called prokaryotes.[10] Algae, in contrast, comprise a different evolutionarily related group, or clade, that was rather late to the sunlight-converting game. Algae, land plants, and animals are all eukaryotes, meaning they are composed of larger and more complex cells.

Cyanobacteria—like algae and land plants—make their own food by photosynthesis, which requires taking in carbon dioxide and water, and then rearranging these molecules into simple sugars, such as glucose. So by making fuel for some tooth-decaying anaerobes, photosynthesizers are part of the supply chain in tooth decay; but let's not go down that slippery slope of blame just yet. Regardless, cyanobacteria normally consist of individual cells, but also can form beaded filaments and colonies.[11] Cyanobacterial cells, however, are distinct from those of algae and plants by what they lack. Like all prokaryotes, they do not have nuclear envelopes holding their DNA in neat little bundles, but instead have their genetic material dispersed throughout each cell.

In 1967, evolutionary biologist Lynn Margulis proposed that

sometime in the past few billion years certain cyanobacteria and other prokaryotes evolved into functioning parts of more complex eukaryote cells.[12] According to Margulis, this symbiosis resulted in the formation of organelles, which as their name implies are like simplified organs in a cell. Examples of organelles include chloroplasts in plants, which make food from sunlight, and mitochondria in animals, which convert food into energy. She and other scientists dubbed this "cell within a cell" hypothesis as endosymbiosis. For a while some biologists ridiculed this hypothesis as speculative claptrap. But when multiple lines of evidence kept supporting it later, it was begrudgingly considered as not impossible, and endosymbiosis eventually became about as certain of a hypothesis as anything can get in biology.[13] A clincher for many was when microbiologists discovered modern cyanobacteria actually residing in eukaryotic cells and making food for them,[14] the microbial equivalent of saying, "We told you so, and by the way, here's breakfast." This intracellularly expressed revelation implies that all modern algae and land plants owe their livelihoods to the true inventors of photosynthesis, cyanobacteria. Not surprisingly, modern cyanobacteria can thrive in nearly every environment, from marine to terrestrial, with only one major requirement: light.

Among the more surprising types of cyanobacteria are those that attack and degrade limestone and other solid objects, such as clams, snails, and corals. These cyanobacteria are euendolithic, with the Greek roots of that unwieldy word corresponding to *eu* + *endo* (= "truly within") and *lithos* (= "rock").[15] The overall effects of such cyanobacteria are astonishing, resulting in countless tiny (less than 10 microns, or 0.0004 inch wide) but distinctive clustered holes or branching tunnels on and in the surfaces of limestones wherever they live. After explaining how these organisms are extremely small and simple organisms, the mere juxtaposition of "cyanobacteria" and "attack rocks" may seem absurd. This premise appears all the more ludicrous when considering that oxygen-producing organisms tend to make their local environments less acidic. And of course, cyanobacteria do not have tiny drill bits, files, or other tools they suddenly whip

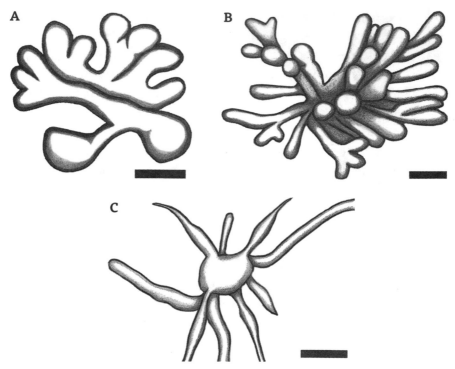

Cyanobacterial, algal, and fungal microborings represented in positive relief. A, Fossil boring similar to the modern cyanobacterium *Hyella* in size and form, in a gastropod shell from the Neogene of Austria; scale = 10 microns. B, Fossil boring similar to those made by modern green alga *Acetabularia*, in bivalve shell from the Eocene of France; scale = 100 microns. C, Boring produced by fungus *Conchyliastrum*, in a modern bivalve shell of Scotland; scale = 30 microns. All examples drawn after photos in Glaub et al., "Microborings and microbial endoliths."

out and start using whenever humans walk away from their microscopes. All of this biochemical strangeness understandably leads to a single-word question: How?

The answer, just like cyanobacteria, is both simple and complex. One of the breakthrough studies of cyanobacteria breaking through rocks came in 2016, when Brandon Guida and Ferran Garcia-Pichel took a close look at the filamentous and euendolithic cyanobacterium *Mastigocoleus testarum*.[16] The goal of their research on *M. testarum* was to figure out how this organism lived in rock without using acids or implements. In short, they discovered enzymes from the cyanobacteria break apart and transport calcium away from rocks and shells, making room for bulbous or thread-like colonies of cyanobacteria in their newly carved-out microscopic homes. Traces of this activity are clusters of holes connected to hollow branching networks, looking like impressions of grapes or roots, but far smaller than those of any plant. For comparison, the finest of human hairs are only slightly less than 20 microns (0.0008 inch) wide, whereas cyanobacteria borings are commonly less than 10 microns (0.0004 inch) wide.[17]

Because euendolithic cyanobacteria leave such distinctive traces, geologists and paleontologists with access to scanning electron microscopes (SEMs) can recognize these enzymatic etchings in ancient marine limestones and mollusk shells. Such trace fossils accordingly lead to two major conclusions. One is that the original environment where the rocks or shells were located was on a seafloor with cyanobacteria living there, a reasonable assumption even if none of their cells were fossilized. Second, these photosynthesizing cyanobacteria were living close enough to the original ocean surface that light reached them.[18] Based on distributions of cyanobacteria in modern oceans, we can then infer they were likely in shallow-marine environments that ranged from less than a meter (3.3 feet) to a maximum of about 200 meters (650 feet) deep. On the other hand, such minuscule traces could never have been formed by cyanobacteria in deep-marine environments, which, similar to movies based on DC comics, were too dark to allow them to live.

The fossil record for euendolithic cyanobacteria extends back to at least 1.5 billion years ago, showing how their adapting to rocky substrates happened a mere 2 billion years after the formation of the first stromatolites. In fact, cyanobacterial borings are documented in 1.5-billion-year-old stromatolites from China, switching from building to destroying.[19] Bioeroding cyanobacteria have persisted in shallow-marine environments since, with their trace fossils reflecting this continuity. Cyanobacteria—whether euendolithic or otherwise—eventually moved from those environments into all parts of the earth's surface with water and light, including soils. Hence these extremely successful single-celled survivors still play a role in atmospheric cycles by taking carbon dioxide out of the air and converting it to oxygen on a globally significant scale.

Nevertheless, one of the seeming geochemical contradictions of rock-dissolving cyanobacteria is that by eroding calcareous minerals in shallow-marine environments throughout the world, they are also like billions of little acid bottles wielded by overeager students in introductory geology labs. Through this collective action, are cyanobacteria releasing enough carbon dioxide to counter their own oxygen production and the effects of their ancestors and descendants? Not really, although their bioeroding effect is significant, and has been for a long time. For example, geologists who studied fossil reefs from the Permian and Triassic Periods (about 250–200 million years ago, or MYA) found that cyanobacterial borings covered as much as 60 percent of these ancient reef surfaces.[20] Studies of cyanobacteria in modern marine environments of the Bahamas show they can erode as much as 210 g/m^2 (about 7.4 ounces/10.8 ft^2), shaving off a half pound of weight per area over time.[21] Multiply this effect for exposed areas of tropical and semitropical limestones in shallow-marine environments worldwide, and the overall effects of these minute photosynthesizers become much more apparent and impressive.

: : :

Mushrooms are wonders of the world for us, whether appreciated for their beautiful forms and colors, consumed as delicious dishes, used for hallucinogenic experiences, or for inflicting near-instantaneous and agonizing death. Yet mushrooms and their kin do not just live or die for us, nor do they always look like beige bulbs on stems, nor live solely in forests. To merely say fungi are diverse is like saying Beyoncé can sing and dance. Granted, a word-association game that uses "fungi" will cause most people to reflexively blurt "mushrooms," but fungi also include yeasts, molds, rusts, and (my favorite) smuts, such as corn smut. Fungi range from single-celled entities to translucent sheaths to colorful shelves on rotten wood to the largest colonial organisms on

Earth, inspiring "humongous fungus among us" rhymes.[22] Some fungi also help plants by enveloping their roots and pulling in phosphorus, nitrogen, and other elements that help these plants grow, and grow well, a symbiosis that has persevered for more than 400 million years.[23] Similar to cyanobacteria, fungi are in most places, but unlike cyanobacteria, they can thrive in darkness. This means fungi can colonize environments from caverns to deep-ocean trenches to forgotten leftovers in the backs of refrigerators.

For those of you who most commonly encounter mushrooms while shopping in a grocery store, you may be shocked to learn that such venues do not follow an accurate biological classification scheme. For instance, putting mushrooms between bell peppers and broccoli and calling them "vegetables" is as inappropriate as seating Banksy between Frida Kahlo and Diego Rivera at a dinner party and introducing all three as artists. You see, despite most fungi living and growing in soil, they do not produce food: they eat food. Cyanobacteria, algae, plants, and other organisms that use light, air, water, and other nutrients to grow are photoautotrophs (where *photo* = "light," *auto* = "self," and *troph* = "food"). In contrast, fungi live off both the living and the dead; hence they derive their food from a variety of sources as heterotrophs (where *hetero* = "different"). Such lifestyles of the dank and moldy reveal how their true deep-time kinship is actually with animals. An evolutionary chart of plants, animals, and fungi shows a closer common ancestor shared between the latter two than with plants, with the divergence of fungi from what would eventually become fun guys happening about a billion years ago.[24] In short, employees of a more evolutionarily informed supermarket should place mushrooms next to meat and seafood.

Despite our common interactions with fungi on land, these organisms originated in the sea; the oldest known body fossils of fungus-like cells are in 2.4-billion-year-old volcanic rocks that formed in marine environments.[25] Other fossils suggest fungi reached land by about 635 million years ago,[26] and by the Devonian Period they had formed gigantic colonies that paleontologists at first mistook for trees.[27] But marine fungi persisted through today, with more than 400 documented species living in oceanic habitats, from abyssal to intertidal. And if you thought fungi were already resilient, then prepare to be more impressed, as they inhabit extreme environments ranging from hydro-thermal vents with superheated water, or waterways polluted by acid-mine drainage.[28] Thus their borings are sometimes the sole indicators of life in places where little else survives. And while living in the sea, some species managed to evolve the ability to break up rocks, shells, and other hard substances. Fungi also do not discriminate when diminishing substrates, putting rocks as varied as granite, basalt, sandstone, limestone, marble, and gneiss on their to-erode checklists. So, if someone feels inclined to modify the handy game of "rock, paper, scissors" to "rock, fungi, scissors," I hope this information proves useful for figur-ing out the proper order.

How these soft-bodied organisms leave their traces in hard substances is similar to cyanobacteria via intertwined physical and chemical processes. Bioeroding fungi are endolithic, either living in preexisting crevices or making their own. For the lat-ter, they send filaments (hyphae) into pores or other openings, which then expand and exert pressure on their hard surround-ings.[29] In calcium carbonate skeletons of corals, clams, snails, or other invertebrates, fungi may augment this physical pressure by exuding enzymes that break down proteins left over in those skeletons, supplying the fungi with food.

Although many marine fungal borings are in the same mi-croscopic size range as cyanobacterial borings, their forms differ enough to distinguish them from their photosynthetic cohorts. For one, fungal hyphae form thin filamentous tunnels that of-ten expand along their lengths that—much like some people's moods—change from fine to swell and back to fine again. Some swellings connect to thicker tunnels that allow spores to exit from whatever is hosting the fungi, letting loose a new generation to colonize new frontiers. In contrast, cyanobacteria and algae tend to stay the same diameter when making their way through a hard substrate. Fungal filaments also tend to branch and taper toward the ends of those branches, which cyanobacteria do not do. Perhaps most importantly, fungal borings don't just break shells, they break rules, as their hyphae cross shell layers within clams or other invertebrates, whereas cyanobacteria and algae stick to easier paths.[30]

So why should ichnologists, paleontologists, or geologists bother to learn the difference between cyanobacterial and algal borings versus those of fungi? Because the difference between their original marine environments is like day and night. To wit, cyanobacterial and algal borings are the works of photosynthesizers, which need light, whereas fungi do not. So, let's say a paleontologist finds fossil clams, examines them under a powerful microscope, and finds abundant microborings in their shells, which she identifies as those created by cyanobacteria and algae. These trace fossils accordingly tell her that these shells were in a shallow-marine environment bathed by sunlight. Alternatively, fungal borings with nary a photosynthesizer trace in sight tell her these clamshells were in deep-water environments, perhaps hundreds or thousands of meters below the water surface, where the sun does not shine.[31] A more intriguing story may come from a mixing of the two suites of traces in the shells, with fungal borings cutting across cyanobacterial borings. Because fungi can live in both shallow and deep water, the shells then may have first lain in shallow water, but were moved by currents from shallow to deep, where fungal eroders did their work.

Sure enough, such hypothetical situations are exemplified by reality, as this is exactly the sort of detective work that ichnologists apply when interpreting ancient marine environments. Once given lots of rocks and body fossils bearing microborings, these scientists can tell which are from specific marine environments, from former coasts to abysses.[32] Distinctive assemblages of such tiny trace fossils hence serve as guideposts for geologists who try to interpret Earth history, applying these minuscule clues to learning about planetary change.

: : :

For lichens, symbiosis is a way of life, as any recipe for lichens calls for algae and fungi working together as one. The "algal" part of a lichen colony can be from cyanobacteria (which we learned earlier are not algae) or green algae.[33] These photosynthesizers, though, are often held captive and used by the fungi, and if you think this sounds like typical heterotrophic behavior, you would be right. What happens in this imbalanced relationship is that the fungi protect the cyanobacteria or algae from being eaten by other heterotrophs—such as animals—but this also means the fungi consume at least some of them. This symbiosis is apparent if you cut through a lichen and look at the cross-section under a microscope. There you can see that lichens are like pita pockets, with: a tough fungal exterior (cortex) serving as the "bread" above and below; an upper layer of cyanobacteria or algae (the algal zone) as the "lettuce"; and a lower layer of fungal filaments (medulla) as the "meat" (or "falafel," if you prefer).[34]

Colorful patches and patterns of lichens are in nearly every terrestrial environment bathed by ample sunlight, including those in polar areas. For instance, despite Antarctica lacking native flora today, Antarctic rocks that are not covered by ice, snow, or penguin poop are adorned by strikingly beautiful expanses of green, yellow, and orange lichens. Indeed, lichens are so extraordinarily hardy and resilient that some are also among the oldest organisms on Earth. A few living examples, such as the Arctic species *Rhizocarpon geographicum*, have been dated at thousands of years old.[35]

As most woods-wandering folks know well, lichens thrive on trees and soils, but also rocks; those that do are called saxicolous. Given that, rock dwellers are then divided by lifestyle. For example, do they live *on* rocks, or *in* rocks? If the former, they are epilithic, if the latter, endolithic, but of course some lichens do both. Epilithic lichens can be further classified through their forms, which are expressed by their thallus, or the main fungal body holding its algae or cyanobacteria. Do they form low-profile crusts? Then they are crustose. Do they make semispherical bodies that look vaguely like fruit to a starving person, or someone who may have ingested mind-altering and nonsymbiotic fungi? Then these are fructitose. Do they form layers like leaves, or pages in a book? Such lichens are foliose.[36] Endolithic lichens

Multiple generations and species of saxicolous (rock-dwelling) lichens eroding a silica-rich metamorphic-rock boulder near Covington, Georgia, with bare patches and pitting caused by previous lichens on the same surface.

include: cryptoendoliths, which hide in caves and other open spaces within rocks; chasmolithic, which sneak into cracks; and euendolithic, which actively penetrate rock.[37] Who knew that lichens had such varied forms or lived such diverse lifestyles? Well, lichenologists did, but now you do, too.

Effects of lichens are both physical and chemical, in that they mechanically break apart and dissolve rocks, respectively. For one, the bottom surfaces of some lichens have tightly packed fungal hyphae that form root-like extensions (rhizines) attaching them to rocks and other solid surfaces.[38] Other hyphae may also invade and widen spaces between sand grains, fractures, or even cleavage planes of minerals. On a larger scale, a lichen thallus can shrink and expand in response to atmospheric moisture; wetter makes it bigger, whereas drier makes it smaller. Similarly, the same tough exteriors of lichens that allow them to live in Arctic and Antarctic environments also mean they live through freeze and thaw cycles. So again they contract and enlarge, exerting force against their host rocks. However, hyphae also allow water to flow more easily into a rock, where freezing conditions lead to icy interiors. Because water expands with freezing, it pushes apart and otherwise weakens its host rock. In some instances, lichens envelop and include mineral bits in their thallus, pulling away these detached parts from rocks.[39] In this sense, lichens are not only weathering rocks, but also eroding them, moving bits and pieces away from their place of origin.

Salt crystals can also form and grow in direct response to chemical reactions caused by lichens and other microbiota in crevices and other spaces.[40] Just like ice wedging, this crystal growth expands those spaces. However, lichen-produced acids have the most direct chemical effects on rocks. While the photosynthesizing parts of lichens make oxygen, the fungal parts make carbon dioxide. When this gas combines with water in and on rock surfaces, it forms carbonic acid, the same acid in carbonated beverages.[41] Although the effects of carbonic acid are relatively minor, lichens enhance their acidic dispositions by also exuding organic acids, such as oxalic acid and citric acid of lemon-juice and paper-cut infamy. The effect is to wear down rock surfaces by prying apart minerals, sediments, and layers, while also chemically changing them.

Paleontologists have documented the bodily remains of fossil lichens, but as far as I know their trace fossils are not yet documented. How would we distinguish such traces from those of fungi working on their own, or physical or chemical weathering? This is such a good question that paleontologists have no idea how to answer it. Yet we can confidently infer that as long as lichens have lived on rocks, they have also degraded those rocks.

: : :

If there ever comes a time for me to sit in a dentist's chair for another root canal, at least I can distract myself by reviewing the other simple life-forms that are tearing down hardened defenses, edifices, and foundations far grander than my teeth. Collective action works, especially when trillions of small organisms are doing the acting every day and everywhere. Through their combined efforts, the rock-destroying effects of anaerobic bacteria, cyanobacteria, fungi, lichens, and other seemingly lowly life have a cumulative effect on the breath of our planet, and by extension on the general well-being of all other life tethered to that cyclicity and breath.

In the oceans, from the deep sea to the intertidal, cyanobacteria and fungi are actively bioeroding rocks and skeletal remains of invertebrates and vertebrates alike. Lichens do their work on land, from ocean shorelines to mountains, stripping rock surfaces. Land-dwelling fungi that do not symbiotically merge with algae nonetheless join them in this diminution, a wholesale slaughter of landscape exteriors that reveals illusions of immutability. The primary impact of this minute bioerosion is the conversion of so much hard material to small-grained versions of itself or its complete dissolution and redistribution of atoms. This is why the bioerosion wrought by bacteria, cyanobacteria, fungi, and lichens actually contributes to an even larger picture, which is its contribution to the sedimentary part of the rock cycle.

The rock cycle—which is the transformation and reuse of earth materials—has been around in one form or another since the Earth began cooling from its early molten and meteorite-bombarded state about 4 billion years ago.[42] Not coincidentally, this cooling led to the formation of early plate tectonics, with rigid lithospheric plates riding atop hot, plastically flowing asthenosphere.[43] Soon after plate tectonics started, and as liquid rock poured onto the early earth surface or cooled below to form igneous rocks, water accumulated in what would become ocean basins. This is where that water and other chemicals started weathering and eroding rocks, taking away pieces as sediment or dissolving elements and compounds and interacting with a newly formed atmosphere.[44] In and just below these watery environments, sediments bound together and formed sedimentary rocks. Later in plate-tectonic history, igneous and sedimentary rocks were heated just short of melting or placed under tremendous pressure by colliding plates, producing metamorphic rocks. Although meteorites and other bolides continued to strike the earth over the next 4 billion years, these extraterrestrial donations of matter and energy were negligible compared to what was already there, and what was already being recycled.[45]

Another momentous development in the history of the earth also happened at the boundaries between these primitive plates, which supplied both a kitchen and a recipe for primordial soup. Although biologists once thought life was birthed in shallow, lightning-sparked seas, they now suspect primitive prokaryotes formed in the deep sea at hydrothermal vents caused by early plate tectonics.[46] These prokaryotes then evolved into anaerobic bacteria, cyanobacteria, and eventually eukaryotes, including fungi. Branches on their evolutionary trees adapted to increasingly more common hard places, and some were naturally selected to eat rocks. As continents grew out of the rock cycle throughout the first few billion years of Earth history, land environments beckoned, offering free real estate for new branches of cyanobacteria, algae, fungi, and other microorganisms. Early in this history of life on land, some cyanobacteria, algae, and fungi encrusted these formerly barren terrains, providing the first foundations for freshwater and terrestrial ecosystems that we take for granted today.[47] Within just a few million years of colonizing the land, sedimentary residues liberated from the bedrock hitched rides in the downward movement of flowing water, with most making their way to the oceans but some dropping short in landward basins. Once bioerosion had evolved at scales from minute to massive, the rock cycle was changed so that geology had become indistinguishable from biology, and vice versa.

Then there is climate. With the advent of cyanobacteria in the Archean Eon about 3.5 billion years ago, the earth's atmosphere slowly shifted from reducing (anaerobic) to oxidizing (aerobic) by about 2.4 billion years ago.[48] Moreover, once these cyanobacteria and green algae spread onto rocky surfaces in shallow seas throughout newly opened ocean basins, they broke down rock while taking in greater volumes of carbon dioxide and exuding more oxygen. Marine fungi did the opposite in water both shallow and deep, consuming oxygen and releasing carbon dioxide as they bored into hard surfaces. When lichens first evolved on land, they generated both carbon dioxide and oxygen, but also had an atmospheric impact through their weathering of landscapes, which took up more carbon dioxide. Overall, these bioeroders changed the world, while affecting climate in a big way.

As greenhouse gases in particular plummeted via these and other biological processes, the earth cooled, and stayed cool for a long time. This 200-million-year span of time during the Proterozoic Eon earned the memorable nickname of "Snowball Earth," in which multicellular life laid low until a thaw better allowed eukaryotes to organize into animals, which moved onto and into substrates and consumed organics, including one another.[49] Fast-forward to today, when we confront a multipronged and cumulative disaster of our own making, global climate change. With increased atmospheric and oceanic temperatures, as well as excess carbon dioxide leading to higher ocean acidity, endolithic fungi will increase in numbers, prevailing over their photosynthetic (and oxygen-producing) rivals in marine ecosystems.[50] This fungal surge will produce more carbon dioxide, adding to the overall effect of global temperature and climate change.

Hence all of this microbial and low-lying bioerosion contributed to a quiet revolution, one absent the crunching, cracking, rasping, or other onomatopoeia we normally associate with the breaking of solid media. Through the aggregate effects of rock-destroying cyanobacteria, algae, fungi, and lichens, the surface of the earth was irrevocably altered, and it will continue to change long after we are gone. For all life today, small bioeroders matter, regardless of whether they reside in our mouths or mountains.

Chapter 3

Rock, Thy Name Is Mud

Chitons are cool. Imagine the sheer delight of college students when they see chitons for the first time, their enchanted utterances resounding over coastal outcrops. This is where they find these seemingly motionless primeval-looking animals on rocky surfaces, places eroded by them, their ancestors, and other species of rock-eating animals. At first glance, these flattish, oval, and segmented animals invite comparisons to long-extinct animals called trilobites. But those animals had legs, eyes, and antennae, all of which chitons lack. Moreover, trilobite bodies can be divided crosswise into three parts, with a definite head and tail between the main body, as well as three lengthwise divisions, or lobes (hence their name, "tri-lobite").[1] Most tragically, trilobites also died out about 250 million years ago. So for any of my students who excitedly blurt, "It's a trilobite!," my first response is "I wish," followed by a series of questions leading them down a different path, one not of the jointed-appendaged, but of the limbless. Eventually they arrive at a conclusion that these wonderfully enigmatic animals are molluscans, related to snails, clams, octopi, and others of their lineage.

Chitons are composed of eight overlapping plates (valves) of the mineral aragonite encircled by a band (girdle) of the same material; these plates also allow chitons to bend as needed.[2] A thin organic layer on these plates provides colors, which in some chitons are dull, whereas others are startlingly variegated and vivid, bordering on brazen. Their sizes range from less than the width of an adult human finger to the enormous gumboot chiton (*Cryptochiton stelleri*), which can be as long as a forearm (35 cm/14 in) long, while also weighing nearly 2 kg (more than 4 lb).[3] The gumboot chiton—common on rocky coastlines of the northern Pacific Ocean—earned its all-weather shoe-wear nickname from its "rubbery" exterior, with tough, purplish flesh covering all eight valves.

Chitons are placed within an impressively syllabic clade (evolutionarily related group) called Polyplacophora, and their primitive appearance actually checks out with their evolutionary history. As mentioned previously, they are molluscans, but likely diverged more than 500 million years ago from their more familiar kin, such as gastropods (snails), bivalves (clams), and cephalopods (octopi, squids, and more).[4] Represented today by a little less than a thousand species and about 400 fossil species, chitons have a worldwide distribution but mostly live in tropical shallow to intertidal marine environments. A few, however, are adapted to colder places, such as Alaska, and some even live in deep-sea environments. All chitons are marine, and although some tolerate staying out of the water between tides, none are terrestrial. This means you should not expect to see chitons scraping algae off sidewalks, patios, or building exteriors.

Students curious enough to try picking up chitons off their rocky places quickly experience one aspect of their lifestyles that is immoveable, which is that they are immoveable. Chitons are so tightly bound to their host rocks that neither my students nor I can pry them off unless with a knife, which I absolutely forbid. Also, to actually see a chiton move would be an exercise in patience worthy of its own meditative practice. So, instead of fruitless wresting and steadfast watching, I tell students that a soft body lies underneath a chiton's armored exterior, one containing a fleshy foot like that of a snail. Other soft parts contain tiny light receptors that act as eyes, gills, a digestive tract, nervous system, reproductive bits, and all the other parts it needs to get through this life and pass on genes to future generations.[5] One of these anatomical parts, though, is not so soft. A chiton's file-like radula, which acts as a rasping tool, not only has aragonite, but also includes iron in the mineral magnetite. The inclusion of iron in its radula better allows chitons to scrape algae or encrusting animals off rocky surfaces, which they ingest and digest as they move.[6] Of course, all good things in digestion come to an end. So, soon after a student finds a chiton and we talk about it for a few minutes, I save my most impactful point for last by pointing to a pile of tiny mud-filled pellets behind each chiton and saying, "See that? That's its poop."

I get to witness my students' polyplacophoran-provoked joy once every two years, which is when I take them to San Salvador Island (Bahamas) for a field course. Bahamians refer to San Salvador as one of the "Out Islands," as it sits atop its own tiny shallow-marine platform on the southeastern edge of an archipelago while surrounded by abyssal depths.[7] This geological isolation means that the journey to San Salvador normally takes more than a day of travel from our university's home base in Atlanta, Georgia. Most people make the trip to San Salvador to stay at exclusive resorts there, where they lie on its white-sand beaches, admire its shallow turquoise-aquamarine waters, or otherwise do what nonscientist humans refer to as "relaxing." Nevertheless, San Salvador also invites more than a thousand people each year to stay at the not-so-luxurious but always comfortable Gerace Research Centre (GRC) there. The namesake of the GRC was a scientist—Don Gerace—who more than 50 years ago leased a former US Navy submarine-listening base for $1 and converted it into what later became a world-renowned scientific field station.[8] Since then, it has hosted generations of biologists and geologists, many of whom bring classes to San Salvador so students can learn directly from its modern and ancient environments and their respective biotas.

Usually, on the first day of this biannual trip to the GRC, another instructor and I gather our students and walk for 10 minutes to an off-white and undulating rocky ridge east of the compound. No matter what time of the year, energetic waves smash against the eastern side of the ridge, hinting at how those impacts, hour by hour, day by day, are at least partly responsible for the jagged coastline below us. While marveling over exposures revealing internal layering diagnostic of coastal dunes and fossil burrows of their former occupants,[9] we also stop to look at the rocky intertidal area where modern-day processes erase this evidence of the geologically recent past. There we see that brute force is not all that wears down these rocks. Instead, the slow, inexorable erosion of chitons and their molluscan compatriots, gastropods, is mostly responsible.

The chitons on San Salvador are West Indian fuzzy chitons (*Acanthopleura granulata*), which are called "fuzzy" because of short projections on their girdles that evoke fringed skirts.[10] They blend easily with their host rocks, with subtle gray to brown colorations, and are about 7 cm (3 in) long. These molluscans are in the upper part of the intertidal zone, just above or within a splash zone that also provides inadvertent field-course baptisms. Among the chitons and likewise scraping the limestone are snails called nerites and periwinkles. The nerites are colorful and rounded, whereas periwinkles are more subdued, pointy, and knobbed along their spirals; both types of snails range from pea- to marble-sized. Among these marine snails are the four-toothed nerite (*Nerita versicolor*), the bloody-toothed nerite (*Nerita peloronta*), the common prickly winkle (*Echinolittorina tuberculate*), and the beaded periwinkle (*Cenchritis muricatus*).[11] Unlike chitons, they are less fixed to the rock and can be picked up for closer examination without harm before placing them back on their homes. They also differ by lacking magnetite; hence their algal scraping is done with radulas only as hard as their host rocks.

How much damage do these other small bioeroders inflict on Bahamian coastlines? In one study done on the rocky coastlines of Andros, the largest Bahamian island, bioerosion rates were estimated at 2 mm (0.08 in)/year, and about 3 kg (6.6 lb)/year from each square meter of rock.[12] These rates were attributed to just a few invertebrates, which included the previously lauded chiton and nerites, but also limpets (snails that seemingly forgot to coil), sponges, and barnacles. Like all animals, chitons, nerites, and periwinkles not only eat, but also excrete. Through these

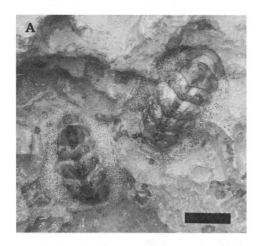

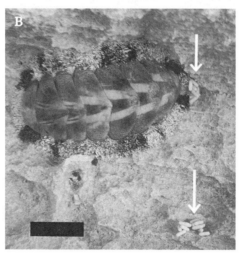

Chitons and gastropods bioeroding coastal limestones on San Salvador Island, Bahamas. A, West Indian fuzzy chitons (*Acanthopleura granulata*) in a rocky intertidal environment. B, West Indian fuzzy chiton caught in the act of changing limestone into mud via fecal pellets (arrows). C, Four-toothed nerites (*Nerita versicolor*, FTN) and common prickly winkle (*Echinolittorina tuberculate*, CPW) eroding the rocky surface. Scale = 1 cm in all photos.

paired biological processes—eating and pooping—and done by the millions, their radula-aided scrapings both diminish Bahamian rocky shorelines while also adding mud to offshore areas. Although I don't tell my students this when we snorkel through the beguiling aquamarine waters around San Salvador, they are also likely swimming through fine particles of waste ejected by molluscans and other invertebrates.

So, let us consider the numerous spineless animals living in shallow seas throughout the world that perform this conversion of the solid into the suspended, changing rock into mud, and let us wonder about when these animals began to carry out such changes. From there, we can ponder how these bioeroding marine invertebrates will fare in the near future as the world's oceans change with climate. Which will suffer, which will thrive? Or will some of these animals pay no heed, and simply bore away?

: : :

However hard it might be to swallow, marine animals have not always eaten rocks, nor did they always move across rocks to scrape up their meals. The oldest clues for animals rasping surfaces of any kind come from trace fossils from the Ediacaran Period (630–542 MYA), and in rocks slightly more than 550 million years old. These trace fossils consist of a series of linear and parallel scratches on fossilized microbial mats (also known as biomats), which grew on Ediacaran seafloors and probably included cyanobacteria and other photosynthetic microbes.[13] The presence of these scratches implies that animals did not just sit at home and wait for ocean currents to deliver suspended organic goodies, which is exactly what suspension feeders do. Instead, the makers of these trace fossils had the gumption to go fetch vittles for themselves, and they had the right bodies to do so, bodies that included something like a radula.

Before these trace fossils were described, though, paleontologists had already identified and named the body fossil later identified as the accused scratch perpetrator, *Kimberella*. This odd fossil was first recognized in Ediacaran rocks from South Australia, then later in rocks of the same age along the White Sea in Russia and in central Iran; once considered rare, thousands of specimens were eventually found.[14] *Kimberella* superficially looks very much like a chiton, as it bears the same low profile, overlaps with the range of chiton sizes, has an oval outline, and is ringed by a girdle. But when you look closer, its anatomy departs radically from chitons. Of these differences, the most obvious is its apparent lack of hard parts and definite segmentation. For one, instead of eight interlocking plates, *Kimberella*'s back had a shell-like covering that was stiff but flexible, acting more like a plastic coating than an armored defense. Underneath this covering were a series of bands interpreted as musculature, which propelled this animal along seafloors like a severely flattened slug.[15] However, one important part is also missing from these fossils: a radula. Its absence does not necessarily mean *Kimberella* lacked this anatomical accoutrement, as some modern molluscans have organic radulas, which are unlikely to be preserved in 500-million-year-old fossils. For paleontologists, what changed *Kimberella* from a head-scratcher of a fossil to a mat-scratcher in the Ediacaran was the direct association of their bodily remains with bundles of linear grooves.[16] This rare combination of body fossils and trace fossils shows that *Kimberella* was grazing, and grazing very much like some modern molluscans, such as chitons and gastropods. Ichnologists even boldly assigned an ichnogenus (scientific name) *Kimberichnus* to trace fossils made by *Kimberella* to distinguish its scratch patterns from similar trace fossils credited to gastropods and chitons, which they call a different ichnogenus, *Radulichnus*.[17]

Nonetheless, before we laud *Kimberella* as the grandmother of animal-led bioerosion, we should note that biomats may have been firm, but they were not hard. Think of these matgrounds as more like cellophane than fiberglass, pliable covers that were easily wrinkled if subjected to shear forces or pressures from above. Below these covers were sediments sealed off from oxygen-rich ocean waters, leading to low-oxygen (anoxic) conditions below that forbade entry to any Ediacaran animals, all of which needed oxygen to live.[18] Biomat thickness and density were thus factors preventing animals from making their way through to the forbidden zone below. Yet once animals did break through biomats, the gates to this underworld were flung open and animals plumbed into these formerly anoxic sediments, irrigating them with oxygen and causing anaerobes to declare the microbial equivalent of "There goes the neighborhood" before perishing. This mixing of seafloor sediments toward the end of the Ediacaran, which radically changed the chemistry of the oceans and eventually the earth's atmosphere, has been dubbed the agronomic revolution.[19] And if you have a tough time imagining slugs and worms as farmers, just think of this activity as an underwater plowing of the fields.

At the same time as *Kimberella* is the oldest trace fossil evidence of animals drilling, breaking into, or otherwise eroding solid substances, and in another Ediacaran fossil, *Cloudina*. These tiny fossils, which look like a stack of Dixie cups, were among the first animals to secrete hard, calcium carbonate skeletons, and were also among the first to have their shells attacked by other animals. Some of their skeletons bear small circular holes in their sides, as if something tried to mock their initial attempts at personal protection.[20] Thus, before the end of the Ediacaran Period and the start of the Cambrian Period about 542 million years ago, at least a few animals had already evolved methods for getting past hard stuff.

: : :

Talking about when animals began eroding rock also requires asking what constitutes a rock. For example, a rock is often defined as an aggregate of a mineral or minerals, and a mineral is a naturally formed, solid, inorganic substance with a definite crystalline structure and chemical composition.[21] Rocks are then classified on the basis of their origin within the rock cycle as igneous (made by cooled magma), metamorphic (end products of other rocks, altered by heat and pressure), or sedimentary (as mentioned previously). Sedimentary rocks are formed from sediments, which get most of their materials from rocks at or just below the earth's surface.

Loose sediment with no cement or matrix—its grains only

separated by air or water—is easily classified as just plain old "sediment." But what if this same sediment is compacted, with spaces between grains reduced so that they lose their individual feel and become more of a cohesive unit? If this is too abstract a concept, think of how a pile of sand is transformed into a sand-castle. Dry sand on its own tends to form piles with a maximum angle of 34°, called its angle of repose.[22] To make vertical walls, ramparts, drawbridges, and other medieval home improvements requires adding both water and pressure, packing the grains tightly together so moisture between those grains makes them stick. In doing so, an aspiring sandcastle architect is effectively changing that sand from soft to firm. Now, if that builder some-how injected epoxy resin or another stealthy cement into that sandcastle to stiffen it, they effectively changed the sand from firm to hard, and soon after can start advertising for knightly renters.

Geologists are well known for their esoteric and often exclu-sionary jargon, but when classifying the spectrum from soft sedi-ments to hard sedimentary rocks, they did just right. The order of sedimentary layers from soft to firm to hard is called (drum-roll): softground to firmground to hardground.[23] Once in a while a pedantic geologist might insist a softground is so thoroughly saturated with water that it is better classified as a "soupground." But such distinctions also provoke arguments about the differ-ences between soups and stews, whether or not chili is a soup, or if real chili includes beans, and thus are best avoided altogether.

For ichnologists, this three-part continuum of softground, firmground, and hardground is a handy way for them to classify animal traces and trace fossils with reference to substrates. What is a substrate? It is the medium that preserves a trace, no mat-ter what that medium might be. A substrate could be the mossy carpet of a forest floor holding the shapes and patterns of mouse footprints. It could be the algae on a pond surface parted by the wake of a swimming alligator. It could be dewdrops smeared across a leaf by a ladybug out for a morning stroll. Of course, all of these traces and their substrates are ephemeral, and their like-liness of entering the fossil record is far lower than winning the lottery. Hence, in most of ichnology, we look to substrates with a better potential for permanently preserving traces, such as sediments, rocks, shells, wood, and bones. Notice that of these substrates, one out of the five can be classified as "soft" or "firm," whereas the other four might be labeled "hard."

This rating of substrates as if they were pillows, mattresses, or other sleep-enabling accessories led paleontologists to wonder when animals began moving into firmer and harder materials. *Kimberella* scratching on biomats and little holes drilled into *Cloudina* toward the end of the Ediacaran were a nice start, but these actions are embarrassingly inadequate when compared to the rock-eroding and shell-eating done today. Given the amount of energy or adaptations involved in penetrating rock, paleontologists rightfully assume that animals first burrowed in softgrounds, then some excavated new niches in firmgrounds, which is still burrowing, just more difficult. Later, some animals evolved from burrowing to scraping and breaking rocks. Based on trace fossils, this momentous development in animal life happened in the Cambrian, at about 500 million years ago.[24] However, most Cambrian borings are simple, small, vertical tubes with little rounded ends that look like old-fashioned bulb thermometers. These trace fossils, which were probably made and occupied by small suspension-feeding animals, were named *Trypanites* by ichnologists.[25] *Trypanites* then in turn became the namesake for trace fossil assemblages in marine hardgrounds since the Cambrian Period.

Yet animals did not get really boring (which is to say, exciting) until the Ordovician Period, starting about 480 million years ago. Most people with even a casual interest in paleontology have heard of the "Cambrian explosion" (mentioned previously), a time of rapid diversification in animal life during the Cambrian (542–488 million years ago).[26] The Cambrian explosion has also been dubbed "the biological big bang," both of which are misleading metaphors because the diversification happened over about 20 million years, making it more of an immensely protracted thud. Regardless, the period that followed this time of renovation, the Ordovician (488–44 million years ago), was a party time for evolution that made the Cambrian explosion seem more like the last sad sound in a microwaved bag of popcorn. This was the time of the Great Ordovician Biological Event, which if that's too much to say, just call it GOBE.[27] Within the GOBE, then, was the OBR, which a few paleontologists (by which I mean, very few) have celebrated as the Ordovician Bioerosion Revolution.[28] The Ordovician is when both numbers and diversity of bioeroding animals took off, in which they sought and colonized rocky places throughout the world's oceans. A variety of animals thus joined cyanobacteria, algae, and fungi in scraping, drilling, and breaking rocks, shells, and other hard surfaces, adding to a biological rock cycle within the larger rock cycle by including new and formidable sources of weathering and erosion.

Still, seeing that both fossilized burrows and borings are preserved in rocks, how do paleontologists and geologists tell the difference between them? The easiest distinction between burrows and borings is that burrows pushed sediment aside, whereas borings cut across sediments, matrix, cement, shells, bones, wood, and anything else that might be in that bioeroding animal's way. Think of the difference between someone poking a finger into your conventional sandcastle, versus using a power drill to assault your epoxy-resin-impregnated sandcastle, in which the drill bit cleaves through sand and resin alike. Of course, geoscientists use other clues to distinguish what were originally softgrounds, firmgrounds, and hardgrounds in sedimentary rocks. But when trying to tell the difference between an animal burrow and an animal boring, only one makes the final cut.

And just who were a few of these pioneering marine invertebrate bioeroders? Are any of their relatives alive today, giving us modern examples of their rock-altering ways? Do these marine invertebrates use different means for scraping, drilling, or otherwise degrading rocks and shells? The answers to the questions are you'll see, yes, and yes.

: : :

In my experience, sponges are often underappreciated, serving as mere side characters in documentaries about coral reefs, or covered in biology classes as the "simplest" animals. So if you are looking to start a Sponge Appreciation Society and need a list enumerating why sponges deserve praise, the first item should state that sponges were very likely the first animals. The evidence for such an audacious claim, however, is based mostly on their modern biology rather than their fossil record, because the first sponges were probably soft-bodied and hence not so amenable to fossilization. The main tool used to estimate sponge origins is through molecular clocks, which are genetic markers in modern animal lineages that we assume changed at certain rates through time.[29] These molecular clocks tell us that sponges should have evolved by about a billion years ago—more than 400 million years before Ediacaran animals like *Kimberella* and *Cloudina*.[30] Geochemists backed up this prediction when they excitedly pointed to 630-million-year-old organic compounds called steranes that were considered as organic compounds diagnostic of animals and specifically of sponges.[31] Alas, this origin story was later discredited when other researchers showed that some algae contained the same compounds.[32] Not to be outdone, though, sponge science got a big boost in 2021 from a study by geologist Elizabeth Turner, when she reported possible 890-million-year-old sponge body fossils from northwestern Canada, close to their predicted molecular-clock age.[33] Sponge body fossils are very slightly more common in 580–542 MYA rocks, but did not become regular participants in the fossil record until well into the Cambrian.

Like other animals, sponges are heterotrophic (they eat), produce gametes (they reproduce), and are multicellular (they do not sing "One Is the Loneliest Number"). Sponges are sometimes described by three general body plans: asconoid, syconoid, and leuconoid.[34] Asconoid sponges are the simplest, looking like a vase or barrel, with inner surfaces lined with special cells called choanocytes, or "collar cells." Syconoid sponges take this one step further by adding pleats to their inner surfaces, which increases surface area and allows them to add many more choanocytes. Leuconoid sponges go completely wild, making vast networks of chambers and tubes, with choanocytes in each chamber. Variations on these three body plans mean that sponges can create a bewildering array of shapes and sizes augmented with bright colors.

So let's say you spot a typical barrel-shaped (asconoid) sponge and peer closely at the cells lining its inner portion. There you would see cup-like structures with little whips inside them, which are its choanocytes. The "collar" (cup) part of a choanocyte is made of microvilli, extensions of cell membranes that help absorb nutrients, like those in our intestines.[35] The "whip" part of a choanoctye is its flagellum, which beats water so it directs nutrients and oxygen to microvilli. Thousands of these cells lining the interior of a sponge, with flagella beating the water around them, cause water with suspended organics to flow toward the cells and thus feed the sponge. Choanocytes in sponges are remarkably similar to single-celled organisms called choanoflagellates, which move to the beat of their own flagella. So biologists reasonably proposed that sponges evolved as collections of cells very much like modern choanoflagellates, but in doing so became the first animals.[36]

Sponges are classified under the clade Porifera, which in turn contains: Calcarea, or calcareous sponges that have calcite or aragonite parts (spicules) supporting them; Hexactinellida, "glass" sponges with silica spicules; and Demospongia, "soft" sponges that are mostly organic but may also have mineralized bits.[37] Of these, the demosponges include clionaids, such as *Cliona celata*. Clionaids cause grief and despair in shell collectors by boring into shells, while also altering entire ecosystems. But considering that clionaids are among the so-called soft sponges, how do they break and enter hard cases ranging from coral reefs to oyster shells? Similar to lichens, they apply a combination of chemical and physical assaults. First, they attach to a calcium carbonate surface (whether aragonite or calcite), and then secrete acids that weaken these minerals. Second, their choanocytes create enough of a current to fracture these weakened spots and pull them away as tiny chips.[38] How tiny? Silt-sized, which means they are smaller than sand grains (< 2 mm/0.08 in wide) and larger than clay particles (> 1/256 mm/0.000008 in), but still count as "mud." These sponge-caused silt chips add up, though, making a significant part of the mud budget in reefs worldwide.

When clionaid sponges bioerode a reef, shell, or other hard substrate, they form extensive networks for their choanocytes. They can bore both laterally and down, and choose directions depending on what else might be living with the sponge. Sponges that bore laterally are cavity forming, sending horizontal tunnels between roundish chambers just underneath the surface of their host substrate. Sponges that bore down are gallery forming, opening a substrate more to the outside world to receive light, which is better for photosynthesizers.[39] Hence, gallery sponges can have algae living with them, producing both food and oxygen, making them ideal roommates.

Once a clionaid sponge dies and its soft body decays, what is left behind to tell us of its former presence? A sponge-eroded shell surface will look as if someone used it for target practice and fired at it repeatedly with a miniature shotgun using millimeter-wide shot. These numerous small holes on the surface are apertures that connected to networks of chambers and galleries below, which is where the bulk of the sponge lived. Yet these networks can be quite shallow. For example, if you find a shell with extensive sponge-inflicted damage on one side, such as one valve of an oyster, flip it over. You might be surprised to see the other side of the shell still intact, with nary a hole in sight. This contrast of bored and non-bored surfaces shows the sponge settled onto whatever side was facing the ocean, whereas the bottom side was protected from etching and chipping activities above.

Because the trace fossils of these borings are so distinctive and ichnologists wanted to communicate their identity, they applied the ichnogenus *Entobia* to them, which is much easier than saying "a bunch of small closely spaced holes that connect to galleries and chambers." *Entobia* as a name actually dates to 1838, long before anyone knew for sure what caused these traces, but the link between them and sponge makers was well established

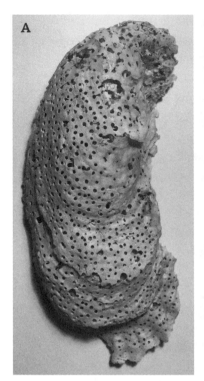

Sponge and bryozoan borings in shells of the same species of bivalve, the eastern oyster (*Crassostrea virginica*). A, Sponge borings, Sapelo Island, Georgia. B, Bryozoan borings, Edisto Island, South Carolina.

by later in the twentieth century.[40] *Entobia* ranges from the Devonian (about 420–359 MYA) to recent, but sponge-caused bioerosion probably stretches back to the Ordovician.[41]

: : :

Despite their soft bodies, marine worms are also capable of boring into shells. Their bioerosion differs from sponges, though, by not leaving mere minute holes in shell surfaces, but instead carving out curving U-shaped grooves and lengthy meandering tunnels. The worms responsible for such shell vandalism are quite varied, but most are from the clades Annelida and Polychaeta, which are segmented worms.

Modern polychaetes are sometimes nicknamed "bristle worms" for their abundant bristles (chaetae *or* setae) on "false legs" (parapodia) projecting from their sides.[42] Although earthworms lack parapodia, they have tiny bristles, attested by anyone who has held an earthworm while digging in a garden, baiting a fishhook, or saving it from sidewalk doom. Worm handlers also may have noticed their squirming friend was segmented along its length. This segmentation, bristles, and other anatomical features show how earthworms are related to polychaetes. However, unlike earthworms, the vast majority of polychaetes—consisting of about 10,000 species—are in oceanic environments ranging from shallow coastal pools to deep-sea trenches.[43] Like chitons, many of these worms are quite colorful, but unlike chitons, some of them are also terrifying, such as the 3 m (10 ft) long *Eunice aphroditois*, which uses hooked jaws to snatch and eat fish that make the fatal mistake of swimming by their burrows.[44]

The most infamous of modern polychaetes that affect hard substrates are those in the genus *Polydora*, nicknamed "clam worms" because they bore into live clams and oysters and cause their shells to "blister."[45] They are not restricted to bivalves, though, but also bore into gastropods, corals, and nonliving limestones.[46] Similar to sponges, these worms use a combination of chemical and physical means to enter hard places. For the chemical part, they have acid-secreting glands that weaken a shell or other hard substance. Once the shell is softened, they use their chaetae to scrape their way in, tunnel down, and turn up, often making a U-shaped structure with the upper parts of the "U" squashed together.[47] When looking down at a shell, coral, or limestone surface with these borings, many closely paired holes hint at the destruction beneath. Yet if that surface is later worn down by other means, hole shapes may range from ellipses to figure-eights to dumbbells, representing cross-sections of the borings from different depths.[48] Rowdier departures from this basic form include meandering or otherwise looping tunnels with multiple pouches, or tunnels that make hairpin turns but lack pouches.

Trace fossils of polychaete borings may go as far back as the Cambrian, with the ichnogenus *Trypanites* often credited to bioeroding worms.[49] However, trace fossils in fossil shells closely matching those of modern *Polydora* traces in shells are also well documented, telling us similar animals made these borings in the geologic past. Ichnologists label the simpler U-shaped borings as the ichnogenus *Caulostrepsis*, dating back to the Devonian, whereas the more expressive meandering, looping, and pouched forms are called *Maeandropolydora*, which are in Jurassic and younger rocks.[50]

: : :

When I was a graduate student and teaching undergraduate paleontology labs, I always felt challenged when talking about bryozoans. This uncertainty was understandable, as these animals differ from ones most people would encounter in everyday life, nature documentaries, or cartoons. Nevertheless, bryozoans are noteworthy animals for their evolutionary longevity and stunning diversity. Most importantly for our purposes, at least a few bryozoans bore into rocks and shells, leaving distinctive traces telling of their former presence.

Bryozoa means "moss animals," not because they gather around immobile rocks, but for superficially resembling moss. They are mostly marine animals represented by almost 6,000 species, and form millimeter-sized and closely spaced pouches (zooecia, plural of zooecium) holding individual animals (zooids) in sedentary colonies.[51] Zooecia are composed of organic material or calcium carbonate, with the latter more likely to fossilize, which they have done in marine environments since the Cambrian (about 480 MYA).[52] Modern bryozoans make up for their tiny individual sizes by forming massive colonies that often encrust on shells, rocks, reefs, or ship hulls. As suspension feeders, they let their food come to them as small organic particles suspended in the water, and like sponges, they use group action to pull in these particles. Instead of flagella, though, zooids are adorned by bundles of tentacles called lophophores that pop out to collect and deliver organic particles suspended in the water to waiting mouths.[53]

Considering how bryozoans settle down and stay in one place with few moving parts other than their lophophores, they lack the means for actively drilling or scraping into shells or rocks. Yet to attach to a hard surface, at least some colonies secrete acids on zooid bottoms to help stick them to those surfaces. This fixating behavior is expressed as thin, shallow, branching, or intersecting tunnels that swell in places along their lengths, rendering "spiderweb" patterns on shells or rocks.[54] Other bryozoan borings make themselves known as a series of small, shallow pits looking like dashed lines that also might branch. Trace fossils of boring bryozoans date back to the Early Ordovician (more than 470 MYA), with paleontologists assigning the ichnogenus name of *Ropalonaria* for the spiderweb-like tunnels. In contrast, shallow pits forming dashed lines were assigned the ichnogenus *Finichnus*, known from the Late Cretaceous (about 70 MYA) through today.[55]

: : :

Barnacles live in many of the same environments as bryozoans but are quite unlike them, in that they do not form colonies and are anatomically more complex. Most barnacles superficially look like corals by possessing hard, conical shells and attaching to solid surfaces. But they also have jointed legs, revealing their arthropod identity. Barnacles are crustaceans belonging to the clade Cirripedia, but share common ancestors with crabs, lobsters, shrimp, and crayfish.[56] Their "shells" are actually a series of overlapping plates like those on a lobster or crayfish, but arranged around their soft parts as protective body armor. Like a superglue commercial gone wrong, barnacle larvae use cement glands in their heads to attach to surfaces, ranging from wharves to whales.[57] Once connected to a surface, barnacles have eight pairs of long, fine legs, and they know how to use them. Functioning much like lophophores, individual barnacles beat their legs to pull in food particles (including live plankton) suspended in the water. Barnacles are divided into two main groups based on their mode of attachment. The most common are acorn barnacles, which cement their shells directly to a surface. In contrast are gooseneck barnacles, so called because of their attaching with fleshy stalks (peduncles), which were then compared to geese necks by sailors who clearly spent too much time at sea.[58]

These unusual crustaceans accordingly attracted much attention from a budding young naturalist named Charles Darwin. Darwin was smitten by barnacles in the 1830s while voyaging on *The Beagle*, and later expressed this obsession in the best way possible by writing a two-volume book about them, with each published in 1851 and 1854.[59] Among his most notable observations of barnacles were off the coast of Chile, when he realized that the stout shell of a living marine snail was "*completely* drilled by the cavities formed by this animal" (emphasis by Darwin, but I would do the same). After describing these parasitic barnacles, he realized they were soft, shell-less acorn barnacles adapted to living in another animal's shells, rather than making their own.

These parasitic barnacles described by Darwin, as well as others that drill into clams, snails, corals, and other calcium carbonate substrates, belong to a clade of acorn barnacles called acrothoracicans.[60] How do they bore? Young barnacles settle onto a surface, and then use calcified projections (studs) on a peduncle to wear away a surface and make a chamber.[61] They grow up and otherwise live inside little cavities, with no need for a protective covering like their outer-life barnacle relatives. The openings to these borings look like slits, half-open eyes, or commas, but widen below into living chambers. The barnacles then poke their legs out the slits to gather food, keeping the rest of their bodies below a rock or shell surface and safe from predators.

Although barnacle body fossils date back to the Cambrian, trace fossils of barnacles—or at least barnacles acting very much like modern ones—date back to at least the Middle Devonian (about 390 MYA).[62] The most common trace fossil attributed to barnacles is *Rogerella*, which matches the sizes and shapes of modern barnacle borings, but other distinctive forms warrant additional ichnogenera, including the intriguingly named *Zapfella*.[63] The value of these trace fossils is that they tell us barnacles were present even if their bodies are no longer there, extending our perception of former marine biodiversity and behaviors.

: : :

Hard times call for drastic adaptations, and at least a few clam lineages rose to the challenge by evolving into full-bodied drills that enabled them to bore deeply into corals, rocks, and even swimming animals. Among the clams best known for boring into rock are those belonging to the genus *Lithophaga*, with Greek roots that literally mean "rock eater" (*lithos* = "rock," and *phagos* = "eat").[64] All of these clams are mussels, but unlike most mussels that attach themselves to water-body bottoms, they actually go into hard bottoms. Other well-known rock-boring clams include piddocks, which are in the clade Pholadidae and sometimes called "angelwings" because their opened shells are reminiscent of heavenly messengers' feathered appendages.[65] Biblically inspired folks then could not help but accuse a few of these innocent clams of unholy intentions, naming one the false angelwing (*Barnea truncata*) and another the fallen angelwing (*Petricola pholadiformis*).

Molluscan damnation aside, a clam surrounding itself with rock has many advantages, but most importantly prevents other animals from eating it as it filter-feeds with its paired siphons. Siphons are like two-way underwater straws, one of which (inhalant siphon) sucks in water with goodies, and the other (exhalant siphon) pumps out water that no longer has goodies. Given the benefits of feeding while ensconced, the question still remains: How do clams drill into rock? Most apply a winning combo of thick, corrugated shells and muscular action, with species of *Lithophaga* using mussel muscles.[66] For those clams that mostly use their bodies, their shells act as roughly textured files abrading against corals, limestone, or other hard substrates. Smoother-shelled bivalves use acids, so these are corroding while eroding.[67] In either case, clams bore by rocking up and down, back and forth, and rotating clockwise and counterclockwise. Movements up, down, and around are enabled by a clam's muscles relaxing and contracting on either side of the shell, with a ligament between the valves acting as a hinge to keep them connected.[68] Regardless of whether clams exude acids or grind away, their exhalant siphons eject fine, clay-sized particles caused by their activities, "dusting" their homes as they excavate.

Bivalve borings have a wide range of sizes that depend on clam species and ages, but their overall shapes are similar. Most borings look like stereotypical cartoon-cavemen clubs (or vases, depending on mood) with circular, oval, or dumbbell-shaped apertures that expand downward and with rounded ends. Other clam borings are long, shallow grooves made by clam bottoms as they adjusted their positions on a hard surface. One happy circumstance for biologists and paleontologists alike is that they sometimes find clamshells in their former homes, which makes linking a trace to its maker much easier.

Trace fossils matching the shapes of club-shaped borings are called *Gastrochaenolites*, whereas groovy borings are dubbed *Petroxestes*; both extend back to the Ordovician (more than 450 MYA).[69] Between then and now, clams have seemingly bored into a variety of items, including rocks, corals, bones, wood, and even coprolites (fossil feces).[70] But perhaps the most audacious example of fossil bivalve borings is from a Late Cretaceous (~70 MYA) sea turtle in Japan. Based on how holes in its carapace (some with their bivalve makers) show signs of healing, the sea turtle was apparently alive and swimming while clams impudently settled onto and drilled into it.[71] As for wood-boring clams, which originated in the Jurassic (about 150 MYA), their significant effects on driftwood and human-made wooden ships warrant a full chapter of their own, which I promise to provide later.

: : :

Patrick, the lovable dunce of the animated TV series *SpongeBob SquarePants*, does much more than make us shake our heads at his sweet dim-wittedness. He also effectively represents our evolutionary connection to the animal lineage that inspired him. As a sea star, Patrick would be classified as an echinoderm of the clade Echinodermata (*echinos* = "spiny," and *derma* = "skin").[72] Yet Patrick's representation by a sea star also alludes to a deeper

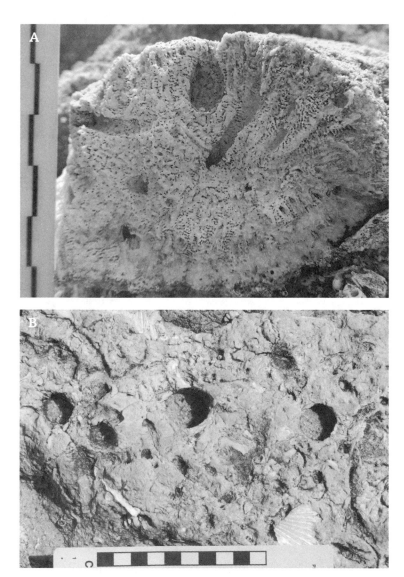

Trace fossils of rock-eating clams. A, Lengthwise section of borings in a fossil coral (*Acropora cervicornis*), Late Pleistocene, Bahamas; scale in centimeters. B, Circular holes made by bioeroding clams in the Oura megasurface (Miocene), Portugal; scale in centimeters.

connection, in which sea stars and all other echinoderms share a common ancestor with humans and other backboned animals.

One of the easiest ways to understand the relationship between you and Patrick is to place images of echinoderm and human embryos early in development next to one another, showing that they are nearly indistinguishable. This striking resemblance is because of their anuses. You see, echinoderms and chordates are deuterostomes, which means the very first opening in their embryonic development (blastopore) becomes an anus rather than a mouth.[73] If you find this relationship between echinoderms and vertebrates still difficult to believe, another way to remind yourself of their kinship to vertebrates is that they wear their skeletons on their insides. Unlike arthropods, molluscans, and all other hardened invertebrates, echinoderms have endoskeletons, made of calcite plates that reinforce soft tissues on their outsides and protect vital organs on their insides.[74]

Along with sea stars (stelleroids), other echinoderms are remarkably diverse and have been evolutionary wonders of marine environments for more than 500 million years, dating back to the Cambrian. Modern examples of their clade include, but are not limited to, sea cucumbers (holothuroideans), brittle stars (ophiuroids), sea lilies (crinoids), sea urchins, and sand dollars (both echinoids). Clades of extinct echinoderms include those that sound like skin disorders (cystoids), recurring skin disorders (cyclocystoids), or video-game characters (blastoids).[75] Notice that I also said echinoderms are evolutionary wonders of marine environments, not freshwater or terrestrial environments. So if you ever find a wayward sea star or other echinoderm near a beach, whatever you do, do not put it into freshwater, as this will surely kill it. The same principle applies to keeping it on a shelf at home, or wearing one as a sheriff's badge, which will quickly become a stinking badge, which you do not need.

Of all echinoderms, the ones most famous for degrading solid rock are echinoids, and specifically certain types of sea urchins. Like sea stars, these rounded but spiny animals have their mouths centrally located on their bottoms so they can feed directly from whatever surface is underneath them. Urchins are mostly grazers, scraping algae off rocks, shells, and corals with five tightly interlocking plates or "teeth." This anatomical feature was nicknamed "Aristotle's lantern" in honor of Greek philosopher and scientist Aristotle (384–322 BCE). Although better known for his philosophizing and Platonic relationships, Aristotle was also keenly interested in animals, and in a moment of enlightenment he compared the shapes of sea urchins to Greek horn-shaped lanterns. However, a bad translation from Greek originally led to biologists referring to just the mouth parts as the "lantern" instead of the entire animal.[76]

Sea urchin teeth, such as those of modern species of *Diadema* and *Echinometra*, are composed of calcite, just like the rest of their endoskeletons. This mineral composition means that most surfaces they scrape are at least as hard as calcite, which should wear them down. Fortunately for these animals, though, their teeth are structurally strong and grow as quickly as they erode.[77] The five-fold arrangement of echinoid teeth also means that when muscles open and close them, they gouge surfaces with each scrape. Repeated rasping thus causes five-point-star patterns corresponding to the symmetry of their mouthparts, with those patterns commonly overlapping. If given enough reason to stay in one place, echinoids also can scratch out circular pits, but if they move along a surface, they may make continuous and sinuous gutters.[78] The larger implication of these animals consuming so much limestone is that they convert rock into mud- and sand-sized sediment. Based on several studies of sea urchin rock munching, they can erode 3–9 kg/m² (0.6–1.8 lb/ft²) from a given area, accounting for significant changes in their rocky environments and even limiting reef growth.[79]

How long have echinoids bored into rocks and other hard marine substrates? At least since the first modern-style sea urchins evolved, which was in the Late Triassic (about 230 MYA). Trace fossils credited to echinoid bioerosion also range from the Triassic to the near present, and ichnologists have assigned ichnogenus names to borings with star-shaped patterns (*Gnathichnus*), circular pits (*Circolites*), and continuous grooves (*Erichichnus*).[80] These distinctive trace fossils also show how urchins have affected reef environments in the past 200+ million years, giving us important clues about their effects on those diverse environments, particularly during times of abrupt climate change.

: : :

For more than 500 million years, but especially since the Ordovician, shallow-marine invertebrate animals have broken down rocks, shells, corals, and other hard materials through physical or chemical means. The global environmental effects of shallow-marine invertebrates today are evident as the wholesale transformation of formerly big, solid objects into fine-grained particles. Indeed, most calcium carbonate mud in shallow-marine environments close to the poles comes from bioerosion.[81] Also, when guided by trace fossils of marine invertebrate bioerosion made by chitons, gastropods, sponges, worms, bryozoans, barnacles, bivalves, and echinoids, we gain an appreciation for the variety of methods used by these animals to wear down hard materials, shaping their worlds over time as their lineages evolved.[82]

Yet we still do not quite know how these modern invertebrate bioeroders or bioerosion in general are affected by rapid climate change, such as that happening now. Warmer oceans will certainly impact marine invertebrate life cycles and growth, meaning some species may be favored, but others may not fare so well. The geologic record is helpful in this respect, as the "Big Five" mass extinctions were all linked to rapid climate change,[83] and lineages of most bioeroding marine invertebrates survived all of those extinctions. Still, the rapid climate change happening now is unique. For one, it is unfolding at a much faster rate than most others, and for another, it is being caused by just one species: us. The ever-upward trend of carbon dioxide produced by our burning fossil fuels since the late eighteenth century is now warming oceans and the atmosphere, causing more frequent (and fierce) tropical storms, ferocious wildfires, severe droughts, and other disasters, all affecting today's ecosystems and more than 7 billion people.[84]

From the perspective of shallow-marine environments, perhaps the most unfortunate consequence of adding megatons of carbon dioxide to the atmosphere is that some of this carbon dioxide mixes with ocean water. This mingling portends one of the most dreadful and insidious partners to climate change, ocean acidification. Remember when we learned earlier that water and carbon dioxide forms carbonic acid? When the world's oceans become a gigantic bottle of acid, and humanity is the fervent introductory-geology student wielding that bottle, a good amount of calcium carbonate substrates—including those composing coral reefs—will bubble. This dissolution and consequent release of yet more carbon dioxide into the oceans may then hasten a positive-feedback process, justifiably provoking panic in marine biologists and oceanographers.[85]

Now one might think that animals that make a living in hard substrates would welcome acid-induced weakening of those same substrates. And sure enough, this is good news for a few animals. For instance, bioeroding sponges are already taking advantage of dead and dying coral reefs, with signs that their diminution is accelerating with ocean acidification.[86] For other animals, though, acid baths are not so helpful. Gastropods, bivalves, barnacles, and echinoids all have calcium carbonate hard parts, meaning they will have problems growing shells and other bits. Even chitons, those resilient and compelling descendants of the Paleozoic Era, may face a reckoning as the world's oceans become warmer and more corrosive, affecting all life, and not just the eroders.

Chapter 4

Your Beach Is Made of Parrotfish Poop

Whenever snorkeling over a Bahamian reef in the early 2000s, I did not just take in the spectacular colors and motion on and around it: I also listened. At first the sounds were elusive, but once my submerged ears attuned to them, they were hard to ignore. They usually came from below, a crunching and popping reminiscent of sugary breakfast cereals meeting milk. But these sounds were those of fish breaking the reef with teeth that could easily eat rocks. And if I was patient enough, taking the time to float motionless above with face mask pointed down for a few minutes, I could visually link the crunches with the fishes responsible for changing the oceanic soundscape.

I am talking about those lovely denizens of tropical reefs in the Atlantic, Pacific, and Indian Oceans, parrotfishes. Granted, parrotfishes already demand our attention owing to their prettiness, but watching and listening to them chew on rock is a synesthesia that delivers both scientific and aesthetic pleasure. The scientific component of this sensation comes from the overlapping observations of sight and sound, a linkage doubly confirming its reality, whereas the aesthetic facet is the experience itself, enhanced by alternately separating and uniting senses.

But for now, let us focus on the visual. Although parrotfish shapes are not so varied, they come in a stunning variety of colors and sizes. A short list of colors applicable to parrotfishes would include cerulean, teal, mint, turquoise, lavender, hot pink, crimson, lemon, and fire, applied as spots, stripes, splashes, and swaths expressed with a brilliance shaming black-lit velvet Elvises.[1] They are streamlined fish, with laterally compressed bodies and tall middles, tail (caudal) fins pointed on top and bottom ends, and prominent pectoral fins located almost directly above much smaller pelvic fins. Their common names are inspired by their mouths, which protrude such that they resemble a parrot's beak. However, this superficial comparison is belied by rows of tightly packed teeth cemented by bone, an arrangement that no self-respecting parrot would have. Nonetheless, a parrotfish in motion is akin to flight, complete with diving, ascending, and hovering. Their acrobatic swimming is powered by beating pectoral fins, with dorsal and anal fins above and below (respectively) acting as subtle rudders, and caudal fins directing quick changes in direction with each flip right or left.

Yet it is when these fishes pause at a reef that I also pause, waiting for them to nip a bit of coral with their impressive teeth. After snorkeling enough times with my students on our biannual trip to San Salvador Island, I have also witnessed the end products of parrotfish-rock consumption, and in volumes far greater than those of rocky intertidal chitons. Follow a parrotfish long enough as it goes about its life, and you will eventually see a white stream of sand flow out of its rear, gently suspend, and then settle to the bottom, joining sandy deposits of former coral left by previous generations of parrotfishes.

Later in the '00s, I happily expanded my parrotfish awareness when teaching a study-abroad program in the coastal city of Townsville, Australia, which gave me the opportunity to scuba dive on the Great Barrier Reef. While taking in the reef's kaleidoscope of colors, patterns, motions, and yes, sounds, I also witnessed Pacific species of parrotfishes dining on parts of it, echoing their Atlantic cousins. Most importantly, though, I got to watch a few fishes complete the cycle of converting rock to sediment as they ejected sand from below their caudal fins. And, just like in the Bahamas, the sandy areas surrounding the elevated parts of the reef gave testament to many fishy deposits made over many sedimentary cycles.

When I later read more about parrotfishes, their feeding habits, and their ecological effects on reefs, I was surprised to learn that not all species ate corals, but also include sediment swallowers or fishes that simply crop and browse algae. I also found out their interactions with reefs help maintain these diverse ecosystems, and that a paucity of parrotfishes oftentimes signals the beginning of the end for reef health. Thus I wondered more about them: not as a fish-studying ichthyologist, but as a fossil-loving ichnologist. For instance, which species of parrotfishes chew rocks, versus those that eat sediments or otherwise have low-rock diets? How far back in time does the mutually beneficial relationship between parrotfishes and reefs go? Do we have any trace fossil evidence of when the ancestors of today's fishes first bit into reefs and started converting rocky environments into sediment? What are the ecosystem-wide impacts of parrotfishes today, ones that actually cause biologists, fishers, fisheries, and policymakers alike to take notice and voice concern? How are these bioeroding fishes affected by climate change and other alterations to their oceans? All of these questions help guide us to better understand this smaller but essential sedimentary rock cycle driven by parrotfishes, while also exemplifying a reason to support their continued biodiversity and existence.

: : :

The evolutionary history of parrotfishes tells of their living with and adapting to warm, tropical waters over millions of years, and of their close ties to coral reefs and related shallow-marine ecosystems. Today, almost a hundred species of parrotfishes are placed in one clade (Scaridae), and they are considered part of the same clade containing wrasses (Labridae).[2] The modern swimming and coral-crunching species I have seen most often, such as the stoplight parrotfish (*Sparisoma viride*) and rainbow parrotfish (*Scarus guacamaia*), swim in shallow waters of the Bahamas and Caribbean of the Atlantic Ocean, but the most varied parrotfishes live in the Pacific and Indian Oceans. Ancestors of these geographically displaced groups were apparently separated from one another by plate tectonics, with either the clos-

Terminal phase male of a stoplight parrotfish (*Sparisoma viride*); photo by Adona9, Wikimedia Commons.

ing of the Mediterranean Sea or the Isthmus of Panama around 20 MYA.[3]

Adult parrotfishes range in size from not much longer than a typical adult author's forearm to more than the same author's height, the latter represented by the green humphead parrotfish (*Bolbometopon muricatum*) of the Pacific.[4] The variety of parrotfish color schemes listed earlier are more than matched by their complex life cycles and behaviors, which include the ability to change sex.[5] Yet another intriguing behavior of some parrotfishes is their ability to exude a mucus cocoon, used to envelop themselves before going to sleep each night. The mucus is secreted from glands near a parrotfish's gills and extruded from its mouth like it is blowing a bubble, but one large enough to surround its body.[6]

When first documented by ichthyologists, the purpose of this cocoon was unclear. For one, it is far too thin to insulate or otherwise act like a slimy sleeping bag. It also does not hide a fish from predators; it's transparent. Ichthyologists surmised that perhaps the cocoon functioned not so much as visual camouflage, but chemical, masking a parrotfish's scent from moray eels and other predators. Now, ichthyologists who have studied the daisy parrotfish (*Chlorurus sordidus*) of the Indian and Pacific Oceans propose that these envelopes are analogous to mosquito netting, providing protection against parasites.[7] The parasites most likely to affect parrotfishes, gnathiid isopods, are small crustaceans distantly related to the oh-so-cute "roly-poly bugs" one might encounter on land. Gnathiid isopods are not so cute to parrotfishes, though, because they attach to fish skins and suck blood, earning the analogy of "the mosquitoes of the sea." Hence the cocoons intercept the gnathiids before they can latch on to a slumbering parrotfish.[8] Still, not all parrotfishes make cocoons, burying themselves overnight in sand instead to avoid parasites or predators alike.

The evolutionary history of the preceding adaptations in parrotfishes is all the more marvelous when considering how modern-style parrotfish body fossils are geologically young, dating only to about 30 MYA in the Oligocene Epoch.[9] However, molecular clocks and other body fossils both show that their lineages started earlier in the Eocene Epoch, with undoubted ancestors in the Middle Eocene (48–38 MYA) of Italy.[10] Not surprisingly, animals that can chew reefs also leave lasting evidence of such reef-eating, telling us a minimum time for when these behaviors may have evolved. Such trace fossils are in Middle Eocene rocks of India and consist of small, paired grooves that were made by either parrotfishes or fishes acting very much like parrotfishes.[11] Are similar scrapings preserved in older rocks? Probably yes, but these trace fossils so far remain either unrecognized or unreported. Still, the much larger traces of parrotfish activities are of modern reefs and their surrounding shallow-marine environments, which have been shaped by millions of years of gnawing and pooping.

: : :

Corals, which share a common ancestor with jellyfishes, have lived in world oceans since the Ordovician Period (about 480 MYA), although the current clade (Scleractina) has only been around since the Middle Triassic (about 240 MYA).[12] These animals, which are also sometimes nicknamed "stony corals" because of their calcareous skeletons, might live alone but most often are in colonies. Those in colonies usually have symbiotic and microscopic algae in their tissues called zooanthellae (zoo = "animal," and *xanthos* = "yellow," referring to algal color), which partly feed corals, provide them with oxygen, and otherwise help them grow.[13] Corals in turn offer protection, as these animals are armed with stinging cells that kill small prey that come too close. Hence corals with zooanthellae tend to grow up and out quickly, forming the mound-like masses of coral reefs, and lend to the beautiful colors of a living reef.

Just how do parrotfishes consume corals? Before asking "how?," two better questions are, *why* do parrotfishes consume corals, and *which* do this? The answer to the first query is that these fishes are not necessarily seeking coral to eat. Instead, they are going for algae growing on and in the coral, including microscopic algae that bore into reefs. Of course, for those parrotfishes that ingest corals, they sometimes eat live corals and other small invertebrates, whether on purpose or not. The parrotfishes most likely to eat animals, though, are younger parrotfishes. However tempting it might be to blame this dietary anomaly on reckless adolescent appetites, it is more related to anatomical limitations, as younger parrotfishes are also smaller parrotfishes, lacking the jaw strength to break rocks.[14] Once these youngsters grow up, their main ecological role is as herbivores, resulting in extensive menus that include algae ranging from the unseen to the prominent.

Parrotfish feeding preferences sort themselves out according to how they eat. For example, do they swim above the reef and stop to nip, crop, or pull algae poking out from the reef surface? If so, these parrotfishes are browsers. Do they swim closer to the reef surface and get their faces in the salad bar (so to speak) and scrape low-lying algae off the reef surface with their teeth? Then these are scrapers. Do they say, screw it, I'm not just eating what's in the salad bar, but also the salad bar itself, tearing off chunks of it with each bite? These overly enthusiastic eaters are excavators, and are the parrotfishes responsible for most bio-

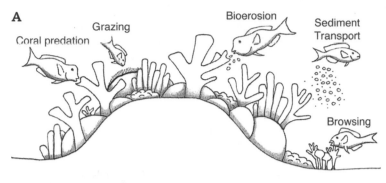

Parrotfishes and their effects on reefs and shoreline ecosystems. A, Parrotfish behaviors on an idealized reef, including coral predation, grazing, bioerosion, browsing, and sediment transport; modified from Bonaldo et al., "The ecosystem roles of parrotfishes on tropical reefs." B, Sandy beach and beachrock on the north shore of New Providence Island (Bahamas), composed of sediments that were produced by parrotfish bioerosion and poop from offshore reefs there.

erosion on a reef.[15] Yet even within the excavator species—such as the stoplight parrotfish in the Atlantic and the green bump-headed wrasse in the Pacific—only the larger adults are ripping into corals. Moreover, the number and sizes of bite traces from excavators and scrapers also reflect the body sizes of the parrotfishes doing the biting, with larger fishes taking more bites and bigger bites.[16] This means that adults of only a few species of parrotfishes are disproportionately responsible for much of the breakage. So when it comes to bioerosion, not all parrotfishes are equal.

As one might imagine, different feeding modes require different abilities and anatomies. Parrotfish "beaks" have teeth that would frighten most dentists, but those of browsers and scrapers are less intimidating than those of excavators. For example, rows of teeth belonging to the daisy parrotfish (C. *sordidus*) of the Indian and Pacific Oceans protrude enough to suggest braces, whereas other species have flatter profiles to their smiles. How do excavator species keep from breaking their impressive chompers on corals? Remember that vertebrate teeth are composed of the mineral apatite, which is harder than the mineral aragonite composing corals, as well as many invertebrate shells (5 versus 3 on the Mohs hardness scale). Furthermore, a 2017 study of parrotfish teeth revealed that these apatite crystals are reinforced as entwined bundles of microstructural fibers.[17] Such teeth combined with impressive jaw levers and muscles thus enable incredible bite forces in relatively small fishes. For instance, given that force applied over a given area is pressure, some parrotfish bites can generate as much as 56 N (newtons) per square millimeter, which correlates to more than 8,000 pounds per square inch.[18] Think of dancing with an elephant wearing heels and it stepping on your foot, and you will get an idea of the forces involved in parrotfish bites.

Nevertheless, these teeth are more for biting, and not for chewing. So how does a parrotfish reduce a chunk of coral before it goes further down a parrotfish? Coral diminution is aided by a second set of "jaws" in a parrotfish gullet (pharynx). This bony arrangement acts as a grinding mill, breaking down pieces of coral into fine sand- and mud-sized particles before moving these bits on to the intestine.[19] Such post-oral processing exposes algae living inside rocky substrates, increasing their nutritional value rather than becoming mere carbonate roughage. Augmenting hard parts are soft parts between the teeth and the pharyngeal apparatus, the gill filaments and a pharyngeal valve.

Gill filaments act like a fine-toothed comb, separating small bits of organic material, whereas the pharyngeal valve secretes sticky mucus that traps more organics, which are passed on to the pharyngeal jaws before going into the esophagus.[20] Given such anatomical attributes, it is perfectly fine to think of parrotfishes as highly efficient fishy food processors, extracting life-sustaining sustenance with every coral-breaking bite.

Although parrotfishes are most famous for their raw and uninhibited coral eating, some species are more inclined to recycle by eating what has already been eaten. With such parrotfishes, sediment that has already passed through other parrotfish guts is swallowed for the algae growing on or in it. In a 1978 study of parrotfish bioerosion in Barbados, researchers compared sediment around reefs there with sediments in parrotfish guts, and found that just one species—the stoplight parrotfish (S. viride)—was making most of the new sediments there, whereas other parrotfish species simply mined algae growing in previously made sediments.[21] Regardless of whether these sediments are fresh out of the gut or put through another alimentary cycle, they become finer, reduced in particle size ever so slightly with each passage. This means that pebble-sized grains become coarse sand, coarse sand becomes medium sand, and so on.

Other questions about parrotfish feeding are not just "What?" and "How?" but also "When?" Unlike some people who prefer eating after sunset and ingesting more calories in the winter, a few parrotfishes are daytime and summertime eaters. In a 2018 study of the daisy parrotfish in the Maldives of the Indian Ocean, researchers discovered it prefers to munch around lunch, eating algae more often in the late morning and early afternoon.[22] On the Great Barrier Reef of Australia, this same species also eats more in the summer than other seasons. In the previously mentioned 1978 study of reefs around Barbados, parrotfishes there showed researchers how frequent eating comes out in the end, as defecations (delicately termed "gut turnovers") happened about eight times a day.[23]

For those parrotfishes that eat not just the coral "salad" but pieces of the salad bar itself, they break apart corals both directly and in concert with other forces. For instance, branching corals—such as the elkhorn coral (*Acropora palmata*) and staghorn coral (*Acropora cervicornis*)—can be weakened by parrotfish bites, which then makes them more vulnerable to breakage from strong waves.[24] Colonies of *Porites* with boring clams, such as *Lithophaga*, also are more likely to be bitten by parrotfishes.[25] And of course, dead or live corals with bioeroding algae inside them also attract attention and are similarly targeted by parrotfishes.[26]

However, an absence of parrotfishes is far worse for reefs, which grow slower without them there. In 2017, marine biologist Katie Cramer and her colleagues reported the long-term consequences of parrotfish by studying former reefs on the Caribbean (eastern) side of Panama.[27] Using sediment cores—which involves drilling into sediments with a metal tube (core) and extracting sediments from the tube—researchers unraveled the past 3,000 years of reef history there. They accomplished this by identifying the relative abundance of fish teeth, coral colonies, and sea urchin parts, with all bits and pieces placed precisely in time. Once all of these data were analyzed, the picture was clear: whenever parrotfishes were abundant, reefs grew. Surprisingly, bioeroding urchins also increased in numbers with parrotfishes, which meant the latter somehow benefited from a parrotfish presence. The opposite situation, though, was that whenever parrotfishes declined in number and variety, reefs and their associated communities suffered, a cautionary tale from the near past we shall revisit later.

: : :

Back in the days before cruise ships became living laboratories in disease ecology, advertisements for their Caribbean destinations often featured white-sand beaches juxtaposed with aquamarine waters, perhaps completing the idyllic fantasy with imagery of adoring couples walking hand-in-hand along said beaches. Yet this idealism is literally grounded in poop, as most of those white-sand beaches passed through parrotfish guts. This depositional reckoning, though, does not imply that parrotfishes purposefully beach themselves to evacuate their bowels. Instead, these sandy beaches represent post-defecation movement of sand by tides, waves, and especially tropical storms. When hurricanes generate high-speed winds, these in turn produce strong waves, which suspend and transport sediments ranging from mud to boulders, and potentially move them long distances. Hence sand that may have been dropped by parrotfishes on the far side of an offshore reef can become part of a secondary dump when carried to the shore by waves and deposited there.

Regardless, the origin story for much of the sand made in these tropical environments starts with parrotfishes. Marine biologists, many of whom are arguably more varied and colorful than parrotfishes, were the first people to make connections between coral reefs, parrotfishes, and the white-sand beaches so beloved by travel agents and tourists alike. Their studies of parrotfish behaviors started in the 1960s and have persisted well into the first two decades of the twenty-first century as these biologists keep learning more about the essential ecosystem services provided by these fishes.[28]

As mentioned earlier, one might think parrotfishes' munching on corals is bad for the coral, but the opposite is mostly true. After all, only scrapers and excavator parrotfish species are directly feeding on live corals, and very small percentages (1–3 percent) at that.[29] Remember, these fishes want algae on and in the coral, not the coral itself. This means the main ecological service provided by parrotfishes to ensuring coral reef health is akin to "weeding." Parrotfishes, whether cropping, browsing, scraping, or excavating, remove marine algae from reefs. If algae are otherwise left unchecked, they occupy settling spaces normally reserved for juvenile corals, as well as other sedentary animals that affix themselves to reefs, such as sponges, bryozoans, and barnacles.[30] Algal overgrowths can then crowd out and overwhelm, changing these environments from coral reefs to former coral reefs. Corals also compete aggressively for space, with colonies of different species sometimes fighting over and dominating parts of a reef. Fortunately, parrotfishes discourage coral bullies by slowing their growth and spread. For example, parrotfish feeding prevents species of the coral *Porites* from taking over too much territory in reefs.[31]

Still, perhaps the most important contribution by parrotfishes to shallow-marine environments around reefs is from what comes out in the end, or rather, ends. When processed carbonate material pops out as fine-grained sand and mud in a parrotfish's feces, it becomes part of the sediment load in those shallow-marine environments. Excavator species are far more effective than scrapers when converting coral into sediment, too. For instance, a 1995 study of bioerosion rates for the heavy-beak parrotfish (*Chlorurus gibbus*) on the Great Barrier Reef estimated that an individual fish could transform slightly more than 1,000 kg (2,200 lb) of coral into sediment each year. This was more than forty times the bioerosion rate of the sediment-swallowing daisy parrotfish (*C. sordidus*).[32] A 2018 study of Maldive parrotfishes and their bioerosion also demonstrated that an excavator species there—the steephead parrotfish (*Chlorurus strongylocephalus*)—produced almost 130 times more sediment than scraper species, with individual steepheads converting more than 450 kg (1,000 lb) of coral into sediment each year.[33] Because parrotfishes swim, they also move their sand loads away from wherever they fed. This combination of scraping, excavating, and milling rock into sand and mud, then transporting and depositing it away from its place of origin, means that parrotfishes are significantly shaping their local environments. Multiply these processes by many fishes over time, and the amount of sediment produced on and around reefs is astonishing.

Perhaps most remarkably, parrotfish poop can even help maintain islands. In a 2015 study around the island of Vakkaru in the Maldives (Indian Ocean), researchers found that parrotfishes produced about 85 percent of the sand around this island.[34] Vakkaru and the 1,200+ other islands of the Maldives are coral reef islands, in that they owe their existence to nearby reefs. Moreover, most of these reefs are atolls, which formed around volcanoes that have long since eroded and collapsed into the seafloor. This volcanic-origin idea, first proposed by a young Charles Darwin when he was not thinking about barnacles, was then later confirmed in the twentieth century.[35] In 2016, other researchers backed up the parrotfishes-making-islands hypothesis, but with a different Maldives island north of Vakkaru, Vabbinfaru.[36]

In an ideal world, all such scientific insights about parrotfishes and their outsized ecological and geological effects should also cultivate a better appreciation for the origins of the much-admired white carbonate-sand beaches of the Bahamas, Caribbean, and many Pacific islands. Indeed, knowledge of their fish-driven genesis argues that truth in advertising for Caribbean cruises should be bolder and mention how those sands got there, highlighting the true romanticism of these beaches and their sedimental journey.

: : :

Given the important roles of parrotfishes in maintaining healthy reefs while producing sandy beaches in multiple oceans, one would think they would be revered, cherished, and protected everywhere. Alas, they are not, and their numbers have diminished significantly over the past few decades through a combination of overfishing, pollution, and the ever-worrying effects of climate change on reef health. With overfishing, people catch parrotfishes and often keep the largest excavator and scraper fishes, a removal that greatly reduces their ecological contributions.[37] Reef declines caused by human overuse, pollution, invasive species, higher ocean temperatures, and ocean acidification are all too obvious as dead or dying reefs in the Atlantic Ocean of the Bahamas and Caribbean, and even in the most massive of reef complexes, the Great Barrier Reef of Australia. This situation means that the future fate of reefs partly depends on our protection of parrotfishes and their environments.

Fortunately, scientists who care about the ocean and its life have studied these problems and worked closely with people who suggest and propose policy and management of tropical reefs. Among the measures taken by local governments and in international agreements are marine protected areas, otherwise known as MPAs.[38] MPAs can function like national parks, in which recreational or commercial fishing is limited or banned outright, along with other restrictions. Yet policies and laws are only as good as the paper (or electrons) codifying them unless incentives and enforcement are applied. Also, MPAs are not necessarily a cure-all for marine problems, as no amount of strictly enforced regulations in these areas will stop global problems, such as ocean acidification, warming seas, or plastic pollution.

Remember the 2017 study done on the 3,000-year history of the Caribbean reefs of Panama? The results showed that some of the most obvious declines in parrotfish abundance—and hence reef growth—happened in just the past few centuries, coinciding with increased fishing and other reef use and abuse from native peoples, European colonizers, and (of course) pirates.[39] Modern-style pollution associated with humans has added yet another stressor, as has the addition of invasive species. For the latter, Pacific Ocean lionfishes—represented by species of *Pterois*—spread into Caribbean and Bahamian waters starting in the 1980s and 1990s,[40] and juvenile parrotfishes were among their many prey items. Interestingly, reclusive parrotfish youngsters are less likely to be eaten by lionfishes, an argument for staying close to cover as a way to ensure survival.[41] Yet shyness around predators also ensures that parrotfishes will swim less and eat less, resulting in less algae eating, and hence less bioerosion.

In short, reefs are in trouble today owing to myriad problems, but bioerosion by organisms other than parrotfishes is related to nearly all of them. Bioerosion on a coral reef is more insidious when boring algae and sponges take over, a problem that is becoming more common with ocean acidification.[42] Unhealthy reefs with relatively few parrotfishes cropping their algae also witness relatively more echinoid bioerosion, with sea urchins overgrazing and damaging reef surfaces.[43] However paradoxical it might sound, parrotfishes help prevent these types of bioeroding by bioeroding, a sort of "fighting fire with fire," albeit underwater.

Knowing that parrotfishes are also essential for sand production in shallow-marine tropical areas, we also must think of these animals as essential workers for the building of entire ecosystems, such as submerged sandy areas around reefs, as well as beaches, and even islands. Thanks to these swimming parrot-fishes, sand made by their teeth and guts is distributed through-out shallow-marine environments behind, in front of, and all around reefs. These sands then serve as habitats for offshore burrowers—such as crustaceans, polychaete worms, burrowing sea urchins, and much more. Upper layers of offshore sandy deposits are not just burrowed, but also later moved by tides and waves.

Storms carry huge loads of these sands onto beaches, where winds sort their grains farther up the beach to form coastal dunes. Once stabilized by plant roots, dunes are protectors, slowing or diverting storm surges that can flood island interiors, especially those with cabanas.[44] Dunes are also their own distinct ecosystems, hosting not just plants, but also onshore burrowing animals, such as ghost crabs, insects, and even small mammals. Beach and dune sands are vital for mother sea turtles, which crawl across beaches to dunes, where they dig out urn-shaped nests and then deposit and bury eggs that incubate in these sands.[45] Many species of shorebirds likewise use sandy areas just behind coastal dunes for nesting, nurturing new generations of shorebirds for the future.[46] An awareness of this connection between offshore sands made by parrotfishes and the survival of endangered sea turtles and shorebirds is yet another reason for their conservation.

Finally, with the ultimate form of flooding—sea-level rise—the more sand made by parrotfishes, the better for small, isolated islands if they and their ecosystems are to stay above these rising waters. Just so long as the parrotfish can avoid threats to their continuation and abundance, and are allowed to eat and poop with wild abandon, these islands may still have a chance to keep up with the escalating waters of the future.

Chapter 5

Jewelry-Amenable Holes of Death

When you're an ichnologist, you never forget your first drillhole. For me, it was when I was in my early 20s, on a sunny Saturday morning in early summer while standing on a country roadside in Indiana. The hole looked like a tiny paper punch had perforated a fossil of what is called a brachiopod, a perfectly circular absence on the side of its otherwise flawless shell. The shell, in turn, was embedded in a layer of limestone lying at the base of a roadside outcrop. Although I was only just then becoming acquainted with brachiopods, I nonetheless knew the hole in the shell of this one was somehow different, and not a part of its original anatomy or lifestyle.

To folks who do not collect fossils, the word "brachiopod" may sound like an exotic new brand of running shoe. An etymologist would likewise find it downright confusing, as its Greek roots translate to "arm foot" (where *brachio* = "arm" and *pod* = "foot"), an anatomical mixing that makes no sense. But to people who live near the state borders shared by Indiana, Ohio, and Kentucky, brachiopods are fossils, and they are such common fossils there that nearly everyone living in this area for more than a few months has encountered one, whether they know it or not. Brachiopods were relatively small animals, ranging from the size of an apple seed to a baseball, and they look like clams because their shells consist of two valves fitted together, a bodily circumstance practically begging for people to call them "bivalves."

Although technically correct, this naming would not pass muster with any dedicated shell collector, who will haughtily inform you that bivalves are mollusks and that brachiopods are not.

An easy way to get past such amateur status is to look more closely at a brachiopod's two parts (valves) making up its shell. Most bivalves have two equally sized valves, each a mirror image of the other. In contrast, brachiopods have one valve noticeably smaller than the other; the lesser half is its brachial valve, whereas the greater half is its pedicle valve.[1] Before these two valves were fossilized, the brachial valve held a soft, feathery appendage called a lophophore (that's the "arm"). This body part, which you may recall is also in bryozoans, circulated seawater (and the food it contained) through a gap between the two valves, the brachiopod equivalent of a mouth. The pedicle valve, on the other hand, is where a fleshy stalk (pedicle) may have once helped it hold on to the seafloor (that's the "foot"). Those brachiopods lacking a pedicle simply relied on flat bodies to keep them on the seafloor surface.

As I held that punctured brachiopod, I thought of its life, but also of how I owed my personal discovery of it and its hole to its former seafloor home. My sole reason for being on an eastern Indiana roadside that Saturday morning was because of the outcrop of grayish limestones and shales there, strata formed from sediments in a shallow inland sea about 450 million years ago. These rocks represented the halcyon days of the Great Ordovician Biological Event (GOBE), when animal diversity rose dramatically as new species spread throughout the world oceans.[2] The very small sample of that biological revolution expressed themselves in the outcrop as neatly stacked layers, one atop the other like a pile of unread books next to my bed. The limestones were harder and less weathered than the shales, so these strata stuck out of the roadcut, whereas the shales were more withdrawn and sandwiched between limestone beds.

All sediments composing these shales and limestones were originally deposited in warm, shallow, tropical marine environments, and their different characters reflected alternating episodes of quietude and terror, respectively. Many of the shales told of vast areas of mud accumulating in bays, perhaps concentrated as sand-sized pellets made by filter-feeding animals like brachiopods, bryozoans, and crinoids. In contrast, many of the limestones were storm deposits, formed by waves generated by Ordovician hurricanes.[3] Today, fossil enthusiasts in the Cincinnati area are thankful for those past storms. Such disturbances instantly buried and concentrated whatever life lived on or in the sea bottom, and happened often enough to make the greater Cincinnati area a world-renowned place for its stunningly abundant and exquisitely preserved Ordovician fossils.

While I examined the outcrop that Saturday morning, the limestones confirmed such fossil treasures: brachiopods of many sizes and shapes; twig-like bryozoans; pieces or entire bodies of trilobites; bits of crinoid stems resembling buttons; the occasional actual bivalve; spiraled gastropods; and other forms I could not identify then, and may still have trouble doing so now. The shales also held some fossils, but not so many. The easiest way to harvest their ancient dead was to look down on the base of the outcrop, where many of the calcified animal remains and broken slabs of limestone had accumulated under the influence of rain and gravity. It was in this fossil scrap heap that I spotted the chunk of limestone holding the brachiopod with its anomalous hole, an oddity that visually separated it from all other grayish pieces.

This outcrop was the first of several I visited that summer in an attempt to study their sequences of layered rocks (strata), body fossils, and trace fossils, and later put these all together to interpret their original environments. These goals were not for mere mental exercise, though, but for completing my master in science (MS) degree at Miami University in Oxford, Ohio. The more specific goal of my research was to discern why some of the fossils—the bivalves and gastropods—were both unusually small and exceedingly abundant in some layered strata, including at this Indiana roadside. Questions I intended to answer included: Did these tiny molluscans (dubbed "micromorphs") have their normal growth stunted by poor environmental conditions? Had evolutionary processes selected for small body sizes to favor their survival and reproduction? Or perhaps they were juveniles that never had the chance to grow up before death and fossilization. These were the questions tested then and later.[4]

Because this outing was in 1983 and well before GPS navigation and self-driving cars, I needed written directions to find the outcrop, a vertical slice of bedrock just across the Ohio-Indiana border and about a 30-minute drive west from Oxford. I also could not afford a car then, so I borrowed a Chevy Suburban from the Department of Geology. It was a behemoth of a vehicle, and I imagined how its low gas mileage likely prompted every major oil company to send annual thank-you cards to the department. Yet it served its purpose, getting me there and back, while also transporting heavy samples of limestone to a geology lab where I cut, polished, dissolved, or otherwise tormented these rocks so they would give up their secrets. That morning I collected my first slabs of limestone, but also took the perforated brachiopod with me.

Despite a focus on my thesis project, I was easily distracted by fossils that seemed unusual, and inexperienced enough to lack jadedness. So I sought out my thesis adviser, John Pope, to show him this brachiopod and its perfectly circular hole. John was an expert on Ordovician brachiopods,[5] and sure enough, he had seen thousands of examples of what I thought was novel, readily identifying the drillhole as the work of a predator. "Probably a gastropod," he said without much drama, as he explained further that some modern marine snails drill into other animals before eating them. Hence this hole was probably the trace fossil of a snail that had done the same. Nevertheless, I found this bit of information perplexing, and my naïve grad-student mind struggled with it. Despite my early childhood interest in natural history, I never knew or imagined snails as killers that crept up on hapless prey and drilled them to death. Instead, the only snails I knew were slow, slimy-bottomed, scum-grazing, and garden-dwelling animals that, yes, were cute, but never ferocious.

Memories of that drillhole came back to me in 1988, which is when I first picked up shells of modern marine clams and snails on Georgia barrier-island beaches. More than a few of these shells had cleanly cut and perfectly circular holes, looking much like the Ordovician hole I had seen in Indiana. Hole diameters varied, ranging from pinpricks to ballpoint-pen pokes, but a closer examination revealed that most of these holes were wider on the outside surface of each shell versus the inside. This gave the holes a beveled appearance, shaped more like a miniature megaphone rather than like the cardboard tube of a sometimes-scarce roll of toilet paper. Later, in the late 1990s and onward, field trips with students led to more discoveries of punctured shells, where each student would ask, "What's this?" and muse how these shells were premade for earrings or necklaces. "Then you would be wearing a necklace of death," I would respond before explaining their grisly origins.

With such search images in mind, the task of finding shell-hole combos on the Georgia coast became easier, and in less than an hour my students or I could fill a collecting bag with shells bearing many variations of these holes. Yet any such sampling trip also posed a few mysteries. Most holes were beveled, but others were straight and even throughout their depth. The majority had round outlines, but a few were ovals. Some shells had incomplete holes that started on the outside but never made their way in, yet a few of those same shells may also have see-through holes. Still, all of these shells and holes told of stories that began with hunting and evasion, but most often were followed by struggle, overcoming, acquiescence, death, and consumption. In a much broader sense, these little mysterious voids also pointed to long-past struggles and evolutionary responses that alternately favored predators and prey, expressed over more than a half billion years of animal life.

And just who were these assassins of the Phanerozoic Eon, and how often did their behaviors manifest? Thanks to millions of holes in millions of shells through the ages, as well as body fossils of potential perpetrators, paleontologists have a good idea how to answer such questions. The easiest response to any given borehole is "snails," but of course that answer leads us to

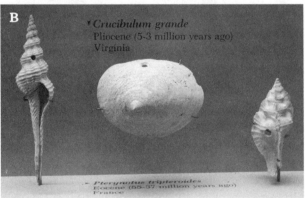

Crucibulum grande
Pliocene (5-3 million years ago)
Virginia

– *Pterynotus tripteroides*
Eocene (55-57 million years ago)
France

Predatory-snail drillholes in clams and snails. *A*, Drillholes in two fossil bivalves made by naticid gastropods in the Waccamaw Formation (~2 mya), North Carolina. *B*, Three different species of drilled fossil gastropods, with two (left and right) bearing muricid drillholes and one (middle) a naticid drillhole; specimens on display in the Smithsonian National Museum of Natural History, Washington, DC. *C*, Beveled drillhole in a modern moon snail (*Neverita duplicata*), a trace of cannibalism.

pose a more complicated query: Which snails, and when? And just how long have these murder mollusks performed their invasive operations? What exactly happens when modern snails take away the lives of other animals? Also, because nature is not always perfect, why do these predators sometimes fail to kill? Yet another concern is how to tell which holes are traces of predatory intent, and which have more innocent origins. Then our inquiries branch off in eight directions, as we depart from the usual suspects to consider alternative drillers that are still molluscan, but not snails. All such mysteries are answerable, and most by simply looking at what is not there, a small absence of shell so clearly defined.

: : :

Killing by drilling evolved at least 500 million years ago, which we know thanks to trace fossils. Fortunately, we have ichnologists, paleontologists, and marine biologists all working together to solve these fossil crimes, a super-team of scientists who have given us a few solutions to holey mysteries while raising more inquiry.

As you might recall, the geologically oldest known candidates for drillholes are in the Ediacaran fossil *Cloudina* from almost 550 million years ago. These tiny, circular holes in the thin mineralized walls of *Cloudina* do not reveal much of their makers or their intentions.[6] For one, we do not even know if they were from predators, parasites, or animals possessing *Cloudina* bodies after they died. Because these trace fossils are from near the very start of abundant mobile animal life, and most of those animals had soft bodies with low chances of fossilization, possible drillers are still unknown and perhaps unknowable. The makers of these holes also may have died with the extinction of *Cloudina* by the end of the Ediacaran. Strangely, despite the "Cambrian explosion" of biodiversity after the Ediacaran, which developed many predatory guilds, this evolutionary burst did not produce a huge suite of new drillers then: though many were there, few were drilling.[7] Otherwise the Cambrian was more of a time for attacking soft-bodied prey, and many of the attackers also might have been soft-bodied.

Enter the Ordovician. Now, *this* was a time for invertebrate life unleashed, for soft, shelled, carapaced, or otherwise-bodied animals to fill all previous niches while inventing a few more along the way.[8] For one, burrowing animals plumbed to unheralded depths and produced burrows of such complexity that we still argue about how some of these burrows were formed, and by whom.[9] For another, bioerosion became an alternative lifestyle for animals in which the harder the substrate, the better.[10] Vertical borings going down into rocky surfaces became more abundant, brachiopod shells began to bear the scars of bryozoans, and other etchings left more lasting impressions than those from mere sediment-shifting burrows. When outside assaults from marine predators grew in number and ferocity, natural selection also led to more shells and other forms of skeletonizing that aimed to defend and deter.

Alas, packing on more mineral matter was for naught, as natural selection also produced animals that pierced those defenses and left calling cards that have lasted well over 400 million years. Although the first artisans of Ordovician bioerosion likely carved out rocky domiciles to protect their soft-bodied makers, drilling represented a novel type of boring that belied protection while also representing an intention to enter and eat bodies. At first, brachiopods were the most commonly drilled animals, but with their tragic waning later in the Paleozoic, drillers switched to other prey.[11] Clams, snails, crustaceans, and echinoderms were thus added to the menus of predators that sliced oh-so-selectively, carrying on a GOBE tradition that shows no sign of abating anytime soon.

As one might expect for animal lineages that have been preying on other animals for a long time, they also coevolved, meaning that drilling did not stay constant, but fluctuated. We know this because diligent paleontologists noted and measured millions of drillholes from the Paleozoic, Mesozoic, and Cenozoic Eras, and recorded their patterns throughout time.[12] Not surprisingly, numbers of drillholes fell considerably after mass extinctions, as both prey and predator succumbed to whatever awful world-altering events happened in each. Yet the dance between the seekers and the sought resumed with each recovery of life, as the survivors either passed on their shell-penetrating abilities to next generations, or these evolved anew in descendants.

Similarly, prey animals kept changing in ways so they could live long enough to pass on behavioral or anatomical defenses through their genes. For example, gastropods and bivalves evolved thicker shells, as well as spines, ribbing, corrugations, and other physical barriers to discourage predators from accessing their innards.[13] Size variations also helped, as prey animals could have either been so small that predators passed them by, or so large that they took more energy to fight than calories gained from a successful kill. But no matter how evolved, drilling predators persisted and prevailed, and will likely continue well into the future.

: : :

Despite this rich fossil record, there is no time like the present when talking about drilling snails. Modern predaceous marine snails that drill into shells are represented by two clades, Naticidae and Muricidae, which collectively consist of almost 2,000 species.[14] Naticid ancestors probably began drilling in the Jurassic at about 200 MYA,[15] but similar holes in many animals from the Paleozoic tell us that previous gastropod lineages also evolved identical means of preying. This is a deadly example of convergent evolution in which similar adaptations arise in different lineages. Although paleontologists are not exactly sure which gastropods made Paleozoic drillholes, they suspect that species of platyceratids—which lived from the Ordovician through the Permian (about 450–250 MYA)—were responsible for boring into brachiopods, crinoids, and other hard-bodied animals.[16]

Let us start with naticids. Naticid shells have two striking aspects: beautiful patterns of stripes, spots, or swirling mixes of green, brown, purple, or yellow in harmony with their coils, and overall roundish shapes. Look closer at that shape, though, and you will see it is slightly squashed in its vertical dimension, like a slightly deflated beach ball. Shell collectors in the southeastern US nickname these gastropods "moon snails" for their circular profiles, whereas others call them "shark eyes," because darker colors concentrate around their tops, lending to a pupil-like appearance.[17] Unlike many other marine snails with shells that coil into high spires or have knobby and spiky ornaments, moon snails are exceedingly smooth. Given their distinctive shapes and textures, once you have held enough moon snails, you can identify them with your fingers. Moon snails range from lentil- to tangerine-sized, but most shells I have seen are marble-sized.

Live moon snails are a sight to behold. Yet they are difficult to see in the wild, as they mostly live in sandy offshore environments or in the lower parts of beaches. They also conceal themselves by burrowing, nearly always keeping a layer of sand between them and the outside world. So the best time to see a living naticid is at low tide, and the easiest way to find one is to look for it at the end of its burrow. Naticid burrows manifest on beaches as curving sand ridges with round mounds at one end, looking like greatly elongated and enlarged commas.[18] Once you have spotted a few of their burrows and developed a proper search image for them, they seem to multiply. Given such experience, I have seen dozens of such burrows during Georgia beach strolls, their widths varying in accordance with the size of their makers. The identity and size of their burrowers is also easy to test by gently pushing fingers down into the sand beside the mounded end of the burrow, curling fingers underneath, and lifting the snail, but with ample amounts of sand around it. Such careful efforts are rewarded with a live moon snail, its smooth shell enveloped by a thinly applied extrusion of its fleshy foot and slimy mucus. This covering by foot and mucus is how the snail keeps its shell smooth and unscratched, supplying protection against quartz sand.[19] The mucus-secreting foot further lubricates the moon snail's way as it extends ahead of the rest of its body, anchors in the sand, and then pulls the rest of the body forward, one expansion and contraction at a time.

Human discoveries of shallowly burrowed moon snails are

made possible because they are often trapped by desperate circumstances. When tides drop and expose tidal flats, moon snails living on them must conceal themselves. Although a few snails seek prey by gliding along tidal flat surfaces under open skies, most of them burrow, a sensible way to deny easy meals to crabs or shorebirds while also protecting precious bodily fluids. Still, whenever these snails burrow, they are also hunting. In fact, my PhD adviser, Robert ("Bob") Frey, out of sheer admiration for moon snails' predatory supremacy in subsurface realms, often referred to them as the "lions of the tidal flat." As a moon snail hunts, it detects potential prey via chemicals or sounds traveling in pore waters between sediment grains.[20] Once these chemical cues or vibrations are detected, the moon snail turns toward its source and keeps moving until it finds something hard in the sand along the way, such as a clam, another snail, or other animals. It then surrounds the encountered object with its foot, wrapping itself around its prey to immobilize it.

With its imminent meal properly enfolded, a moon snail secretes acid in a small spot on the shell to soften it.[21] What it does next is what it does best, which is to drill, an action accomplished by its radula. Like chitons and many other gastropods, each moon snail has a mineralized radula attached to a proboscis, which helpfully includes its mouth and esophagus. But unlike most other mollusks, moon snails do not use their radulae to scrape algae or otherwise graze from horizontal surfaces. Instead, its radula is its personal file, an anatomical tool used to scrape against a specific spot, down and past its prey's shell or carapace and inside to where the good stuff resides. For this purpose, it is applied precisely, turning right, left, right, and back again, but in tight semicircular patterns reminiscent of "wax on, wax off" hand movements. The rotational scraping of the radula is applied at a high angle, producing a hole wider on the outer surface of the attacked animal and narrower on its inner surface. This breaching of shell defenses—which may only take a few hours—is soon followed by the proboscis, which the moon snail inserts through the hole. The prey is then eaten from the inside

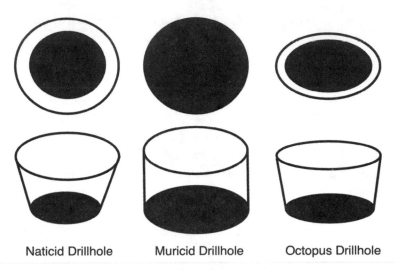

Naticid Drillhole   Muricid Drillhole   Octopus Drillhole

Differences between shapes of naticid, muricid, and octopus drillholes in shells, with external view (above) and perspective view (below), and black indicating drillhole bottom. Drillholes are not shown at the same scale, as octopus holes are often significantly smaller.

while still alive, until it is not.[22] The moon snail uses its radula to rasp the interior, a macabre way of cleaning its plate that leaves behind an empty shell bearing a single hole.

Although naticids are often associated with holes that people find in clam and snail shells, these predators also expand their palates to crustaceans, such as crabs; even ostracods (tiny bean-like crustaceans) have suffered the same fate as their legless companions by dying at the feet of naticids.[23] How do we know? All animals bear the same characteristic beveled trace, an instant spoiler alert ruining any suspense building to a reveal of the perpetrator. Any good ichnological detective can thus pick up a modern shell with such a hole, glance at it, and say confidently, "Yup, that's from a moon snail." Ichnologists assign the ichnogenus name *Oichnus* to such trace fossils, and when cone-like in form, they are more specifically called *O. parabaloides*.[24]

Alongside naticids in shallow-marine environments throughout much of the world are muricids, sometimes called "rock snails," "murex snails," or far more revealing appellations, such as "oyster drills." The majority of these predatory gastropods are high-spired, making them taller than wide. Most are also far more ornate than smooth-shelled naticids, sporting prominent knobby or spiky projections. For example, one species of tropical muricid (*Murex pecten*) has more than a hundred long, straight to curving spikes along its length, inspiring the nickname "Venus comb."[25] Such *Mad Max*–style accoutrements are clearly for protection against other predators (including other muricids), but also keep these snails above muddy or sandy surfaces; unlike naticids, most muricids do not burrow when hunting. At first, naming these snails was far too easy, as nearly all of them were assigned to the genus *Murex*. Biologists later sorted out other genera, though, identifying more than 1,500 modern species; paleontologists have so far also identified more than 1,000 fossil species.[26] Their clade (Muricidae) started in the Cretaceous about 120 MYA, leaving plenty of shells during their evolutionary history, but also plenty of holes differing from those of naticids.[27]

Like naticids, muricid gastropods also strike radula-induced terror in their prey by latching on to living targets and drilling in. Their holes, though, are not beveled, but maintain an even width throughout their length, looking more like cylinders than cones.[28] This shape shows the angle of attack by a muricid is more vertical (up and down) than that of a naticid, implying that the radula moves in a tighter circle. Muricids also are quite selective, asking themselves "to drill, or not to drill?," as some may find slight gaps in shells that allow entry. After all, why pierce a wall if a door is already open? Poisons sometimes aid entry, with muricids releasing toxins that relax a prey enough to let down their defenses.[29] This chemical warfare hence results in less drilling, meaning that muricid kills do not leave as much lasting evidence as naticids.

Muricids' distinctive drillholes show that they definitely prefer eating other molluscans (bivalves and gastropods), but also go after barnacles and echinoids.[30] So a jointly owned naticid-muricid restaurant would include a long list of appetizers and entrées. Nonetheless, any muricid or naticid dining in such a restaurant for the first time might feel a wee bit uncomfortable once they see their own name on the menu, and then notice other diners of their species either staring hungrily at them or gliding across the floor toward their table, radulas ready.

: : :

Cannibalism in our own species is fortunately quite rare, although it happens in other animals more often. Black widow spiders (various species of *Latrodectus*) and praying mantises (more than 2,000 species) are perhaps the most infamous examples in the animal world of eating their own, with females inviting males over for mating and dinner, followed by the females dining alone. Yet some vertebrates also partake in same-species courses, including many species of fishes, as well as African lions (*Panthera leo*), in which male lions kill and eat the offspring of rival males.[31]

Add moon snails to lists of invertebrate cannibals, a reminder announced by every moon snail shell that also bears the hole of another moon snail. Granted, these gastropods are gastronomically broader when consuming their species, not limiting themselves to one sex: after all, when you are a hermaphrodite and a cannibal, everyone can be a mate and a meal. Still, this mode for satisfying multiple urges that also kills within your own species seems contrary to evolutionary theory. A commonsense perspective is that if a species regularly eats its own, it also may be contributing to the demise of that species. This is not how evolution works, though, because as long as the number of individuals eaten in a population is relatively low and its reproduction rates are high, a cannibalistic species will persist and perhaps even thrive.[32] The fossil record of naticid drillholes within naticid shells—extending back to the Cretaceous—also supports the notion that cannibalism has been a part of their evolutionary history for quite a while.[33]

So, although cannibalism in naticid and muricid gastropods is natural, it nevertheless provokes two questions. First of all, why would a moon snail eat its closest relatives? Second, how could a predator succeed in attacking and eating the very same predator? Partial answers to both questions emerged from a 2014 article about the tiger moon snail (*Notocochlis [Natica] tigrina*) of coastal India. In this study, done by paleontologist Devapriya Chattopadhyay and her colleagues, they combined lab-bound experiments with field observations to test how moon snail populations and food resources might influence cannibalism.[34] What they discovered was that even when bivalve prey was plentiful, some moon snails still drilled into other moon snails. However, these attacks were selective, as smaller moon snails were far more likely to drill into and eat bivalves than each other, whereas bigger moon snails eschewed clams and instead chewed their smaller relatives. The researchers made both of these deductions based on traces, as the smallest holes in the moon snail shells were notably wider than holes in bivalve shells. Because drill-hole width correlates with body size, the researchers concluded which moon snails were eating other prey, and which ones were keeping it in the family. The authors proposed that population density was another factor in whether cannibalism happened, as plentiful younger moon snails (i.e., "Millennials") meant more were available as food for less abundant and older gastropods ("Boomers").

Another study on tiger moon snails (*N. tigrina*) published in 2016 confirmed that yes, cannibalism happened, but more often than previously thought, including in places where alternative prey—such as other gastropods and bivalves—was available. The researchers, led by paleontologist Arijit Pahari, collected clam and snail shells from the Chandipur tidal flat of eastern India over three years (2013–2015), and calculated a "drilling frequency" based on number of drilled specimens versus not-drilled for each species.[35] Based on this frequency, they found that almost 60 percent of *N. tigrina* shells had evidence of being killed by fellow *N. tigrine*. Why? Pahari and his colleagues proposed that these moon snails were eliminating their competition while also gaining more sustenance from their own species than from others. Such gruesome results led to the scientists opening their paper with a memorable sentence: "Chandipur intertidal flat in eastern coast of India is a killing field." Indeed it was, and still is. Naticid drillholes thus help evolutionarily minded folks to reconsider the supposed rarity of cannibalism and its disadvantages, while also continuing to ask why this seemingly self-destructive behavior persists in some species.

: : :

Snails are not the only molluscans that snuff out lives via drilling, and few animals, whether invertebrate or vertebrate, are as clever with their killing as octopuses. Octopuses are represented by about 300 species, most of which live in warm, shallow-marine environments throughout much of the world. They belong to one evolutionary branch of molluscans, Cephalopoda, which also includes nautiloids, ammonoids, cuttlefishes, and squid.[36] Cephalopods in general have a long lineage beginning in the Cambrian, but octopuses did not make themselves known as fossils until the Late Carboniferous Period, at about 300 MYA.[37] Octopus- or squid-like animals are the stuff of human legends, ranging from seafaring tales of krakens—which seemingly achieve peak destructiveness upon release—to Cthulhu, the Great Old One created by famed racist (and author) H. P. Lovecraft, to Japanese erotica.

Understandably, many books have been written about actual octopuses, octopodes, octopi, or whatever your preferred plural might be for these marvelous eight-armed animals. But perhaps we should ask them which label they prefer. After all, octopi are regarded as the smartest of invertebrates, easily solving whatever problems we primates assign them. Among their many feats are ingenious escapes, such as when an octopus dramatically left an aquarium for a nearby ocean.[38] Octopi also can learn (and remember) how to navigate mazes, unscrew jars, use tools, and apparently predict World Cup matches. Examples of octopus tool use include their arranging shells as barricades around their dens, or placing their bodies in coconut shells as they casually walk away.[39] If the latter sounds more like play than defense, you may be right, as biologists have also tested for and found compelling evidence of octopus frivolity. Most recently, a 2021 study provided intriguing evidence of octopus dreaming, in which the researchers documented color changes that lasted nearly a minute during octopus sleep cycles.[40]

Given their reputations, an octopus drilling into another animal for food might seem like just another item in a long list of skills. Yet this ability is often used as a last resort, applied only when previous methods fail to subdue their prey and leave an octopus no choice but to puncture a shell or carapace. For example, let us say a hungry octopus encounters a luscious bivalve on the seafloor and decides it will do nicely for lunch. It embraces the clam by wrapping eight sucker-laden arms around its shell and trying to pull apart its two valves. If the clam tires and a gap appears between the valves, the octopus slips inside and begins eating. But if the clam is too resistant to this tug-of-war, the octopus has a backup plan: although it has a beak for feeding, it also has a radula for drilling. So the octopus manipulates the clam's body to place the shell near the octopus's radula. The radula grinds against the shell, but in a small oval spot that is typically less than a millimeter wide at its top and narrower at its bottom.[41] After a few hours, the octopus reaches the inside surface, stops drilling, and injects poison through the hole. The poison, which comes from a salivary gland conveniently located near the radula, causes paralysis and eventually death in the bivalve.[42] Either way, the clam relaxes and opens up its valves, involuntarily inviting the octopus in for a fresh meal.

Among the prey items drilled and poisoned by octopi are crustaceans (crabs and barnacles), bivalves (clams and oysters), gastropods, and—in a twist of cephalopod betrayal—nautiloids.[43] The only trace that any of these animals were attacked and killed by an octopus is a tiny oval to roundish hole, or two, or more in their shells. The locations of these holes over inner muscles of their prey also speak to a precision unseen in most other drillers. The distinctive sizes, shapes, positions, and greater numbers of these holes on shells thus distinguish these holes from those of predatory gastropods. And unlike those gastropods, octopi are scientists, so multiple drillholes are not signs of ineptness, but of experimentation and learning from their mistakes. Using their sensitive arms as extensions of their nervous systems and memory, they find the right place to drill, and presumably do the same for the rest of their lives, which may be only a year or two.

Ichnologist and paleontologist Richard Bromley—who studied octopus drillholes in modern and fossil shells—thought these drillholes were distinctive enough to assign their own ichnospecies name, *Oichnus ovalis*.[44] This splitting of the ichnogenus *Oichnus* was meant to help paleontologists distinguish between drillholes that were probably made by naticids, muricids, or octopi.[45] By applying these criteria to fossil shells, we now know octopi have been drilling into clams, snails, and other animals since at least the Late Cretaceous (about 75 MYA)[46] and well into the Cenozoic Era, including now.

Bromley also referred to octopi as "taphonomic losers," which was not said out of scorn but from recognizing how these soft-bodied animals are unlikely to fossilize.[47] Sure enough, their body fossils are extremely rare, but their trace fossils are far more common, giving us the only clues of an octopus presence for much of the fossil record. Nowadays, octopus drillholes in shells can also be used in biogeographic studies, helping marine biologists detect an octopus presence where none have been since previous times of global warming.[48] Indeed, as climate change continues, biologists anticipate that octopus drillholes will help them map range extensions of their makers into warmer waters.

Did Cretaceous octopi also dream, and if so, of what? Did they visualize opening gigantic clams and feasting on their interiors, or did they experience nightmares of attacks by huge marine reptiles? We might never know, but their drillholes reveal octopus thoughts and decisions made from their evolutionary past that continue through today.

: : :

One of the more common misconceptions about evolutionary theory is that natural selection results in perfectly tuned adaptations that never result in errors, leading to clichés that predators are "killing machines." Yet even the most touted of top predators make mistakes that would encourage death by embarrassment if these animals felt such shame. Also, evolutionary processes have molded responses in both prey and predators over time, with defenses matching tactics and vice versa.

As mentioned previously, drillholes left by naticid, muricid, and octopod predation are not just random, but also inform us of tactics evolved over millions of generations. In clams, the holes are most often positioned close to where their two valves hinge, a bump called an umbo. Not coincidentally, this is also where muscles hold the two valves together.[49] Similarly, holes in gastropods are more often toward the top spire of gastropods, meaning that these predators also found a place that attacked musculature and helped open up shells to allow entry.[50] Paleozoic brachiopod shells show a similar distribution, with most clustering near the hinge that held their two valves together.[51] How do we know this? Because paleontologists have mapped the distributions of hundreds of thousands of holes in hundreds of thousands of shells, generating robust statistical probabilities showing where attacks were most effective.[52] Similarly, incomplete drillholes, which reflect failed attacks, are most often farthest away from the umbo in bivalves, and half holes in gastropods are closer to their lips.[53]

One sign these predators make mistakes is when they fail to follow the schoolyard advice of "pick on someone your own size," an admonition that especially applies to moon snails preying on their own. As mentioned before, drillhole width is also directly correlated with driller size, so we can directly measure body size of the prey and infer the size of its predator in the same specimen.[54] In these instances, most complete drillholes indicate their makers were larger than those they killed, but the opposite when drillholes were incomplete. However, if a prey is *too* small for a hungry snail to bother eating, then small size may actually help it avoid attack.[55] Competition is also a factor, such as when a snail fights another snail for its prey, stopping it mid-drill and leading to *thanatos interruptus*. Regardless of cause, though, these holes inspire feelings of triumph on behalf of their former molluscan owners for escaping certain death, but complete holes in the same shell inform of later predatory success.

Perhaps the most self-owning of mistakes a predatory snail

could make, though, is to not just drill into something already dead, but also into something unrelated to any of its prey. Yes, sometimes naticids and muricids go after nonliving materials and mistake them for food. Evidence of these predator failures comes from gastropod shells with more than one complete drillhole. Two or more complete holes tell us that either the dead came back to life only to be killed again (a horror movie cliché), or a predatory snail mistook an already-dead snail for a live one. But surely the silliest of predation errors among moon snails was when multiple snails tried eating fish ear bones. In these bones—long since separated from their original owners—are the distinctive drillholes of moon snails that must have mistaken their shapes for those of small clams.[56] To the moon snails' credit, the drillholes only went partly into the bones, meaning they stopped drilling once they realized they were just exercising. Still, one can almost imagine them afterward slinking away, hoping none of their relatives saw them: I mean, cannibalism is fine, but fish ear bones?

Illustrations of shell outlines in which researchers mapped complete drillhole locations thus inform us of optimal killing tactics used by drilling snails since the Mesozoic. But such maps also remind me of survivorship bias in humans. Survivorship bias is what happens when the survivors of a calamity are the ones telling tales about it, whereas those who did not survive cannot say what went wrong, because their voices have been silenced. Of course, scientists mostly study the remains of animals that did *not* survive, and nearly every hole that goes completely through a shell is a trace of death. Even incomplete muricid drillholes can represent fatalities, as those gastropods may have released toxins and stopped drilling once their prey stopped resisting. Yet the mapping of complete drillholes more often tells of terminal efficiency, similar to the military expression "one shot, one kill."

∴ ∴ ∴

Among the arguments arising from the study of drillholes in shells is how to separate predator-inflicted damage from non-predator-inflicted damage. For instance, if I am with my students on a field trip to the Georgia coast and one of them finds the shell of a drilled clam, I will inevitably launch into gleeful descriptions of the animal that did it and the gruesome methods involved in drilling. Yet sometimes an unimpressed student will respond to my tale with "How do you know it isn't just a hole from something else?" This is a quite reasonable inquiry, and I thank that student for practicing skepticism while also providing an opportunity for hypothesis testing, as all good scientists should do.

So we will now consider other possibilities for holes in shells. For one, instead of predation, how about ordinary wear and tear caused by waves moving already-dead shells in the surf? Sure enough, surf-inflicted damage can cause holes to develop in clamshell umbos, next to where the two clam halves were hinged together.[57] Add a mineralogical factor, such as the difference in hardness between aragonite shells (a mere 3 on the Mohs hardness scale) and quartz-rich sands (hardness of 7), and such erosion is more likely, similar to vigorously rubbing sandpaper across those shells. Still, when considering the tremendous variability of wave energy, time exposed, shell thickness, and many other factors, these holes should have a broad range of shapes and sizes, bearing little resemblance to honest-to-gosh drillholes.

Another possibility is that the holes are of biological origin, but made by animals that excavated cylindrical homes in already-dead shells. This alternative explanation was applied to at least a few beguiling holes in Ordovician brachiopods that penetrated their shells and were about the right size for predatory drillholes.[58] The clue showing these borings were postmortem rather than dealing death, though, was how they passed through shells and continued into the underlying rock, which does seem like overkill. The non-predatory nature of these holes is further supported by their random locations on brachiopod shells, whereas lethal attacks would have been concentrated in specific places.

Yet another biological cause of a drillhole is not from predation, but from parasitism. The difference between these two behaviors is literally life and death, as parasites allow their hosts to live just a bit longer, however miserably. Parasitism in fossil or modern shells is indicated by repeated and similarly sized drillholes on the same victim, especially if the holes show signs of healing, meaning the animal was alive after each attack.[59] Think of how numerous bites by chiggers, ticks, or mosquitoes on humans usually heal, and unless these parasites are carrying diseases, they do not kill their hosts. So if skeletal tissues show signs of growing around holes, then these traces together are more likely the handiwork of parasites that settled in for a prolonged meal, rather than a quick drill and dine.[60]

Of course, we also think of humans, the most anthropocentric species of all, by asking ourselves, "Could it have been us?" The disbelief expressed by my students that these holes came from something as simple as a snail is also linked to human conceit, that only we could have carved such perfectly circular or oval perforations in bivalve or gastropod shells. This reasoning is further encouraged by the placement of the holes, so aptly suited for the passage of strings or other cordage and allowing them to be worn on necklaces. Surely these relate directly to us, and to our culture. Archaeologists have indeed documented that our modern penchant for personally adorning ourselves with the remains of dead molluscans is nothing new for *Homo sapiens*, and for at least two other species of humans, *H. neanderthalensis* and *H. erectus*.[61] But we also know all too well that holes in shells could have been prefabricated, with humans exploiting unpaid invertebrate labor before stringing them.

In 2017, zoologist Anna Maria Kubicka and her colleagues published an exhaustive survey of reports from archaeological sites in Europe, Africa, and Asia of molluscan shells with holes, which were presumably used for human adornment.[62] The researchers then compared archaeologists' descriptions of these holes and their locations on shells to those made by predatory molluscans, as described by biologists and paleontologists. In this study, they tested the archaeological assumption that human-caused holes were far more regularly placed than those of predatory mollusks, which archaeologists assumed were random. Kubicka and her colleagues found that the placement of human-caused holes on shells and types of shells was actually similar to those of predator-inflicted holes.[63] For example, although piercings made by humans sometimes coincided with the molluscan species of predatory snails, most snail drillholes were concentrated on the umbo areas of bivalves, or high on the spires of gastropods. How to tell them apart? Humans more often used a tool to strike a shell, rather than the precise and methodical drilling of a radula.

Given a greater awareness of how both archaeologists and paleontologists can be wrong when trying to identify drillholes, and when armed with a checklist of criteria, we can be fairly sure how to discern predatory drillholes from wear marks, non-predatory borings, parasitic traces, and human alterations. Accordingly, paleontologists can further attest that the majority of single circular to oval holes in fossils from the Paleozoic, Mesozoic, and Cenozoic Eras represent instant fatalities, rather than other types of before-death and after-death experiences.

: : :

Let's say you're a marine snail and you've been killed slowly by a naticid or muricid gastropod, or more quickly by an octopus. You die, and depending on your life choices and invertebrate beliefs, you either go to gastropod heaven, hell, or just the fossil record. Before that, though, your dead shell may become reanimated: not in a zombie sort of way, but it does involve walking, as the usurper of your shell has legs. Congratulations, shell: you are now the proud mobile home of a hermit crab.

Still, hermit crabs can be picky. In fact, they may very well decide not to move into a home with a drillhole, and instead pick shells with no holes in their walls or roof. Are these hermit crabs superstitious, refusing to live in a home that may still harbor the ghost of its former owner? Does this discriminatory behavior also mean that some hermit crabs are actually ichnologists, detecting predatory traces and passing on the word to other hermit crabs as real-estate advice? The answer is actually more complicated, and related to natural selection.

Hermit crabs have a long, fascinating evolutionary history, starting early in the Jurassic (about 200 MYA) and continuing through the Cretaceous, with at least some surviving the mass extinction at the end of that period (66 MYA).[64] Fossil hermits' proclivity for occupying dead and abandoned molluscan shells also began in the Jurassic, with one specimen found in the coiled remains of an ammonite.[65] Modern hermit crabs—which are not "true" crabs but are related to them—are represented by nearly 1,000 species.[66] Almost all live in marine environments, but also include some terrestrial species, including the heavyweight champion of modern land invertebrates, the coconut crab (*Birgus latro*). Owing to their soft and vulnerable bodies, most hermit crabs depend on dead gastropod shells for protection. These shells they switch out often as they outgrow them; they may also, for other reasons, lead to fierce competitions and fights for the "right" shells that make the people on *House Hunters* look like amateurs.[67]

Given such shell-inspired antagonism, one might think that a hermit crab would be fine with picking a gastropod shell with a drillhole in it: after all, it's just a little hole. Also, shell-drilling predators might leave such shells alone, sensing that they contain scrawny hermit crabs rather than delectable molluscan flesh. But evolution had something to say about this, particularly in the coevolution of hermit crabs with other predators. These predators did not drill, but instead took advantage of the structural weaknesses in shells imparted by drillholes. Such shells were crushed, killing their squatters and any living relatives of the shell's original owner. This in turn led to the hermit crabs of today that tend not to choose shells with drillholes, no matter how well they fit.[68]

So just who were these shell crushers, which are far more brutal than drillers in terrifying their prey? Some were crabs, and some were not, but what they all had in common was an appetite for shell destruction.

Chapter 6

Super Colossal Shell-Crushing Fury!

Few animals were as lucky as the whelk in my hand. For one, despite its lifelessness and hollow interior, the weight and size of its shell—spanning across the length of my fingers and palm and projecting its pointed bottom atop my wrist—told of many years lived. When looked at from its top, I could see that life history spiraling away from an apex that spoke of its beginning. From these first few whorls, when it was a mere protoconch, the whelk grew out and around, its shell accreting clockwise as its soft innards enlarged in tandem with this protective covering. Its interior was bright orange, a color repeated in alternating patches down the length of its exterior. These and other traits helped identify it as a knobbed whelk (*Busycon carica*), the official state shell of Georgia, which was geographically apt, as I was on a Georgia coast beach of Sapelo Island.

Still, the whelk's life was not free of drama, with moments of peril recorded in its shell. A ragged, raised line ran along its length on one of the outer whorls, an interruption of its smooth exterior that reflected a violent encounter with another species in its past. Just past this demarcation was a second, and a third beyond that. After the third irregularity was the whelk's lip, the last growth of its shell before it died. Although I could not tell how it passed, these three lines were irrefutable testimony that it was attacked thrice, yet escaped each time to heal, grow, and live longer. In between those assaults, the whelk caused its own

Shells with repair scars from failed attacks. A, Lightning whelk (*Sinistrofulgur perversum*) with three repair scars (arrows) from attacks; Sapelo Island, Georgia. B, Knobbed whelk (*Busycon carica*) with prominent repair scar that was probably inflicted by a crab and smaller borings from sponges and polychaetes made after it died; Edisto Island, South Carolina, scale in centimeters.

breakage on various clams, using the outer lip of its shell to pry them open and consume their soft interiors. But these were mere chips taken off the edges of shells, and self-inflicted breaks along the whelk's lip did not record nearly as much damage as those from outside attacks. The rough, crooked lines running down the length of its outer whorls spoke of an attacker that grasped the whelk's shell and tore at it like an eager child opening a wrapped gift. So many similar patterns recorded on the shells of whelks and other gastropods told of commonplace horrors, life-and-death struggles we might never witness. Also, whatever did this, it had claws.

On that same beach, and seen most often during low tides at dawn, were parallel sets of four-by-four impressions, a stitching made by eight pointy feet in the wet sand there. Backtracking took me to the origins of these patterns, circular burrows in dunes just above the beach summit, some with freshly heaped sand outside. Shallow digs interrupted the trackways in the recently submerged sand; these little pits sometimes held broken multicolored bits of dwarf surf clams (*Mulinia lateralis*). With each lowered tide, these fingernail-sized and thin-shelled clams burrow down into the sand to hide themselves from predators, a tactic that failed against this particular one. Whatever did this, it likewise had claws, and it could dig.

Repeated visits to Sapelo revealed other attacks on shell-bearing animals. For instance, whenever the intertidal sand flats of Sapelo Island are exposed at low tide, broad, circular to oval depressions interrupt these otherwise rippled surfaces. The part of my memory occupied by dinosaur tracks often equates these depressions with sauropod-dinosaur footprints, especially if they align and alternate in diagonal patterns. Yet these were excavations made by animals without legs or shovels. The signs included finely ground shells, further showing how their makers were searching for burrowing clams and other invertebrate foodstuff, then shot high-pressure streams of water into the sand to uncover and chew their prey. Whatever did this, it could spit, and it could crunch.

On that same Georgia-coast island, more broken shells informed of takeout meals that involved molluscan flight, with both gastropods and bivalves briefly falling from above and losing a bet with gravity. In some instances I found the broken shells of giant Atlantic cockles (*Dinocardium robustum*), a stout, thick-shelled clam resistant to outside forces but with a fleshy interior that dared other animals to test it. Other snail and clam deaths did not involve gravity, though, but percussion, in which pointed or bladed implements hammered the sides of snails or clams until stress limits were exceeded and shells shattered. Whatever did this, it could fly, and it could peck.

Although all of these modern molluscans had shells as self-protection and were connected to lineages reaching back to the Cambrian that likewise had shells, predators between then and now found ways to literally shatter those defenses. Nothing about these traces were as subtle as a small drillhole, but instead told of sudden violations of once-safe bodies.

: : :

Shell-crushing behaviors evolved very soon after the advent of shells in the Cambrian (about 500 MYA), manifesting as yet another strategy during the "marine predation revolution" of the Paleozoic that continued well into the Mesozoic and to now.[1] Among the first shell crushers were marine arthropods; later in the Ordovician, vertebrates applied for membership in this shell-breaking club. These new crushers were eventually represented by an array of fishes, reptiles, birds, and mammals.

One way to understand how Cambrian animals broke shells is to study living animals with the most primeval of visages, horseshoe crabs. Horseshoe crabs, also known as limulids, are modern animals, but ones that would not look terribly out of place on a Cambrian seafloor. For those unaware of horseshoe crabs, these marine arthropods have been around since the Ordovician (more than 450 MYA) and are represented by four modern species, three in Asia and the American horseshoe crab (*Limulus polyphemus*) of North America.[2] Bearing broad head shields reminiscent of *Star Wars* bounty hunters, these animals live in shallow-marine waters and eat small invertebrates, including clams and snails. Limulids are also the only marine invertebrates that have sex on the beach, and by the millions. These orgies take place in May–June on the eastern coast of the US, as larger females crawl out of the water to lay their eggs in beach sands and smaller males also crawl up to fertilize these eggs, or at least try.[3] Have I actually watched this mass mating in action? Oh yes, and in the steamiest place possible for marine invertebrate voyeurism: Delaware.

Like many amorously adventurous animals, horseshoe crabs eat with their legs. Underneath a horseshoe-crab head shield (opisthosoma) and rear end (prosoma) bearing a pointy tail (telson) are its gills, legs, and mouth, where they do their breathing and feeding. Of their six pairs of legs, the rear five pairs in limulids (a pair of pedipalps and four pairs of ambulatory legs) are primarily for walking. But the first pair of legs (chelicera) grabs and moves food toward the mouth. Most of the walking legs then supply chelicera with already-broken food. This breaking is made possible by gnathobases, short and stiff bristles high up on their legs that shred any small living things in the sediments below them.[4] To better understand this process, imagine getting down on your hands and knees and attaching scrubbing brushes to the insides of your legs. With these brushes, you may use them to break down potato chips, crackers, or other hard food by rubbing your thighs together, followed by scooping up the broken foodstuff with your hands and shoving it into your mouth. And just so long as you do this gnathobasic activity in the privacy of your home, why not?

Biologists have known about horseshoe-crab feeding for quite a while, but paleontologists only recently started applying these lessons to Cambrian animals, such as trilobites and other marine arthropods. Once these scientists began looking closely at the fossilized legs of these arthropods, they found bristles up high on these appendages and close to their mouths. But because paleontologists cannot reanimate the dead (yet), they had to study the functional worth of these gnathobases with computer models.[5] For this, paleontologist Russell Bicknell and his colleagues used computer modeling to examine the legs of three Cambrian arthropods and modern L. polyphemus in two separate but complementary studies published in 2018 and 2021.[6] The researchers inferred that Cambrian arthropods' gnathobases responded differently to stresses, which implied they ate different foods. For example, the trilobite Redlichia rex and crustacean-like Sidneyia inexpectans had short spines on their legs similar to those of modern horseshoe crabs, which meant they ate hard foods, such as small, shelled animals. In contrast, the Cambrian trilobite Olenoides serratus had long spines on its legs that were too weak to break shells, so this animal must have been on soft-food-only diets.

Cambrian arthropods also supported these computer models' conclusions by providing actual fossil evidence of Cambrian shell crushers. Among the clues were trace fossils, like gut contents with shell bits in Sidneyia.[7] Sidneyia and Redlichia were not the only animals doing crunches, though, as some trilobites were bitten. We know trilobite parts went missing while they were still alive, and not from postmortem breakdowns, because some wounds were healed. This showed their owners recovered from these injuries, while also leaving impressive scars.[8]

From these leggy beginnings in the Cambrian, shell crushing changed over the next 500 million years or so to include what became the most commonly used implements: claws and jaws. Claws evolved from the forward-most walking legs in certain crustacean lineages, whereas jaws came from gill arches in the jawless ancestors of all modern vertebrates, discussed later.[9] Both claws and jaws also represent wonderful examples of levers with fulcrums, gratifying biologists and paleontologists alike who embrace their inner physicists.

For those of us whose high school physics classes are but fuzzy memories of vectors, scalars, particles, and waves, a review of levers is warranted. Levers are simple machines that perform work, but they do not necessarily need batteries or solar panels. Yet they do need three items: effort, fulcrum, and load.[10] Effort is the force applied to a lever, a fulcrum is a pivot point on the lever, and a load is the object moved or otherwise affected by a lever. Levers are split into three classes, with each based on where the fulcrum is located relative to the effort, handily listed as first, second, and third.[11] For instance, if you ever had a trebuchet exchange with a neighbor living in the castle next to yours, then you used first-class levers, as trebuchets or other catapult devices have a fulcrum between their efforts and loads. In contrast, a nutcracker is a fine example of a second-class lever because its fulcrum is the hinge between its two handles, whereas the load (the nut) is placed between it and the hands squeezing its handles. Then there are tongs, which superficially look like nutcrackers, but their loads are at their ends and hands are placed between the load and the fulcrum. Hence these salad-clutching devices are third-class levers.

Given this basic leverage, arthropod claws are better understood as second-class levers (crushing), although they can also work as third-class levers (grasping), but not so often as first-class levers (flinging). These sorts of actions do not use gnathobases, but instead apply great forces on shells, thin to thick and small to large. Although invertebrate claws reinvented shell breaking, vertebrate jaws joined in later. But first, let us talk about when the Mesozoic marine predation revolution saw the birth of a splinter group that literally splintered molluscan shells and other invertebrates' carapaces. These were the crabs.

: : :

Probably the most famous of shell crushers are crabs, feared and admired worldwide for their big, meaty claws. True crabs and hermit crabs share a common ancestor, having split from a decapod lineage early in the Mesozoic.[12] Crab-like body fossils from the Early Jurassic (~185 MYA) of Great Britain and crab trackways in the Middle Jurassic (~160 MYA) of Portugal show how they were well on their way to scuttling grandeur before the Cretaceous.[13] Also, if imitation is flattery in evolution, many other animal lineages complimented crabs by evolving crab-like body plans, a process nicknamed carcinization.[14] Crab cosplayers today include crab spiders and crab lice, but humans also cannot help but see crab-like forms manifest in other ways, from constellations to nebulae.

Have you ever just stopped to look at a crab? I have, and you should, too. Unlike their decapod kin, such as shrimp, lobsters, and crayfishes, a crab's main body (carapace) is composed of a fused head and middle (cephalothorax) and rear (abdomen) that make it wider than long. Eight jointed walking legs (pereiopods) project from this broad body, with four on each side, lending to their signature sideways motion.[15] Because these legs are pointed, most crabs walk in a tippy-toed manner, although the rear-most legs of swimming crabs have flat paddles. Crab eyes are on stalks that can rise above the body or stay flush against it in recessed areas. Also at a crab's front end and between its eyes are antennae and antennules; just below these is its mouth. A crab mouth is complicated, consisting of paired parts called maxillulae and mandibles that tear apart food before it goes down the crab gullet.[16]

But by far the most prominent parts of crabs that interact with the unwary are two claws. Crab claws (chelipeds or chelae) today are multipurpose, whether used for burrowing, wielding against prey or rival crabs, or in self-defense against much larger animals, such as fishes, birds, or curious children. Claws are composed of a fixed part (propodus) and movable "finger" (dactylus), which have short, barbed "teeth" (denticles) that grip whatever load is caught inside the lever.[17] Claws also may differ in size and hence function, with the bigger claw (superior cheliped) used more for crushing and the smaller claw (inferior cheliped) for holding or slashing. Unlike a *Tyrannosaurus*'s arms, a crab's claws actually reach its mouth, introducing a perhaps still-quivering meal to the start of its digestive passage.

As mentioned earlier, a crab claw acts as a lever, but one encased in armor and powered by muscles. Think of the hinge between the propodus and dactylus on a crab claw as a fulcrum, whereas muscles in the claw supply the effort, and a load can be a snail or clam. If a crab grabs a hapless molluscan with just the tips of one or both of its claws, this initial grab is more like using tongs, or a third-class lever. Yet as soon as the crab manipulates a shell closer to the claw fulcrum and squeezes, it switches to a second-class lever, or "nutcracker" mode. Denticles in a claw also concentrate forces into these small points, like spike heels pressing against the top of a foot.

Before breaking a shell, a crab grabs, holding its prey with both claws. Depending on the size and thickness of the shell, a crab may spin it around to find weak spots between valves or an operculum. If so, it inserts a claw tip and slices away. But if this soft approach does not work, then the crab may simply place the shell in its superior claw and squeeze. A clam or snail held closest to the hinge between the two parts of a claw is subjected to more force and pressure. When a clamshell breaks, fractures are in the middle of the valves and look spectacular, but with a smaller snail, a crab may simply snip its uppermost few whorls, giving new meaning to the phrase "take a little off the top."[18]

Nevertheless, if crabs find their shell-crushing attempts thwarted, they take a more devious approach by holding on to a shell edge where it might be thinner and start clipping. If a determined crab decides to earn its meal by cutting a shell edge, then breakage is less dramatic, but with persistence it becomes lethal. In this case, crabs grab a snail with both claws, but use one to stabilize it and the other claw to grasp the snail's lip before pulling and twisting, breaking off a small piece of shell, and another, and another.[19] Although biologists and paleontologists refer to this action as "shell peeling," it is not like peeling a banana, but more like breaking a tortilla chip into smaller parts. This bit-by-bit procedure continues until a snail's soft parts are exposed and more easily pulled out of a shell by one or both claws. If a claw is positioned at a right angle to the shell, it inflicts a V-shaped excision that leaves little doubt of predatory intention. Nevertheless, at least a few gastropods subjected to this invasive procedure manage to escape crab clutches. If so, peeling is followed by healing, with the shell building anew beyond this excision and forestalling a sense of impending doom: until the next crab attack, that is.[20]

Trace fossils of crab shell-crushing and peeling are recorded in the bodies of gastropods past, manifested as shells with ragged scars interrupting normal shell growth.[21] Likewise, fossil crab claws sometimes show they were well suited to break or peel shells. In a 2015 study, paleontologist Emily Stafford and her colleagues assigned the ichnogenus *Caedichnus* to traces in gastropod shells made by crustaceans that busted their lips.[22] Such trace fossils nearly match the history of crabs, with Cretaceous examples soon following their Jurassic origin and subsequent evolution.

Remember the whelk that lived from the start of this chapter? Considering its large size and thick shell, the list of its possible assailants is actually quite short, and only a few animals in the murky waters of the Georgia coast had both the claws and the strength to inflict such damage on its shell. The most likely candidate for whelk cracking is the Florida stone crab (*Menippe mercenaria*).[23] Stone crab claws are much stouter and hence more powerful than those of most other crabs, so much so that people who harvest these crabs for food take just one claw, then throw the still-live but annoyed stone crab back into the water. (Unlike many animals, this limb grows back.) Although other crab species live in the nearby salt marshes, their claws would never have the opportunity to grasp a whelk, which stays put in the open-ocean side of an island.

Remember the dwarf surf clams that did *not* live? These bivalves were excavated, crushed, and eaten by Atlantic ghost crabs (*Ocypode quadrata*). Nighttime visits on Georgia beaches at low tide are the best times to directly observe ghost-crab behaviors linked to these feeding traces. When the sun has set and the tide is out, ghost crabs leave their coastal-dune burrows, walk down the beach to newly exposed wet sand, stop, and dig. Many dwarf surf clams—buried just under the sandy surface—then reward this diligence. Once these little and thin-shelled clams are in their grasp, ghost crabs break them with their larger claw and use their smaller claw to deliver fresh fleshy bits to their mouthparts.[24] I have also seen empty gastropod shells outside of ghost-crab burrows representing different meals: not of snails, but of hermit crabs. Worn patches on shell bottoms show how their most recent occupants dragged the calcareous shells across harder quartz sand, but wandering ghost crabs interrupted these walking shells by snatching and extracting their occupiers.[25]

The ecological expansion of crabs since the Jurassic into marine environments from nearshore to abyssal, as well as in freshwater and on land, means that wherever crabs lived, shelled or shell-borrowing animals were in danger. So, what to do when you are a tasty treat surrounded by a shell? The solution is evolution. As crabs diversified, natural selection led to bigger clams and snails, with some also growing thicker shells with ridges, knobs, and spikes.[26] After many generations of predation, some snail species also evolved smaller apertures and thicker lips, each discouraging shell peeling. Perhaps most surprising, though, is that some snails grow thicker shells and lips within their lifetimes if they even smell crabs nearby. For instance, invasive crab species, such as the European green crab (*Carcinus maenas*) on the US East Coast, trigger shell-thickening responses in native gastropods.[27] Such rapid reactions to shell-crushing threats likewise hint at past evolutionary escalations between predators and prey.

: : :

There is more than one way to crack a shell, and mantis shrimp arguably represent a pinnacle of achievement in this respect. As fierce and fabulous crustaceans that evolved in the Mesozoic,[28] mantis shrimp totally deserve our respect and awe. Are they revered because of their incredible underwater vision, capable of discerning not only the full human spectrum from red to violet, but also ultraviolet and polarized light?[29] Are they adored for their technicolor carapaces, capable of driving Joseph and all other prophets to wail, gnash their teeth, and rend their clothing with envy? Sure, these are pertinent points, but mantis shrimp—which are not true shrimp, but stomatopods—are most famous for their ability to stun prey and fracture shells with high-speed and powerful punches.

Whereas the two front claws of crabs, lobsters, crayfishes, hermit crabs, and other crustaceans are clinically called chelipeds, the same appendages in mantis shrimp earned their own special name that leaves little doubt of their function: raptorials. The "mantis" part of a mantis shrimp's common name comes from the visual similarity between these limbs and the folded forelegs of praying mantises. However, the death-dealing activities of mantis shrimp are much more varied than those of their insect imitators, which mostly use their forelegs for grasping and holding. Instead, mantis shrimp raptorials are adapted for either spearing or smashing.[30] In this method, raptorial muscles contract, but then body parts (sclerites) act as latches, holding arms in place until this stored elastic energy is released.

So for a shell-smashing mantis shrimp—which might be only about 15 cm (6 in) long—think of its raptorials like crossbows, but loaded with hammers instead of arrows, and with hammers attached to their bodies, and there are two of them. Raptorial releases are often described with terms more appropriate for science fiction: ballistic strikes, cavitation bubbles, and implosions. The "ballistic" part comes from the speed (23 m/s, or 70 km/h) and force (~1,500 N) of the strike.[31] Mantis shrimp also cause two impacts with each punch. The second impact comes from a cavitation bubble, a low-pressure air space that appears behind the first strike because water is displaced so quickly. The implosion is from the near-instant collapse of that bubble, which in turn generates sun-surface temperatures, light, sound, and force.[32] All of these actions take place in milliseconds, from release of the raptorials to implosion, and because both raptorials are involved, four impacts happen in those milliseconds. This is all very bad news for the prey, shelled or not.

Despite these astonishing numbers, though, mantis shrimp do not simply throw one "knockout" punch, bask in glory, and dine like royalty. Instead, they hit shelled targets repeatedly to weaken and crack a shell. In a 2018 study, researchers found that mantis shrimp needed an average of 73 hits to break shells, ranging from as few as 7 to as many as 460 strikes.[33] This repetition was related to the time needed for mantis shrimp to prepare their strikes, recover, and reconsider their targeting if the first and each successive strike do not break a shell. In contrast, crabs apply force over longer periods of time and hence more easily adjust their shell-breaking strategy while still handling their prey. Experimental studies show that mantis shrimp, when given different gastropods, also change their target according to shell shape. For instance, mantis shrimp tend to strike roundish gastropods (such as moon snails) at their apertures, whereas high-spired gastropods have their top spires knocked off.[34] Perhaps the most adorable of predation studies, though, was on the punching abilities of baby mantis shrimp. These researchers discovered that larval shrimp punch at the same magnitude as adults, making their parents proud.[35]

Amazingly, we have trace fossil evidence that stomatopods were punching above their weight long before primates became smart enough to organize mantis-shrimp fan clubs. Trace fossils matching mantis shrimp shell-breaking behavior were identified in gastropod shells from the Middle Miocene (16–12 MYA) of the Netherlands and given the ichnogenus name *Belichnus*.[36] Similar holes in shells from before then likewise hint at unstoppable stomatopods of the geologic past.

: : :

Any large aquarium gives visitors a chance to admire its confined fauna, with a few stars standing out. In most instances, sharks of all kinds generate the most excitement, from barely bipedal children to Baby Boomers rendering *duh-duh-duh* imitations of the theme music from *Jaws* (1975), a film that featured the improbable scenario of an elected official ignoring a mortal threat to his citizens. Yet if you stand back and observe aquarium onlookers for a while, watch their faces when a stingray swims by, its broad, flattening body undulating as it "flies" in slow motion, and watch their appreciation flip from bloodlust to bliss. And if their eyes follow these stingrays long enough, they may even witness these fishes turn up and swim along the wall of glass, evoking gasps as people see a "face" underneath their sleek exteriors.[37]

Despite their ethereal ways of being, stingrays and other rays are active predators and share a common ancestry with sharks. All rays and sharks have skeletons supported by cartilage, rather than bone, and belong to the clade Elasmobranchia; rays in turn belong to the clade Batoidea.[38] This name is easier to remember than most for fishes, as their wide, flat pectoral fins evoke a sense of "bat wings," a feeling further encouraged by their slow-motion flapping as they glide through the water. In contrast to most sharks, rays prey lower on the food chain by going more for invertebrates—such as bivalves, gastropods, and crustaceans—rather than other fishes. But finding these invertebrates is not easy, as many of them burrow into the seafloor, hiding in sediments below where rays swim.

What to do when you lack digging limbs or tools for revealing, extracting, and consuming your favorite morsels from their hiding spots? Rays use water-powered jets. A ray on the hunt swims along a sediment surface until its electroreceptive organs, called ampullae of Lorenzini, sense prey underneath in sediment.[39] Once alerted, the ray stops swimming, positions its mouth over the right spot, and forcefully shoots high-speed jets of water into the sediment. Think of each jet like a garden hose set to its narrowest setting and creating a high-pressure stream that instantly detaches mud from boots, bicycles, or grateful dogs. Now visualize the same process, but with a ray using its mouth and underwater. Once the finer-grained sediments are blown away, their denser inhabitants—such as clams, snails, and crustaceans—lose their game of hide-and-seek and are exposed to the ray, which descends upon and promptly eats them.[40]

Nevertheless, one might reasonably wonder how rays eat shellfish in the first place, considering their skeletons, including their jaws, are supported by mere cartilage, rather than bone. Fortunately for rays and unfortunately for their prey, cartilage does just fine by providing a stiff support for robust musculature, powering the effort needed to break open shells.[41] Jaws and muscles are aided by hard and interlocking teeth, which form flat surfaces (dental plates) that meet to crush shells. However, both jaws and teeth can be arranged into different shapes depending on their prey, exemplifying a "you are what you eat" principle. For example, cownose rays (*Rhinoptera*) eat mostly bivalves but also gastropods and crustaceans. In contrast, eagle rays (*Aetobatus*) opt more for gastropods while mixing in smaller proportions of crustaceans and bivalves. Their jaw-tooth arrangements reflect these dietary differences, with cownose ray jaws forming oval outlines when viewed from the front, whereas eagle-ray jaws are more rectangular.[42] What a ray consumes also depends on body size, as bigger rays can more easily break larger and thicker-shelled clams, whereas smaller rays are limited to easier-to-chew foods, such as juvenile bivalves and crustaceans. Feeding variations corresponding with body size also manifest in local populations of the same species, such as the cownose ray *Rhinoptera bonasus*. In this species, East Coast rays are noticeably bigger than Gulf Coast rays, which translates into East Coast rays eating more clams and Gulf Coast rays chowing down on crustaceans.[43]

Traces left by this jet-propelled feeding are elongate and shallow depressions rimmed by loosened sediment. These depressions often interrupt rippled sandy surfaces and align from rays that swam along submerged surfaces during high tides, while also stopping to scour and feed. After more than twenty years of studying the Georgia coast, I have seen perhaps thousands of ray traces,

evident as "potholes" on beaches at low tide. Most holes are probably from southern stingrays (*Dasyatis americana*), but some also
might be from the Atlantic ray (*D. sabina*) or bluntnose stingray
(*D. sayi*). Feeding pits on the Georgia coast range from about 50 cm
(20 in) to almost 2 m (6.6 ft) wide, and as deep as 30 cm (12 in), with
sediment piles on one side from rays shooting their jets.[44]

But my most memorable experience with ray feeding traces
by far happened while on a 2005 ichnology field trip to New
Zealand. During that trip, we walked on muddy tidal flats of the
North Island to marvel over freshly made ray feeding pits, some
directly associated with the ghostly outlines of ray bodies. Many
of these pits also contained broken bits of clams and snails, expertly masticated and left as refuse in the same spot where the
ray had lain. So, what could have been better for an ichnologist
than seeing perfectly outlined ray feeding traces? How about
trace fossils showing the same behavior, but made by rays about
20 million years preceding our field trip? Just before we went to
the New Zealand tidal flat, the trip leaders took us to an outcrop
with a vertical sequence of Miocene-age strata bearing cross-
sections of ray feeding pits. The pits were no longer empty, but
had been filled with coarse-grained sediment soon after their
makers jetted the seafloor.[45] Similar trace fossils credited to rays,
given the ichnogenus name *Pisichnus*, are in rocks from the Cretaceous in the western US and Spain.[46] Interestingly, the ones
from Spain were mistaken for both dinosaur tracks and dinosaur
nests until ichnologists figured out their ray origins.

As for other fishes, shell crushing or other forms of breaking are not limited just to rays and extend well into the Paleozoic
and Mesozoic. The division between the two eras was marked
by the worst mass extinction in the history of life, the end of the
Permian (about 252 MYA).[47] With 95 percent of species wiped
out, the 5 percent of species left behind were like disaster capital-
ists, moving into devastated and nearly empty ecosystems and
quickly adapting to new conditions, including radical changes in
resources. Bivalves and gastropods were among the marine and
freshwater invertebrates that made it into the Triassic; bivalves
in particular became abundant, taking over suspension-feeding

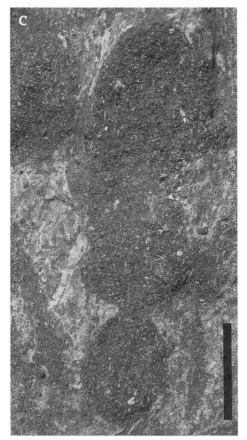

Stingray feeding pits both modern and fossil, showing where these fishes hunted and crunched molluscans. A, Feeding pit, probably made by a southern stingray; Sapelo Island, Georgia. B, Two aligned feeding pits that were probably made by the same eagle ray, with its body outlined by the second pit, Whangateau Estuary, New Zealand; scale = 10 cm. C, Overlapping fossil feeding pits (*Piscichnus*) filled with coarse sand, in the Waitemata Group (Miocene ~20 mya) at Mathesons Bay, New Zealand; scale = 10 cm.

niches once filled by brachiopods, which were almost wiped out.[48] Biomats made a brief comeback in shallow-marine environments for the first time since the Ediacaran, giving grazing-snail survivors plenty to eat.[49] With much food and few predators at its start, populations of Early Triassic mollusks spread and diversified throughout the rest of the Triassic. That is, until fishes and other mollusk-eating animals started putting a dent in both their numbers and their shells.

Mesozoic fishes from cartilaginous to bony ate clams, but one clade of bony fishes, Pycnodontiformes, was probably the most obvious about its clam-consuming ways. These fishes, which originated in the Late Triassic but went extinct in the Eocene (from about 200–50 MYA), had flattened or rounded teeth clearly adapted for crushing.[50] Trace fossil evidence of mollusk-munching fishes also comes from bite traces in clam bodies matching tooth patterns of fish perpetrators, and bromalites, which are fossilized materials related to digestion, from puke (regurgitalites) to gut contents (enterolites) to poop (coprolites).[51]

Do some modern bony fishes also crush shells with their teeth, similar to how parrotfishes break down corals? Yes, and these fishes live in both freshwater and marine environments, preying on clams and snails, with various adaptations in their teeth and jaws for processing shelled prey.[52] Still, some fishes take a different approach in preparing their food by using tools. Yes, you read that right: when certain fishes yearn for a hard-shelled molluscan meal, they go to the nearest underwater hardware store they can find, which is to say, they find hardware. At least three species, the blackspotted tuskfish (*Choerodon schoenleinii*), orange-dotted tuskfish (*C. anchorago*), and graphic tuskfish (*C. graphicus*), are documented as using their surroundings to break open clams.[53] These tuskfishes—which are distantly related to parrotfishes—pick up a clam with their mouths, swim to the nearest coral or other kind of rock, and slam that clam until it goes bam.

: : :

Reptiles evolved alongside shell-eating fishes early in the Triassic, showing how convergent evolution sometimes makes for strange bedfellows, or in this case, strange swim partners. Some of these lineages of shell-crunching reptiles vanished later in the Mesozoic, but a few hung on, with relatives still around today. The first egg-laying descendants of modern reptiles originated in the Carboniferous around 320 million years, and although most continued to stay on land, some adapted to freshwater environments soon afterward.[54]

However, reptiles apparently did not segue to marine environments until the Triassic, an abrupt shift to the wide-open spaces of world oceans that was likely prompted by the end-Permian extinction. Of the few reptilian lineages that survived, some must have lived close to shallow oceanic environs, with their Early Triassic fossils not mere accidents of carcasses getting washed out to sea, but of their living there. Perhaps the best examples of reptiles doing more than just dipping their toes into Triassic seas were placodonts. These stout, long-tailed reptiles belong to the clade Sauropterygia, with its Greek roots (= "lizard flippers") revealing how these reptiles were adapted for the life aquatic.[55] Placodonts shared a common ancestry with long-necked plesiosaurs and shorter necked pliosaurs, which were quite successful marine predators throughout the Jurassic and Cretaceous.[56]

Placodonts, such as *Palatodonta* and *Placodus*, were distinctive from most other aquatic reptiles because of their teeth. To better appreciate the oddness of placodont teeth, touch the front roof of your mouth with your tongue. What you just felt were paired bones composing your hard palate, which helpfully separates food from your nose. Did you feel any teeth on your hard palate? If so, you might be a placodont, but probably not, because they went extinct by the end of the Triassic. Like many other reptiles, placodonts had teeth on their lower jaws (dentary) and upper jaws (maxilla), but also evolved teeth on the roof of their mouths.[57] Other than their chisel-shaped front teeth, which they used to pluck clams off sea bottoms, all other placodont teeth were thick and flat, perfect for crunching clams.

Another important clade of shell-eating marine reptiles that also arose during the Triassic and overlapped with placodonts was Ichthyosauria. Ichthyosaurs were so fully adapted to oceanic environments that their streamlined bodies looked much like those of modern-day dolphins. Ichthyosaurs also gave live birth at sea, which is again more like dolphins and other whales than most other reptiles.[58] Although ichthyosaur diets were quite varied, ranging from squid to fish, more than half of Triassic species had teeth and jaws suited for breaking shells, an evolutionary innovation that developed at least five times in that period.[59]

Other shell-crushing marine reptiles included the lizard-like and often-giant mosasaurs, which originated in the Cretaceous at about 100 MYA. Mosasaurs swam in Cretaceous oceans with ichthyosaurs and plesiosaurs for millions of years, but when both of those reptiles went extinct, mosasaurs were top marine predators until their demise at the end of the Cretaceous.[60] You might think that mosasaurs would not deign to eat lowly clams or snails when nearly every other animal was available to eat, and you would be right. Nonetheless, they did chomp on ammo-

Late Cretaceous ammonite bearing a row of tooth traces in its shell, courtesy of a chomping mosasaur; specimen displayed at Smithsonian National Museum of Natural History, Washington, DC.

nites and other shelled cephalopods, which we know from tooth traces left in their shells identical to the shapes and sizes of mosasaur teeth.[61]

The end-Cretaceous mass extinction took out almost all marine reptiles, except for snakes, sea turtles, and a few crocodilians.[62] Of these three groups, some sea turtles may have eaten hard-shelled mollusks, but most modern sea turtles go for softer food, such as jellyfishes. As for crocodilians, tooth traces in freshwater clams from Cretaceous rocks of Spain are linked to them.[63] Also, at least one crocodilian, *Gnathusuchus* of the Miocene (~13 MYA) in Peru, had peg-like teeth perfectly suited for breaking clams, snails, and anything else with a shell.[64]

: : :

An enduring early childhood memory of mine is of my parents packing at least four siblings and me into a car for a road trip to Manitowoc, Wisconsin, where we visited my mother's beloved sister, my aunt Eileen. It was one of the few times as a kid that I traveled more than an hour out of my hometown of Terre Haute, Indiana, and it was marvelous. Although I did not see an ocean until my early 20s, Lake Michigan—a body of water with no end as it met the horizon—was impressive enough to a Midwestern child of the 1960s. With no dated photos of this trip or living parents, I have no way of knowing my age then, but guess that I was eight years old.

Anyway, on the way to Manitowoc we stopped in Chicago, the first big city I had ever seen. Do I remember its tall buildings, people-filled streets, fabulous museums, or deep-dish pizzas? No, what I remember was getting sprayed by a walrus there. This pinniped outpouring took place at one of the Chicago zoos, where I finally saw real, living examples of exotic animals: gorillas, lions, giraffes, and other animals of Africa that had only existed before in my mind as gray-scale images on the TV show *Mutual of Omaha's Wild Kingdom*[65] or glossy photos in *National Geographic*.

Of course, urban zoos in the 1960s were far more effective at preventing their animals from directly interacting with humans than the fictional *Jurassic Park* of the 1990s. Yet a walrus, like life, found a way. As I eagerly stood behind a railing at the walrus pool while admiring its occupant's prominent tusks, whiskers, flippered limbs, and enormous body, it looked at other nearby bipeds and me, took aim, and spat a stream of water at us. From that distance, the stream arced and broke up into droplets by the time it reached us. Still, my shirt was soaked, and the Martin family had a funny tale long afterward that always started with the distinctive sentence "Remember when that walrus in Chicago sprayed Tony?"

Many decades later, I read scientific articles about walruses (*Odobenus rosmarus*) and was glad to learn their spitting water was not just for inspiring family travel stories, but also for gathering food. Although walruses are huge animals, with some weighing more than a ton, their diet is mostly invertebrates, such as crustaceans, polychaete worms, and especially clams, which they can eat by the thousands per day.[66] In their Arctic marine environments, walruses dive down to shallow bottoms to look for clams and a wide variety of other foods.

To eat a clam, first a walrus blows, then it sucks. Once on a seafloor, they use their prominent and sensitive whiskers to detect buried prey; their tusks are not used for digging, but they may drag them.[67] Like rays, they deliver high-pressure jets of water with their mouths to loosen sediment and expose clams and other food. Unlike rays, though, walruses do not chew clams with their teeth. Instead, they hold clams with thick and tough lips and create enormous negative pressure in their oral cavities, pulling the bivalves' soft bodies out of their shells.[68] This force is also powerful enough to cause a shell to collapse: in short, walruses suck so hard, they break shells.

Walruses are well represented in the fossil record by skeletal material, but their vacuum-affected clams have not yet been interpreted as trace fossils. However, Pleistocene rocks of the Olympic Peninsula in the Pacific Northwest preserve the excavations of walrus water jets, similar to those made by rays. The sizes and abundance of these trace fossils imply that great herds of walruses dined on clams in this area then, perhaps tens of thousands of years before humans were there.[69] And once humans showed up in that area, I hope after such a long and arduous journey that they were treated to walrus-provided showers.

: : :

Given their smooth and slinky bodies, loping gaits, and playful antics, otters are among the most charismatic of mammals, earning adoration both in zoos and in the wild. Of course, like many mammals with "cute and cuddly" reputations, otters are fierce carnivores that often kill their food, a fact that becomes more obvious when naming otters' relatives, such as weasels, badgers, and wolverines.[70] Because all thirteen living species of otters are well adapted for moving in and out of water bodies, the animals they eat are mostly fishes and aquatic invertebrates, including clams. Indeed, mussels and other bivalves are favored snacks of sea otters (*Enhydra lutris*) on the Pacific coast of North America, which they gather by diving to shallow sea bottoms and bringing them up to the surface to consume.[71] Unlike most women's clothing, sea otters have pockets, folds of skin under their arms that hold clams, sea urchins, or other food items during their ascent to the surface.

Sea otters are the heaviest otters in the world, with adult males approaching 40–45 kg (88–100 lb).[72] Still, some mussels and thick-shelled clams present an obstacle to even the brawniest of otters. When challenged, this is when they resort to geological allies: rocks. Otters use rocks as three types of tools: pry bars, hammers, and anvils.[73] For the first, they may wedge a rock under a bivalve and push on it, operating it as a first-class lever to detach the clam from the sea bottom. For the second, they gather clams and rocks in their pockets, lie on their back at the surface, put the clam on their chest, and slam the rock onto it like a hammer. Alternatively, the otter may put the rock on its chest, hold the clam with its paws, and bring it down onto the rock as an anvil. A variation on the anvil theme is where an otter finds a stationary rock, such as a large boulder the size of a small boulder, and strikes the clam against it.[74] All of these methods result in broken shells and exposing fresh molluscan flesh.

Would evidence of otter tool use actually show up in the fossil record? Yes, according to some researchers. In a 10-year study of shells broken by sea otters and otter-used rocks on the California coast, researchers likened such evidence as "archaeological" (although we all know it was ichnological).[75] For example, shells broken by otters typically had an entire valve hinged to just a fragment of the other valve. Favorite stationary rocks that otters used as anvils also developed obvious wear spots from otter pounding. Based on current rates of feeding and past shell middens, these same researchers calculated that between 44,000 and 132,000 mussels were broken in just one small area of coastline over 23 years.[76] Multiply the impact of these impacts over the entire Pacific coast range of sea otters, and that's a lot of shattered shells.

: : :

Speaking of impacts, some clams rise to great heights before their death and consumption. In coastal areas and during low tides, gulls may fly over exposed sandflats and mudflats, searching for the outlines of clams buried below the surface. When clams are spotted, these avian ichnologists swoop down, land, pick up a clam with their beak, and take off.[77] Then, once high enough, they drop them, serving up instant raw clam on the half (or quarter, or eighth) shell. Typically all that is left of this killing is a jigsaw puzzle of clamshell pieces and the gull culprit's footprints.

Despite dozens of visits to the coast of Georgia between 1998 and 2015, I had never actually witnessed gulls dropping clams. Instead I only performed postmortem forensics that told tales of murder most fowl. Yet the clues were all there like a murder mystery: the outline of the mollusk in its former burrow, gull tracks leading up to the clam, take-off tracks, impact marks on the sand with shattered shell bits, landing tracks by the gull, and lots of overlapping standing tracks as it ate fresh clam.

Such foreshadowing meant I was more than overdue to get visual confirmation of gulls killing clams, which was finally granted in March 2015 on Jekyll Island, Georgia. While sitting on

Clams shattered by gulls dropping them onto beaches and wood. *A*, Broken Atlantic cockle (*Dinocardium robustum*) with impact site (arrow) and tracks of the gull that did it; Sapelo Island, Georgia. *B*, Carnage of quahogs (*Mercenaria mercenaria*) left by gulls on a wooden pier, Jekyll Island, Georgia; scale = sandal, 8½ men's.

a deck on the west side of the island and looking at a mudflat (between swatting sand gnats), my wife Ruth and I noticed a laughing gull (*Leucophaeus atricilla*) flying about 10 m (>30 ft) above a wooden pier. At one point, it paused its ascent, and an object fell from it and down toward the pier. *Thunk!* We heard the collision of the object correlate with what we saw, and with much excitement realized we had both just witnessed gull clam-cracking for the first time.

When we hurried to the dock to see what had happened, a pier littered with broken shells greeted us. But what was most surprising about this broken-shell assemblage was that only one species of clam was represented, the southern quahog (*Mercenaria mercenaria*). These thick-shelled clams are common in sparsely vegetated muddy areas of salt marshes, burrowing into the mud and connecting dual siphons to the surface for filtering suspended goodies in the water during high tides. Yet during low tides, they become vulnerable to avian predation. Despite being "hidden" in the mud, somehow the gulls spotted them from the air, landed next to them on the mudflat, and pulled them out. They then used the nearby pier as an anvil, with the clams' hard and thick shells unwittingly becoming their own hammers after landing from a fatal height.

So now it's time to think about broken clams and deep time. If we found such an assemblage of broken shells of the same species of thick-shelled clams in a geologic deposit, how would we interpret it? Would we recognize these pieces as predation traces, let alone ones made by birds? Which also prompts the question, when did gulls or other shorebirds start using flight and hard surfaces to open clams and snails? All of these are good questions paleontologists should ask whenever they look at a concentration of broken fossil bivalves overlapping with the known geologic range of shorebirds. In short, these may not be "just broken shells," but evidence of birds using gravity-assisted killing as part of their predation portfolio.

Other birds break shells without the assistance of gravity, using just their beak. Among these are the appropriately named oystercatchers, which peck into oyster, clam, and snail shells, applying their beaks with considerable force.[78] This action, however, requires the evolution of beaks stout enough to chip, pierce, or otherwise shatter molluscan skeletons. Apparently other birds skipped evolving stout beaks, and instead take advantage of hard places in their surroundings to do the job.

Do birds ever drop animals other than molluscans, and animals equally unlikely of otherwise flying? Yes to both of those, as a few birds are known to grab small tortoises, ascend with their terrapin payloads, and let go of them above rock outcrops to break their shells. In at least one instance, this tortoise-dropping method was also blamed for the death of a human, the Greek playwright Aeschylus (~525–456 BCE), who is frequently credited as the "father of tragedy." According to a possibly fictitious story describing his death, an eagle picked up a tortoise, but misidentified Aeschylus's bald noggin as a rock and dropped the tortoise

on him instead: so, more of a tragicomic ending.[79] The real tragedy of this account, though, is no mention of whether or not the
tortoise survived, or if the eagle succeeded and dined on the tortoise beside the recently deceased playwright.

: : :

Picture a 20-ton sauropod dinosaur walking majestically across a
Late Jurassic tidal flat. Her long neck sweeps back and forth well
above the sandy shore as she scans for predators or others of her
kind, perhaps trying to rejoin a herd lost earlier in the day. Her
tail whips in a different pattern from her neck and more chaoti-
cally, like an enormous grass blade in the wind rather than the
metronome of her head.

Now imagine yourself as a clam exposed on that shore, either swashed onto this surface by waves or stranded by a high
tide. Much like artists and writers, you know that exposure only
brings loneliness and suffering. So you do what your ancestors
did to survive the twin perils of desiccation and predation, and
you burrow. Only you did not burrow deeply enough to avoid the
stress of a walking sauropod's rear foot as it pressed at least part
of its 20-ton bulk onto the sand. The foot did not even have to
touch you directly to kill. The squeezing of water from between
the surrounding sediment brings those grains closer together
and smashes you, a clam that was just trying to make it through
one more tidal cycle until a behemoth brought you to an end.

Envisaging dinosaurs stepping on hapless molluscans is one
thing, but seeing a crushed shell in a dinosaur track more than
fulfills that mental image. This reality check happened while I
was on a 2003 ichnology field trip in Switzerland, visiting one of
many sauropod tracksites there. The trackways recorded in the
limestone were from the Late Jurassic, and because these limestones were made in shallow-marine environments, we ichnologists could tell the sauropods walked along shorelines then.[80]
The oblong tracks from the sauropods' rear feet were big, more
than twice the length of my field boots (size 8½ men's), and reg-
istered just behind their smaller, bean-shaped front-foot tracks.
Other than their enormity, though, what I remember best about

this field-trip stop was how one rear-foot track had a little anomaly in its hindmost center: the coiled shell of a marine gastropod. A closer look at the snail revealed that its topmost spires were missing, perhaps pulverized by the weighty animal's foot.

Nevertheless, this and a few other cases of sauropods and other dinosaurs stepping on molluscan shells do not necessarily mean these shells' owners were alive when they unwittingly became experiments in load bearing. For example, a fossil turtle that was also in Jurassic tidal flat sediments of Switzerland was fractured in such a way that paleontologists proposed a dinosaur as the likely shell breaker.[81] Strong circumstantial evidence pointing to a sauropod as the stomper were thousands of sauropod tracks in an overlying stratum, but the paleontologists concluded the turtle was already dead when trampled. Still, one place where sauropods probably did kill molluscs by smashing them and other invertebrates underfoot is in Late Jurassic lake deposits near La Junta, Colorado.[82] There the clams lie shattered within the dinosaurs' footprints, mute testimony to lakeshore lives ended quickly.

Meanwhile, as sauropods and other weighty dinosaurs sauntered through their landscapes during the Jurassic, they surely walked by or interacted indirectly with forests. Within those forests were trees that in turn hosted animals much smaller than dinosaurs, but far more numerous. These animals tunneled into solid tissues and used tree interiors as homes, grocery stores, nurseries, and much more. Much later, certain dinosaurs evolved to destroy wood and eat these animals. Welcome to the inner wooden lives of insects and the most bioeroding of insect-eating birds, the woodpeckers.

Chapter 7

Woodworking at Home

The ponderosa pine had been torn apart by hand. Deep, parallel gouges scored some parts, whereas other places on the trunk had been punched, impacts denoted by craters with splinters radiating up and out along their rims. A closer examination revealed how the gouges translated to a hand slightly wider than any of ours, bearing five blunt claws that ripped into the rotting, recumbent log.

The year was 2002 and I was in Idaho, sharing the Frank Church–River of No Return Wilderness Area with about thirty other people from across the US, all united by an interest in tracking wolves (*Canis lupus*). Our joint venture was organized by an environmental education organization, the Wilderness Awareness School, which offered weeklong wolf-tracking courses in this federally managed wilderness. In these courses, participants learned how to identify the tracks and other signs of wolves, as well as document their whereabouts via these traces. As for the wolves, they were part of a then-new reintroduction project done cooperatively with the Nez Percé people, the previous land managers there before colonizers ejected both wolves and native peoples.[1] Another integral part of the tracking courses for us was to recognize and record traces of other animals sharing the same ecosystems as the wolves. Every day started with five-person teams leaving their tents before dawn to track. On some days we did not track wolves at all, but elk, because wherever you found elk, you were more likely to find wolves. Give a wolf a menu with a long list of prey items and, in most situations, they will order the elk.[2]

Like many ecosystems, the boreal forest around us—dominated by firs, pines, and larches, but also accompanied by aspens, birches, and cottonwoods—held more than one species of large carnivore, causing wolves not to just seek elk, but also avoid potential competitors. For instance, the largest predatory cats of North America, cougars (*Puma concolor*), shared this area and similarly maintained broad ranges that likely overlapped with those of the wolves. Other significant meat-eaters were black bears (*Ursus americanus*), one of which was responsible for eviscerating the ponderosa pine before us. Although smaller on average and far less aggressive than grizzly bears (*Ursus horribilis*), black bears are nonetheless massive, with males regularly exceeding 200 kg (440 lb) and females 150 kg (330 lb).[3] Big mammal bodies need lots of fuel, and luckily for black bears, they are among the least picky of omnivores, eating a variety of plants, fungi, and animals. For animal-based meals, black bears can also do quite well with scavenging, sometimes by scaring away medium-sized carnivores—like coyotes and wolves—from their kills. This habit we likewise documented when we found a black-bear feeding spot adorned with the bones and skins of at least two dead elk and fresh, pinkish scat containing digested meat and bone.

The ripped and pummeled pine log before us was from a black bear not intent on eating its wood, but rather the live animals inside it. The deepest excavations reached nearly to the log's center, exposing extensive hollow, anastomosing, and interconnected tunnels, their diameters ranging from those of vermicelli to pencils. This is how our tracking team made the connection between this bear and wood-boring insects, such as the beetles, termites, and ants that had made most of the tunnels. Sadly, our team did not stay long enough to identify which tunnels belonged to which insects: remember, large carnivores were in the area. Still, this ecological lesson stuck with me, that so-called dead trees were so filled with invertebrates that they could supply sustenance for 200+ kg (over 400 lb) animals.

More than twenty years have passed since that bear-imparted lesson, but I still think of it whenever I walk by a former tree, whether standing in stark contrast to its green-garbed companions or lying on its side. I often imagine such trees stuffed with wood-chewing insects helping return the tree's essence to its descendants and other life-forms in the forest. And as a paleontologist, I go back even further into the geological past to ask myself how the evidence for such interactions would manifest as fossils. Much later, land-dwelling vertebrates, and especially woodpeckers, evolved to seek and consume these wood-boring insects, and ecosystems changed with interactions between eaters of wood and eaters of the wood-eaters. A quick stroll through the history of wood, as well as the insects and woodpeckers that have bored into it, pulls together many seemingly disconnected themes, all linked by bioerosion.

: : :

For nearly as long as woody tissues have existed, something has been eating them. Knowing about this relationship between wood and its consumers accordingly inspires inquiry into wood origins. Although we often associate wood with land plants, this coincidence was not always true. For instance, the earliest land plants, which probably started during the Ordovician (about 470 MYA), did not have woody tissues like those we find in most trees today. These primitive plants were nonvascular, which meant they were more like modern mosses or liverworts, hugging the ground rather than reaching into the skies.[4] Their low-lying life habits reflected the physiological limits of nonvascular plants, which could not efficiently circulate water, nutrients, and food in broad networks. Such restrictions also meant these plants were confined to wet environments hugging edges of oceans, rivers, or lakes.

Nevertheless, once vascular plants evolved from their predecessors in the Silurian at about 425 MYA, they ascended, and literally so, as both woody tissues and water conducted in those tissues helped these plants partly overcome the oppressive effects of gravity.[5] Given stiffer support, stems gave plants height, helping them gather more light for photosynthesis while also creating the first plant-related shade. After stems came roots, which penetrated primordial soils and extended a plant's reach for water and nutrients, like phosphorus and nitrogen, a quest augmented by symbiotic fungi.[6] The growth and movement of roots also helped weather bedrock, which further developed soils. We humans took advantage of vascular-plant solidity by combining wood with stone or metals for tools, weapons, and homes.[7] Much later, people also used wood for shipbuilding, furniture, picture frames, and art. With the advent of writing in some cultures, wood was then used as pulp for paper, composing non-digital versions of this and many other books.

When thinking of wood, it is probably easy to first visualize the roots, trunk, and branches of a tree from which it came. This visualization becomes more challenging, though, if you move from the tree's outer bark surface to its inside. This is where organic fibers are built around one another to make it solid, but while also leaving spaces between those fibers. The fibers are made of cellulose, which in turn are held together by lignin. Cellulose and lignin are different organic polymers that work very well together to make wood a hard substance that bends under stress but is also difficult to compact. Incidentally, cellulose and lignin are among the most abundant organic compounds on Earth, as these are in all living and most recently dead vascular plants.

Other than ensuring a tree gets above ground level, cellulose and lignin also form xylem. Xylem is one of two major transport tissues in plants; in trees, it moves water and nutrients from the roots through a stem or trunk and into the branches. Xylem is accompanied by another vascular tissue, phloem, which moves food produced by photosynthesis (mostly glucose) to parts of the tree, which converts it into energy.[8] So in terms of functions, one might think of xylem as plumbing and phloem as catering.

In modern trees, foresters also use a more general division of below-bark woods as sapwood and heartwood, which are often distinguished as different hues in a tree cross-section. Sapwood is the outermost portion of wood under the bark, and although composed of about 90 percent dead cells, it still works as xylem. In contrast, heartwood composes most of the inner wood, which in turn surrounds an innermost pith.[9] Heartwood and pith are made of older and nonfunctional xylem; their main job, then, is to keep a tree upright. If you ever see the stump of a recently cut tree, look for these wood divisions, but also for growth rings. These concentric lines mark annual changes in growth that typically corresponded with seasonal cycles, testifying to that tree's life.[10]

Given the start of wood in the Silurian, when did animals start chewing it? It turns out this behavior also began in the Silurian at about 420 MYA, or only a few million years after "first wood."[11] All land animals in the Silurian were invertebrates, and many were arthropods.[12] So because arthropods also have mouthparts and some are well suited for breaking up organic tissues, we suspect these animals began woodworking.

We know land-dwelling arthropods ate land plants not from direct evidence of arthropods fossilized with their jaws fiercely clamped on roots and stems, but from their trace fossils. In a 1995 study, paleontologists reported coprolites (fossil feces) in Silurian rocks from Wales that contained spores and other bits of vascular plants.[13] Since then, other paleontologists have found more such evidence pointing toward either mites (small arachnids related to spiders) or millipede-like arthropods as the most likely poop producers.[14] In rocks from the Middle Devonian (393–383 MYA) of upstate New York, tiny tunnels in fossil wood filled with coprolites also helped paleontologists better make the connection between wood going in one end of an arthropod and coming out the other.[15] Such trace fossils also show a very important step in the ecology of the earth, as they mark the start of herbivory, and more specifically, xylophagy (where *xylo* = "wood" and *phagos* = "eat"). Once this animal-based cycling of plant-based tissues in land ecosystems began, it never stopped, and it will continue as long as land plants and animals coexist. As a bonus, such ancient coprolites also show that animals had vegan lifestyles about 400 million years before it was cool.

Either just before or soon after herbivory began, the first insects descended from land-dwelling crustaceans.[16] But exactly when these six-legged Earth-conquering arthropods showed up is still a matter of vigorous debate. Disagreement arises partly because the early fossil record of insects is spotty, with only a few possible insect parts coming from the earliest part of the Devonian (about 400 MYA) and no definite insect trace fossils from then, either.[17] Yet molecular biologists project insect origins way back into the Ordovician, stating that molecular clocks indicate insects diverged from crustacean ancestors nearly 480 MYA.[18] If they are right, then insects have an 80-million-year gap in their fossil record. Regardless of their origins, though, insects continued to evolve since and eventually became integral parts of all continental ecosystems.

However, insects eating live plant tissues prompted evolutionary responses from plants, in which they developed chemical defenses to prevent mere nibbling from escalating into all-you-can-eat salad bars.[19] So began a coevolutionary dance between plants and insects evident in the fossil record via a combination of plant body fossils and insect (or other arthropod) trace fossils in those plant fossils. Some of these trace fossils include bite traces in leaves, wood that was chewed but not swallowed (frass), coprolites with plant remains, tunnels in wood, and much more.[20]

Following the birth of herbivory, these fossil combos tell us of broad and important phases in plant-insect relationships through time. These phases included: the Late Carboniferous (323–299 MYA), when insects ate plants more than ever before; the Permian (about 299–252 MYA), when plant-eating insects spread into new habitats, such as deserts; the Late Triassic (237–201 MYA), when plants and insects recovered and expanded after the horrific mass extinction at the end of the Permian; and the post-Cretaceous, with a similar recovery of plants and insects after a big space rock hit the earth 66 MYA.[21] Then, about 10 million years later, this biological mending was followed by one of the hottest times in Earth's history, the Paleocene-Eocene Thermal Maximum (PETM).[22] In each of these biologically and geologically dramatic events, plants and insects picked up where they last left off, with their coevolutionary stories told mostly by plant body parts and insect traces in those parts. For the sake of our thematic look at life breaking down hard stuff, though, we will focus on how insects have affected wood specifically, as well as how some birds likewise penetrated and otherwise affected wood since their origin.

: : :

Much to the chagrin of entomologists who study anything else, 40 percent of insect species are beetles, and about a quarter of *all* animal species are beetles.[23] Beetles are placed in the clade Coleoptera, but given the incredible diversity within Coleoptera and our continuing theme, we will look just at those that chew wood. Beetles with such lifestyles are in the clade Scolytinae and known as "bark beetles": not for their sounds, but for their bites.[24] Bark beetles are represented by more than 5,000 modern species and a good number of fossil species, with the oldest example, *Cylindrobrotus pectinatus*, from the Early Cretaceous (about 130 MYA) of Lebanon.[25]

Like most insects, beetles have a life cycle that goes from egg to larva to pupa to adult, called holometabolous. Bark beetles go through this life cycle too, but mostly while encased in wood. Bark beetles are often divided into two broad categories, pine beetles and ambrosia beetles, based on wood preferences and other factors.[26] Pine beetles do indeed live in pine trees and other conifers, whereas ambrosia beetles dwell in a variety of trees. Ambrosia beetles further differ from pine beetles because of their interdependence on fungi, discussed soon. However, both pine beetles and ambrosia beetles make their way through wood the same way, which is with mandibles that tear apart cellulose and lignin fibers as they create tunnels and other spaces.

Pine beetles have received more attention for their wood-eating ways lately because of their laying waste to wide swaths of forest, leaving thousands of dead pines in the wake of their mandibular prowess.[27] Nonetheless, within this destruction are motes of beauty, and no traces of pine beetles are as visually com-

Insect traces in wood, modern and fossil. A, Snag of formerly living loblolly pine (*Pinus taeda*) with meandering and branching tunnels made by pine beetles, accompanied by large holes made by woodpeckers hunting for these and other insects; St. Catherines Island, Georgia. B, Fossil termite galleries in mangrove (*Avicennia*) filled with fecal pellets; specimen from the Miocene (~22 mya) Puketi Formation, New Zealand, in University of Auckland collections.

pelling as those made by a mother and her baby beetles. First an expectant female beetle bores a single, straight, and open tunnel in the wood, in which she lays her eggs. The tunnel is a nursery for larval beetles, which hatch from their eggs and commence chewing their way out of this brooding chamber. Individual larvae pick a direction and move away from the chamber, spreading and diverging, with no tunnels intersecting. This marvelous movement results in tunnels forming spokes around the original birthing tunnel.[28] Even better, these tunnels widen along their lengths. Why? Because the larvae grow as they eat, hence the tunnels become a record of their growth, similar to how parents mark the heights of their children on a wall. At tunnel ends, larvae go into their next stage in life, pupation, which is also where they spend the winter.[29] With spring warming, adults hatch from the pupal chambers and resume chewing, making vertical shafts leading up and away from the larval tunnels. These beautiful patterns are just underneath the bark of a dead pine and follow the circumference of the tree, often revealed with one pull of that bark. Adult beetles also can form S-shaped tunnels, which often cross one another to create braided-river patterns filled with woody feces and frass.

As hinted at earlier, ambrosia beetles differ from pine beetles, and the main difference is their relationship with fungi. Rather than using wood directly for sustenance, ambrosia beetles make tunnels in the wood for growing and harvesting fungi, making them among the few animals that cultivate their own food.[30]

Beetles probably started boring in wood soon after their evolutionary debut in the earliest part of the Permian (~295 MYA), but modern clades of wood-eating beetles did not show up until much later, in the Jurassic and Cretaceous.[31] Thus beetle trace fossils in fossil wood from the Permian and Triassic that look much like modern beetle borings were probably made by previous clades. Not surprisingly, wood-boring trace fossils linked to beetles were first discovered in a place world famous for its fossil wood, Petrified Forest National Park in Arizona. In 1938, the geologist M. V. Walker recognized tunnels in fossil tree trunks there as the probable work of beetles.[32]

Almost 75 years later, in 2012, other trace fossils in fossil wood of Petrified Forest National Park were also credited to wood-boring beetles. Clusters of these pouch-like, U-shaped borings were first identified in the 1990s as bee brooding chambers.[33] But considering that bees did not evolve until 100 million years later, this interpretation was a bit problematic. The mystery was solved when two paleontologists—Leif Tapanila and Eric Roberts—reexamined the borings and found that each had separate openings (entrance and exit), with frass packed into the entrance side.[34] These traits ruled out wood-boring bees as the makers, as these insects make a single entrance-exit hole and are not known for frass packing. Chamber sizes also matched the sizes of adult Triassic beetles, pointing toward these as brooding chambers for beetle larvae. Hence Tapanila and Roberts identified the chambers as the oldest known trace fossils of holometabolous insects in the geologic record, dating at about 210 MYA, and assigned them the ichnogenus name *Xylokrypta* (= "wood hidden").[35] In 2017, paleontologist Zhuo Feng and his colleagues pushed that oldest known date back another 40 million years with probable beetle borings and brooding structures in Late Permian (~253 MYA) fossil wood of China.[36] These and other trace fossils show how the more than 250 million years of shared history between trees and wood-boring beetles is still revealing itself to us, with more evidence likely on the way.

: : :

Of all insects ranked as the least favorite of homeowners, roaches and termites are always sure to make the short list. So if homeowners also believe in evolutionary conspiracy theories, the relatedness of roaches and termites may not come as a surprise. Both groups of insects are within the same clade (Blattodea), which in turn holds another clade (Tutricablattae) that encompasses modern wood roaches and termites. The divergence of termites from their roach ancestors likely happened during the Early Jurassic, at about 190 MYA.[37]

Wood roaches, which live in forested areas of North America and eastern Asia, are best represented by species belonging to the genus *Cryptocercus*.[38] Although wood roach and termite behaviors are quite different, wood roaches are more closely related to modern termites than all other roaches. Wood roaches also live in and eat wood as food, just like termites. Both wood roaches and termites are hemimetabolous, meaning their life cycles lack larval and pupal stages, and instead are born in a nymph stage, looking more like shrunken adult roaches or termites than grubs. Later, nymphs molt with each growth stage until adulthood. Another link between wood roaches and termites is their using the same gut microbes. A difference, though, is that termites have specialized castes within each species—queens, soldiers, and workers—whereas wood roaches simply hang out together in rotten wood.[39]

Given that wood roaches hide in decayed wood, scientists understandably took longer to document their behaviors, but eventually gained some remarkable insights. One notable behavior of eastern wood roaches (*Cryptocercus punctulatus*) is monogamy, with mated pairs making and occupying mating chambers in wood and the male defending them. But if another male wood roach invades this chamber and wins a fight against its resident male, then he also wins the resident female and chamber, evicting the loser and promptly moving in.[40] Such roach feuds and sexy times all take place in complex roach-made spaces beneath wood surfaces.

Despite the best efforts of pine beetles and other wood-boring insects, termites are the insects most often associated with and loathed for their wood-eating activities. Before delving into termites' effects on living and dead wood, though, an overview of termite biology is needed. Termites (clade Isoptera) are eusocial. This means they form colonies that function by having individuals with distinct body plans (polymorphism) reflecting different castes.[41] Castes in a typical termite colony include: primary reproductives, such as a queen and king; soldiers, which defend the colony against attackers, such as invading ants; and workers, which cooperatively build a home for a termite colony and feed the other castes. Although eusocial ants, bees, and wasps have similar caste systems, those insects are holometabolous.[42]

Termite-worker mandibles are adapted for grasping and tearing apart wood fibers, which they do *en masse* to carve out spaces for the colony. Hence any termite traces identified in either modern or fossil examples are assumed to be the handiwork of workers. How to tell termite nests from those of other wood-eating insects? Look for deep, pervasive borings that cross sapwood and heartwood alike; in contrast, most non-termite insects' borings are shallower and simpler.[43] Termite nests in wood owe their complexity to interconnected tunnels and galleries. If the word "galleries" inspires thoughts of large rooms containing framed works of art, simply subtract the art to get their entomological meaning, which refers to spacious interiors. Galleries normally serve as nurseries, where the queen lays her eggs and workers feed nymphs. Tunnels and galleries are usually open, but can get partly filled by frass. Termite feces are also identifiable as appropriately sized cylinders with hexagonal cross-sections.[44]

Molecular clocks, as well as wood roach and termite body fossils in amber, point toward termite origins in the Early Cretaceous, 140–120 MYA.[45] Moreover, early termite body fossils show they already had different body shapes depending on whether they were breeders, workers, or soldiers. Thanks to their relationship with wood and other factors, termites may have been the first eusocial insects, leading the way for other insects to form huge colonies with specialized tasks within each species.[46]

Termite trace fossils in fossil wood seem rare, but are documented from Late Cretaceous and younger rocks. For example, coprolites in Cretaceous amber from France are the right size and shape for those of wood-eating termites.[47] Both coprolites and borings in Cretaceous fossil wood from the Isle of Wight (UK) and the far-off land of Texas are likewise credited to termites.[48] Some fossil termite nests from Tanzania show that termites used fungal gardening toward the start of the Miocene (about 25 MYA), joining an exclusive club of insect farmers, such as ambrosia beetles and leafcutter ants.[49]

As homeowners and foresters know well, insects other than beetles, roaches, and termites likewise evolved to tear apart wood for food, shelter, and/or nesting. Among those insects are some Hymenoptera, which includes wasps, bees, and ants, with their lineages also extending back to the Mesozoic. Wood-altering ways of a few hymenopterans are revealed by their common names, such as sawflies, carpenter bees, and carpenter ants, whereas paper wasps, yellow jackets, and hornets are wood scrapers that chew and reshape wood fibers into paper nests. Given these multifaceted lifestyles over the past 100+ million years, it is not surprising that these insects have fulfilled varying roles of home builders, wreckers, and renovators in wood. How they will change with vascular plants in the future is anyone's guess, but we can be assured that they will continue to fascinate us with their adaptive persistence and resilience.

: : :

Just like with the breakdown of rocks and shells, the effects of insects degrading wood for hundreds of millions of years led to their essential roles in cycling matter and energy in forests. But many of these insects have not done this work on their own, as they were assisted by little internal companions. For example, ambrosia beetles depend on a specific fungus for food, and other bark beetles spread fungi with their tunneling. This interdependency qualifies as a "win-win" for beetles and fungi that mutually benefit from their relationship. However, this mutualism is also a "lose-lose" situation for trees that try to defend against it. For instance, if beetles and their fungi are invasive species, native trees often cannot adapt quickly enough to repel these invaders, so they die, and die quickly.[50]

A second and more widespread wood-eating problem is happening with native bark beetles wiping out pine forests in parts of North America. How could such a biological catastrophe come from native species? The game changer was climate change. Forest scientists now think that warmer temperatures throughout North America over the past few decades have accelerated pine beetle life cycles.[51] More egg-to-adult transitions within a given year mean more bark beetles per year, chewing their way through slow-growing trees that cannot defend against these attacks.

A third insidious problem related to the biodegradation of wood and the influence of climate change comes from the smallest of organisms living in tiny guts. Despite their wood-eating habits, wood roaches and termites by themselves cannot digest cellulose in that wood. To do this, bacteria and single-celled protozoans in their tiny digestive tracts digest the cellulose for them.[52] However, one of the waste products of cellulose digesters is methane, which is a greenhouse gas and a much more effective one than its more famous atmospheric companion, carbon dioxide. By some estimates, termite flatulence produces 1–3 percent of global methane,[53] a volume that will increase if deforestation leads to more dead wood available for termites to eat. In short, if methane-producing beetles and termites are all turbocharged by climate change, then we will enter a new phase in the history of wood bioerosion unlike any from the earth's past. However, given knowledge of past wood chewers and their ecological impacts, we may be better informed as we adapt to and mitigate whatever happens next.

: : :

Whenever Karen Chin talks about fossil poop, people listen. After all, she has studied dinosaur dung since the mid-1990s and is widely acknowledged as the reigning queen of coprolites. Relevant to our wooden interests is one of her earliest and most important scientific discoveries on wood-eating behaviors she inferred in dinosaurs. The 1996 study, coauthored with entomologist Bruce Gill, was related to 75 MYA dinosaur coprolites in northwestern Montana that were mostly composed of fossilized bits of conifers.[54] Based on their abundance, huge volumes (as much as 7 L/2 gal), and location close to nests and bones of the large (9 m/30 ft long) herbivorous dinosaur *Maiasaura*, Chin surmised they were the perpetrators, or what she would call the "poop-a-trators." Regardless of their origin, these fossil feces showed that at least some dinosaurs ate wood. More intriguingly, Chin and Gill found burrows within the coprolites closely resembling those of modern dung beetles. Dung beetles today use wastes from a variety of animals to provision brooding chambers, which they dig in soil or the dung itself. From this evidence, the researchers sketched out a Cretaceous food web, in which dinosaurs ate conifers and pooped, and then dung beetles used this bounty to feed their children.[55]

Still, one part of this explanation bothered Chin, and it had to do with the wood. Although insects have consumed wood as food since the Paleozoic, fresh conifer wood in itself should not have appealed to large dinosaurs, as it does not supply nearly as much nutrition as leaves or other plant parts. So there had to be something more to this wood. After carefully studying fossil

Dinosaur coprolite composed of wood fragments that were likely eaten by the Late Cretaceous (~75 mya) hadrosaur *Maiasaura*, from the Two Medicine Formation, Montana.

wood fragments from the coprolites, Chin realized they lacked lignin, a loss more typical of decayed wood. These and other clues led Chin in 2007 to propose that the dinosaurs ate rotten wood, which for a dinosaur would have been more nutritious than fresh wood because of what it held, such as fungi and wood-chewing insects.[56] Despite her persuasive argument, though, the coprolites did not hold any remains of insects or other arthropods; hence it remained speculative.

Ten years passed before Chin and other paleontologists were vindicated by further support for the audacious idea that some large "herbivorous" dinosaurs did not just eat plants, but also ate invertebrates as a normal part of their diets. In a 2017 paper, she and her colleagues described dinosaur coprolites from southern Utah similar to and about the same age (80–75 MYA) as those from Montana.[57] These coprolites likewise contained fungi-damaged wood fragments, showing how their dinosaurian dumpers ate rotten wood. Mixed in with the woody bits, though, were fossil crustacean parts. Although these body parts were not whole enough to identify their original owners, they were probably land crustaceans that lived in decayed wood. So Chin and her colleagues concluded these dinosaurs chewed wood that also included invertebrates, which enriched their diet with animal protein and calcium in the animals' exoskeletons.[58]

So were these dinosaurs like black bears that rip ponderosa pines, or were they the first "woodpeckers," in which they mined wood for its squirming arthropod goodies? To answer this question, we must go dinosaur watching and see what these animals do to woody substrates and the life within them.

: : :

The staccato burst came from above and behind us. Recognizing this sound, I whipped around and put binoculars to my eyes. The source of the sonic disturbance, clinging on to the vertical surface of a mature longleaf pine (*Pinus palustris*), seemed too small compared to its auditory impact. So we watched it for a moment before it resumed hammering, rapidly slamming its beak against the trunk of that tree.

My wife Ruth and I were in the Piedmont National Wildlife Refuge, managed by the US Fish and Wildlife Service and less than a two-hour drive south of Atlanta, Georgia.[59] We were there mostly to see living examples of the red-cockaded woodpecker (*Leuconotopicus borealis*). These birds are about the size of an American robin (*Turdus migratorius*), with wingspans barely surpassing the length of a standard knuckle-striking 12-inch ruler. Despite "red" as their main color descriptor, red-cockaded woodpeckers are mostly black and white, although males are distinguished from females by a red brushstroke on the sides of their otherwise white heads bounded by black straps.[60] Black-and-white horizontal banding on their backs is more obvious while watching them through binoculars from afar as they deftly scramble up and down tree trunks. Their feet are configured like grappling hooks, with two clawed toes pointing forward and two back, making an "X" pattern, enabling such Batman-like moves.

A quick scan of the forest revealed plenty of testimonials to this woodpecker's most impressive behavioral feat, which is to make round holes in longleaf pines with their heads. Their 5–7 cm (2–3 in) wide holes are located about 10 m (33 ft) off the ground and well below any branches.[61] Such heights also demonstrate that woodpeckers choose larger and more mature trees, and preferably ones with fungi-softened sapwood and heartwood. Each hole marks the entrance of a cavity large enough to hold at least one or more adult woodpeckers and deep enough to reach a tree's heartwood. Depending on volume, cavities are used as roosting places for individual birds and mated pairs, or for nests.

Making a new cavity may require a few years of pecking, a laborious but essential method for ensuring woodpeckers survive long enough to create future generations. The most recently excavated of such holes were denoted by freshly bled sap along edges, which formed a whitish sheet joined by streaks from individual drillholes above and below. This resin shows how woodpeckers pick live trees for their cavity building; the sap also deters tree-climbing snakes that might fancy eggs for breakfast, chicks for lunch, or parents for dinner. If longleaf pines are not so common, woodpeckers may use other species, but longleaf pines are the best for their needs, especially if they are nearly a hundred years old.[62] Interestingly (to me, anyway), red-cockaded woodpeckers are the only woodpeckers that make cavities in live pine trees. Holes also tend to cluster in relatively small areas, with dozens of hole-bearing trees reflecting group behaviors.

As for dying or dead trees (snags), red-cockaded woodpeckers find whatever life is within by striking them with beaks or peeling off bark with feet and beaks to unveil formerly hidden insects. In this and other pine forests, snags accumulate chunky aprons of peeled bark at their bottoms. Once exposed, pine beetles (larvae and adults), termites, wood roaches, carpenter ants, and other insects are either grasped with a long, forked tongue or snapped up by beak. Foraging is often divided vertically between the sexes, as males tend to seek insects on branches and

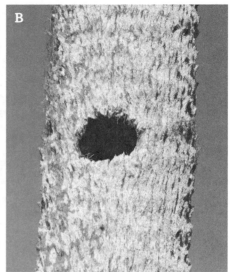

Woodpecker nest holes. A, Red-cockaded woodpecker (*Picoides borealis*) nest holes and "starter homes" (smaller diameter holes) in longleaf pine, Ichauway (Jones Ecological Center), southwest Georgia. B, Red-bellied woodpecker (*Melanerpes carolinus*) nest hole in dead cabbage palm, Ossabaw Island, Georgia. C, Pileated woodpecker (*Dryocopus pileatus*) nest holes in loblolly pine, Sapelo Island, Georgia.

the trunk in higher parts of a tree, whereas females mine the lower trunk for treats.[63] This division of labor makes sense—like a couple grocery shopping at different bodegas to prevent fighting over the same food items—and is especially helpful during winters when insects may be less active.

As one can tell from just these few observations of red-cockaded woodpeckers, they are extraordinary in many ways, but became more so as their populations decreased throughout the twentieth century. In a seeming paradox, humans that now admire this woodpecker were also responsible for reducing its numbers. It turns out that when you wipe out a bird's food source, potential homes, and nurseries, it has a tough time surviving. And in this instance, most of this bird's needs depend on one species of tree, the longleaf pine. Longleaf pines and their associated ecosystems used to cover most of the southeastern corner of North America, from east Texas to Virginia, and likely hosted millions of woodpeckers. But now, in the aftermath of European colonization and American exploitation of these trees for lumber and paper, as well as clearing forests for agriculture, less than 5 percent of that area has enough longleaf pines worthy of the word "forest."[64]

In the southeastern US and in many parts of the world, woodpeckers are both present and very much active participants in forests rural and urban, hinting at their adaptability and ecological effects. Moreover, wherever woodpeckers live, they have an outsized impact on wood-boring insects while also adding their own borings to trees and other plants in their environments. Through their bioerosion, they have also helped shape these environments in ways that greatly benefit their ecosystems.

: : :

Woodpeckers are birds, and birds are dinosaurs. I have been in paleontology long enough to remember when the statement "birds are dinosaurs" was greeted with incredulity, but it is now the stuff of five-year-old kids who eagerly point at a sparrow and say, "Look, Mom. A dinosaur!" Based on fossils and molecular clocks, we know birds descended from flightless (but feathered) theropod dinosaurs by the Middle Jurassic, at about 170–60 MYA.[65] The "oldest bird" honor is still bestowed on *Archaeopteryx*, a Late Jurassic (~150 MYA) raven-sized fossil represented by bones and feathers from Germany, with the first nearly complete skeleton found in 1861.[66] Yet answering the seemingly simple question "What is a bird?" became even more difficult with the discovery of Early Cretaceous (~130 MYA) feathered non-avian dinosaurs in China.[67] Among these non-bird animals were those that could probably climb trees and at least glide to other trees or the ground. As a result, paleontologists use a checklist of anatomical traits to decide whether any given small dinosaur was non-avian or avian.

These early birds did not just get worms, but probably ate a wide variety of items, including insects. Direct evidence of diet for Cretaceous birds comes from fossilized gut contents, with a few specimens containing plant remains, fishes, and lizards. Many Cretaceous birds also had teeth, and for at least one species of Early Cretaceous bird from China (*Sulcavis geeorum*), its teeth looked more suited for cracking insect or crustacean exoskeletons.[68] The foot of a Cretaceous bird from Burma (*Elektorornis chenguangi*) had an unusually long middle digit, prompting paleontologists to surmise it was used to probe for wood-boring insects in trees.[69] Of all Cretaceous birds, though, this foot is seemingly the only piece of evidence suggesting that any acted like a woodpecker.

Toothless lineages of birds belonging to the clade Neornithes evolved near the end of the Cretaceous, and birds from this clade were the only survivors of the mass extinction that killed all non-avian and most avian dinosaurs at 66 MYA.[70] Within the Neornithes are two clades, Paleognathae and Neognathae. Paleognaths included the ancestors of modern flightless birds like kiwis, rheas, emus, ostriches, and (my favorites) cassowaries, whereas neognaths were the ancestors of all other birds. For birds that made it through the catastrophic end of the Cretaceous, neognaths took off both literally and figuratively during the Cenozoic. Aided by flight, they spread throughout the world while filling and inventing new ecological niches.[71] Birds today are represented by about 10,000 species, twice as many as mammals, leading one to wonder if the mammals calling the Cenozoic the "Age of Mammals" might be a little biased.

When did woodpeckers emerge as part of this grand diversification of birds? Learning about their evolutionary tree helps us better understand how most of them evolved with trees. As neognaths, woodpeckers belong to the clade Cavitates, more generally known as "cavity nesters" for their habits of roosting or nesting in tree cavities.[72] Within Cavitates are Coraciiformes, consisting of kingfishers and bee-eaters, and Piciformes, which include woodpeckers, toucans, and related birds. Piciformes fossils do not show up in the fossil record until the Oligocene-Miocene transition, about 23 MYA. One of these fossils is a leg bone from the Early Miocene (21 MYA) of central France, assigned the species *Piculoides saulcetensis*.[73] This fossil closely resembles other bones from the Late Oligocene (25 MYA) from Germany, hinting that some of the earliest woodpeckers lived in this part of Europe then.[74]

Given this fragmentary fossil record and the genetic relatedness of modern woodpeckers, scientists proposed that woodpeckers originated in Africa, Asia, and Europe before moving on to the Americas.[75] However, Early Miocene woodpecker feathers preserved in Caribbean amber show these birds had already spread to the Americas.[76] How did these paleontologists know the feathers came from woodpeckers? Because woodpecker flight feathers are distinctive from those of most other birds, which is related to their habitual beak slamming. For instance, woodpecker tail feathers are stiffer than those of most birds, helping stabilize them against a tree trunk as their heads move rapidly back and forth.

Based on current behaviors, we can reasonably assume that the ancestors of these woodpeckers followed wood and the insects that lived in this wood. So considering the wealth of traces made by woodpeckers, one would think ichnologists have ample evidence of their existence aside from mere bones, and that they accordingly recreated detailed records of woodpecker behaviors over time. Alas, we do not, and we have not. So far only a few possible fossil nest holes have been interpreted from the fossil record,[77] and no undoubted woodpecker tracks, pecking traces, coprolites, or other such evidence are known.

Woodpeckers today are in most of the Americas, sub-Saharan Africa, Europe, and Asia, with greatest diversities in tropical rain forests of South America and southeastern Asia. In a way that makes perfect sense when a bird has the word "wood" attached to its common name, these birds live in places where cellulose and lignin get together to make tissues that also appeal to insects, which most often are in forests or other woodlands. Given such a broad distribution and more than 200 living species, woodpeckers are quite varied in their behaviors and diets, but they are united by their common origins and banging their beaks into hard plant tissues. How do woodpeckers manage to do this every day, year after year, but without hurting themselves? Answering such a question requires studying them through the perspectives of anatomy, physics, and (of course) evolution.

: : :

If you ever had good reason to purposefully pound your head against a wall, you probably regretted it later. This remorse is likely related to how human skulls are poorly adapted to traumas caused by colliding with solid objects, which can cause concussions, neck injuries, jaw problems, and more. A growing awareness of these effects belatedly led to the requirement of helmets for American football and cycling, and other headgear that helps cushion our crania and protect our occasionally valuable brains.

Of course, woodpeckers do not have the luxury of strapping on a helmet every time they need to thump their beaks into trees. Instead they relied on natural selection, which through a combination of genes, selection pressures, and luck resulted in astonishing adaptations that allow them to avoid knocking themselves unconscious every time they get hungry. This ability to withstand such forces becomes more impressive when looking at the numbers: they can move their heads at more than twenty-five pecks per second; at head speeds of 7 m (23 ft) per second; and with more than a thousand times the force of gravity (or g-force) when their beaks hit a hard wooden surface.[78] How do they do this and survive?

To help visualize a woodpecker's needs, imagine the required planning if you were to imitate one by climbing a tree trunk and striking its surface rapidly with a pick mounted on the front of your face. First of all, you must make sure your feet are firmly anchored to the surface below you, preferably gripping the trunk: think of wearing crampons rather than your weekend-special high heels. Second, your butt must maintain an absolutely perfect non-twerking state of motionlessness so that your hips can act more as a pivot point. Third, the pick needs to be as close as possible to your face, with no room for shock absorption when the pick slams into the tree. If you've gotten this far, then trying to peck fifteen times or more per second with your head is ill advised, seeing that the world's fastest drummers can only do this with their hands and drumsticks.[79]

In contrast, woodpeckers' bodies are adapted to address all of these requirements, and admirably so. Remember how their feet have four toes with claws and resemble an "X"? Feet with such shapes, called zygodactyl, are standard for woodpeckers that walk up, down, and across vertical surfaces, which comes in handy for moving about on tree trunks with minimal effort.[80] Such feet also help prevent these birds from knocking themselves off a trunk. Also, woodpeckers do not have tails, but they do have tail feathers, which (as mentioned before) are stiff enough to counter the back-and-forth movement of their heads.

Most importantly, woodpecker skulls are marvels of all-natural engineering. For instance, woodpecker skull bones are more mineralized than their other bones but also thin, making them both strong and resilient.[81] Their brains are tight against their skulls with almost no fluids between, which prevents thinking matter from sloshing around braincases. Strong muscles in the back of a woodpecker neck contract with each strike, an action helped by a woodpecker beak, which is integrated with a woodpecker skull so that it acts more like a rigid hammer than a shock absorber.[82] Then there is the matter of a woodpecker tongue, which laps up wood-boring insects. It is attached to the inside of the skull just below the eyes, runs through its right nostril, and splits in two as it winds up and around the back of the skull, before joining underneath and sticking out the beak. A bone, the hyoid, supports the tongue, with compact bone in its center and more porous (spongy) bone on the outside.[83] Another trick woodpeckers use when they drill is to close their eyes. That simple action ensures they retain vision afterward, because if they didn't, their eyes would pop out of their head. Such an unsightly event is also prevented by a third eyelid underneath two standard eyelids.[84]

Given such specialized anatomical traits, one can quite properly applaud woodpeckers for having evolved them in a mere 25–30 million years, while also keeping up with the continual evolution of wood-boring insects and the trees hosting these insects. Woodpeckers' current challenges, though, are related to human-caused deforestation, whether by cutting down trees for wood products, clearing land for agriculture, or too-frequent and too-ferocious fires related to climate change. Some woodpecker species may still make it through the next few hundred years, but their conservation will require paying attention to their behaviors and ecological effects.

: : :

I am always delighted to hear the raucous call of a pileated woodpecker, a war cry seemingly meant to strike fear in the circulatory systems of wood-boring insects everywhere. Yet these calls are actually meant to communicate with others of their species, whether to announce that they and they alone will eat insects from a given area, or to inform mates (potential or actual) they are nearby. Just in case that message is not clear, the rapid-fire hammering of a beak against a tree trunk seconds after a shout-out adds emphasis, much like an exclamation mark at the end of a declarative sentence. This woodpecker is not just meekly saying, "I am here," but more like "I am *exactly* where I belong!"

Woodpeckers and other birds already use a variety of vocalizations to communicate to others of their own species, such as: contact calls ("I am here, where are you?"); flirting ("Say, nice tail feathers!"); scolding ("Get away from my food, mate, and children!"); alarms ("Predator nearby!"); and begging ("Feed me, Mom and Dad!"). Yet bird calls can take on meaning even for those outside their species. For example, if I hear the frenzied calls of blue jays (*Cyanocitta cristata*) or American crows (*Corvus brachyrhynchos*) announcing the presence of a red-tailed hawk (*Buteo jamaicensis*), I do not have to see or hear the hawk to know it is nearby. Other bird species often react to such alarms and spread the word quickly throughout an entire forest as the original viral tweets. Nonhuman animals, including other mammals, are also well attuned to bird sounds around them, picking up both gossip and warnings while passing the news on to members of their species.[85]

For woodpeckers, striking their beaks against solid objects is not just for acquiring food or making nest holes, but also for augmenting their voices, like singers who also play drums. Similar to good drummers, woodpeckers use a blend of beats per second, variations of beats within a series, and silent pauses between each series.[86] These combos result in distinctive patterns recognized by woodpeckers within each species, while also letting those members know that they are ready to defend their territory or they are available for mating. You know, just like some drummers.

Woodpeckers normally use trees for their drumming, but as percussionists might tell you, not all drums are alike. For example, a woodpecker pounding its beak against a living tree with dense wood will not sound the same as one hitting a snag with hollow spaces caused by wood-chewing insects and fungal rot; the latter tree has more resonance, helping spread that woodpecker's message farther.[87] Woodpeckers also use fallen logs and stumps for creating a similar timbre with timber. Woodpeckers living in urban areas may even expand their repertoire by using human objects for drumming, such as utility poles, roofs, metal gutters, and more.

As woodpeckers evolved in the past 20 million years, so did their sounds. But drumming methods as species-specific communication also hold clues of common ancestry. In a 2020 study published by biologist Maxime Garcia and his colleagues, they analyzed the acoustic "signatures" of woodpecker species to test how these drumming sounds differed with genetic distance.[88] In other words, did the drumming patterns of more closely related species sound more similar than those of distant relatives? From their recordings of ninety-two woodpecker species' drumming, they recognized six distinct styles: steady slow, acceleration, steady fast, double knock, irregular sequence, and regular sequence. The researchers also verified that drumming was indeed a reliable way to distinguish species, and that more closely related species fit into the same acoustic categories.

Amazingly, these scientists pinpointed when in the geologic past these clades of woodpeckers split in their musical tastes. According to their calculations, the "slow and steady" drummers diverged from the "acceleration" drummers about 13 million years ago, whereas the "steady fast" drummers diverged from the common ancestor of both the "slow and steady" and "acceleration" drummers closer to 15 million years ago.[89] The researchers further proposed that drumming was an evolutionary outcome of pecking used originally for foraging that later also applied to communication. This is an excellent example of exaptation, in which an adaptation used for one purpose is later co-opted for another. In this case, the sounds originally produced by pecking on wood for food also became useful for defending territory and finding mates, while also changing over time as woodpeckers diversified in their forested environments. Still, this research showed that species-specific signaling via drumming is real, and that woodpeckers looking to mate will not fly to the beat of a different drummer.

Knowing about woodpecker drumming implies that ichnologists might look at modern trees and discern beak traces caused by woodpeckers that were seeking food versus those simply communicating. After all, feeding should cause more removal of bark and other actions that plumb farther below a woody surface, whereas drumming is more superficial. Could ichnologists likewise recognize trace fossils of either behavior? Maybe, but like most borings, deeper is better for preservation potential. Nonetheless, ichnologists can dream and otherwise imagine finding a series of holes in a Miocene petrified log that correspond with a rat-a-tat-tat message from the ancient past.

: : :

Previously we learned that some beetles carry fungal spores with them and use fungi either to soften wood or to cultivate fungal gardens for food. This mutualism benefits both the beetles and the fungi, as beetles get fed and fungi get spread. However, such symbiosis is often detrimental to living trees affected by such a dual assault, particularly if these beetle-fungi pairings are both invasive species. Climate change also makes matters worse, as native species of beetles and fungi can reproduce more quickly with warmer annual temperatures.

Woodpeckers in other symbiotic situations and ecological factors can become even more complex. For instance, flighted birds in general are well known for inadvertently (and rapidly) transporting smaller animals from one place to another, or aiding in their reproduction.[90] Woodpeckers likewise disperse other species, but because they are commonly associated with forest and adjacent ecosystems, what they spread reflects their communities.[91] Also, woodpeckers sometimes benefit themselves by carrying passengers. For example, red-cockaded woodpeckers make their lives a little easier by giving rides to red heart fungus (*Phellinus pini*) and other fungi. Woodpeckers transfer fungal spores from tree to tree, which infect the heartwood of longleaf pines and other trees. These infections soften the wood, better enabling the woodpeckers to make nest cavities.

An experimental study of this symbiosis, published by biologist Michelle Jusino and her colleagues in 2016, confirmed that red-cockaded woodpeckers were indeed responsible for spreading red heart fungus and other fungi.[92] Over the course of the study, holes accessible to woodpeckers contained the same fungi detected on the birds, whereas holes screened to keep woodpeckers out showed far fewer signs of fungi. Hence this mutualism reflects a balancing act between three species—the woodpecker, the red heart fungus, and the longleaf pine—but one in which the tree comes out holding the short end of the trunk. The study took more than two years to complete and was conducted in longleaf-pine forests on US Marine Corps Base Camp Lejeune in North Carolina, thus involving an unusual triad of woodpeckers, scientists, and Marines.

Another form of mutualism that works out great for a woodpecker species but not so well for another species is exemplified by the rufous woodpecker (*Micropternus brachyurus*) of Southeast Asia. This reddish-brown woodpecker loves eating acrobat ants (*Crematogaster* spp.), which build nests on branches or trunks of live or dead trees.[93] These nests are nicknamed "carton" nests because ants chew wood and paste it together to construct large multichambered structures. Rufous woodpeckers either attack acrobat-ant nests by themselves or as male-female pairs. Once on a nest, they stab vigorously with their beaks to expose its innards, while also agitating thousands of soldiers that march to their deaths. Rufous woodpeckers do not discriminate on the basis of life stages or castes, and readily eat eggs, larvae, and pupae, as well as adult workers, soldiers, and queens.[94] These woodpeckers also may be joined by other bird species that wait for them to start breaking apart a nest and swoop in afterward for an ant buffet. But the real coup de grâce delivered by rufous woodpeckers to acrobat ants is their nest-in-nest brooding. These birds regularly repurpose abandoned acrobat-ant nests to serve as their nests, cutting out hollow spaces where they lay eggs and raise their chicks.[95] In short, rufous woodpeckers eat acrobat ants out of house and home, and then take over their houses and homes.

Considering that rufous woodpeckers are already reusing and recycling, one might not be surprised to learn they also support sustainable agriculture. And not just any agriculture, but one ensuring that science, art, literature, and other creative human endeavors happen: coffee. Acrobat-ant nests protect and harbor mealy bugs (scale insects), which secrete a tasty substance ("honeydew") that the ants eat, analogous to how humans raise cows or goats for milk.[96] Mealy bugs eat coffee trees, and hence are considered pests on coffee plantations. However, when well-meaning coffee planters apply pesticides to ant nests to kill mealy bugs, they also kill ants, and thus take away the food, homes, and nurseries of rufous woodpeckers, the ants' main predator. The most logical solution, then, is to just let the woodpeckers be woodpeckers, and maintain trees with ant nests for shade-grown coffee plantations. This case argues for the intrinsic value of natural history, in which knowledge of ecological relationships also benefits important economic resources, and especially caffeinated ones.

: : :

For woodpeckers, their roosting and nesting cavities in trees are not only homes for the woodpeckers that make them, but also for future generations of their species. At some point, though, the usefulness of these spaces wanes, especially if parasites and other bothersome roommates become too abundant, or the host tree becomes less stable. Nevertheless, cavities do not suddenly become useless, but instead become available for other species, even as a live tree segues into snag.

A variety of insects, amphibians, reptiles, small mammals, and other birds are among the animals found living in former woodpecker homes. For example, birds that live in former red-cockaded woodpecker holes include tufted titmice (*Baeolophus bicolor*), eastern bluebirds (*Sialia sialis*), and even other species of woodpeckers.[97] If a larger woodpecker takes over a cavity and remodels it to her or his liking, then other big animals may hop in later, such as ducks, owls, and raccoons.[98]

Similar ecological echoes from woodpecker-made cavities happen in habitats wherever woodpeckers live, whether in boreal, temperate, or tropical forests. For instance, much like how cavities in your teeth multiply as one leads to more, woodpecker cavities in trees increase after outbreaks of wood-boring insects. What is different about this cavity making, though, is that these insect-prompted cavities also boost biodiversity later. In a 2015 study by conservation biologists Kristina Cockle and Kathy Martin, they found that an infestation of mountain pine beetles (*Dendroctonus ponderosae*) in pine forests of British Columbia attracted larger numbers of woodpeckers that ate these beetles.[99] They further discovered that woodpeckers produced more roosting and nesting cavities than normal, which in turn created more spaces in later years for other birds and small mammals.

The important message coming from such cavity studies is that trees with nest holes no longer occupied by woodpeckers still play an essential role in maintaining biodiversity in their ecosystems. Just as fallen trees or snags in a forest provide many species of fungi and insects with food and shelter, these same so-called dead trees with woodpecker-made cavities serve as homes for many animals long past the lives of their makers. This means that humans should be aware of such microhabitats as they contemplate clearing forests perceived as in decline: save the older living trees, but also save the snags.

: : :

In the summer of 2002, during the same two weeks I spent in central Idaho tracking wolves but also learning that black bears were major insectivores, the effects of another wood-altering species there caught my attention. One afternoon my tracking team encountered a ponderosa pine (*Pinus ponderosa*) that doubled as a food-storage unit. From near its base to as far as we could look up, its bark was filled with acorns, all stuffed snugly into acorn-sized holes. When I examined the holes at eye level, each were ringed with rough edges, speaking of a purposeful chiseling by beaked beings.

These impressive caches were the handiwork of the all-too-aptly named acorn woodpeckers (*Melanerpes formicivorus*), a western US species of woodpecker that stores massive amounts of food for winter. Acorn woodpeckers are gorgeous birds, with wide eyes, black beaks, mottled undersides, black backs, and black-and-white heads topped by bright-red caps.[100] These features lead some people to describe them as "clownish," an unflattering comparison, but one they nevertheless encourage by emitting calls that sound like *waka-waka-waka*. Still, their industriousness is no laughing matter, as ponderosa pines or other

trees may contain tens of thousands of holes created by generations of woodpeckers. In recognition of their seedy abundance,
these silos of the forest are called granaries.

Similar to red-cockaded woodpeckers, acorn woodpeckers
work in groups, typically composed of overlapping family units.
These groups cooperate in drilling and filling new holes, while
also popping nuts in and out of older empty holes, and over generations.[101] A tight fit is important for success, both to prevent
nuts from falling out, as well as to discourage squirrels and other
birds from stealing their food. Accordingly, woodpeckers may
adjust the sizes of holes or experiment with differently sized
nuts; despite their common names, they are also fans of hazel-
nuts, walnuts, and piñon-pine nuts. Amazingly, granary trees
are not hurt by this collective action, as the woodpeckers only go
"skin deep," pitting dead bark and mostly avoiding the cambium
beneath.

Similar to how other woodpeckers do not respect unnatural
boundaries, acorn woodpeckers also drill into and place acorns
into utility poles, wooden homes, fencing, and Sasquatch sculptures.[102] Such creative uses of woody substrates will no doubt
become more common as forested areas and urban landscapes
increasingly overlap.

: : :

Not all wood bored by insects, bears, or woodpeckers stays where
it falls. Some trees go on epic journeys, floating down rivers and
out to sea, where they drift along with currents of oceanic im-
port. What happens to these trees when they take their unwit-
ting sojourns? Surely their cellulose, lignin, annual rings, insect
borings, woodpecker nest holes, and other products of life do not
just stop giving, but also encounter other beings while contributing to their watery environments, too. This is where traces of
lands past and present meet traces of oceans past and present, as
wood takes on new lives and becomes bored once more.

Chapter 8

Driftwood and Woodgrounds

Of all the years to be a marine bivalve, 1588 CE was one of the best. This was when a certain faction of clams took sides in a human war, with their contributions leading to the downfall of one of the greatest navies in history. The victors were the combined forces of the English Navy and species of wood-boring clams like *Teredo navili*, whereas the losers were Spanish sailors of the "Great and Most Fortunate Navy" (*Grande y Felicísima Armada*),[1] a claim falsified by this marine-molluscan flotilla. In 1588, the Spanish fleet represented the then-empire of Spain and had been fighting England for decades, but lost a series of battles with them at sea. One of the deciding factors in their defeat was the weakened hulls of the Spanish ships, perforated by the clams.[2] Centuries later, American schoolchildren learned that this succession of superpowers somehow related to the manifest destiny of England and its descendants in former colonies of North America, but without giving due credit to their invertebrate allies.

This is but one example of how wood-boring clams played a role in shaping human history, in which an evolutionary legacy from the Mesozoic influenced our recent past. For instance, just short of a hundred years before the defeat of the Armada, an Italian navigator hired by Spanish royalty sailed an ocean of varying hues before encountering an island he thought was on the outskirts of India. He was wrong, but ultimately that did not matter. The island—named Guanahani by the Lucayan people who discovered it hundreds of years before the Italian navigator arrived—was on the southeastern edge of an island archipelago in the western Atlantic Ocean.[3] The Spanish renamed the island San Salvador; later, the group of islands west and south of it—with their rocky intertidal zones bioeroded by chitons and gastropods and beaches filled with parrotfish feces—was called the Bahamas.[4] The navigator and his crew enslaved several Lucayans from Guanahani, took them to Spain, and told tall tales to the royalty there about further riches that awaited their empire in the "New World." This sales pitch worked, leading to three more voyages by the navigator. However, the wood-boring clams hampered his last trip across the Atlantic Ocean, riddling the hulls of two ships in his fleet so thoroughly that their crews had to abandon ship and stay in Jamaica for a year.[5] Regardless of such troubles, other wooden sailing ships carrying soldiers, metal swords, cannons and other firearms, surveyors, horses, pigs, rats, and lethal diseases regularly crossed the Atlantic Ocean over the next few decades. Ships from England, Italy, Portugal, France, Belgium, and the Netherlands soon joined Spanish ships in carrying colonizers to the Americas, Africa, and beyond.[6]

Meanwhile, *Teredo* clams and other wood-boring bivalves were colonizing, too. Nicknamed "shipworms" because of their long, fleshy bodies and attraction to wooden sailing ships, these clams started their journeys as floating larvae that settled onto and attached to vessel hulls and wooden docks. Once they grew large enough to do so, these molluscans of mass destruction drilled into hulls, docks, dikes, pilings, and wharves. The damage they wrought was epic, ranging from the collapse of dikes in the Netherlands in 1731 to piers and wharves near San Francisco in 1919–1921.[7] Ships assaulted by these clams included famous ones, such as Sir Francis Drake's *The Golden Hind*, and *The Essex*, a whaling ship so debilitated by shipworms that when a large sperm whale slammed into it, it broke apart and sank.[8] A young novelist named Herman Melville wrote a book inspired by this whale attack, but did not acknowledge the crucial role of the shipworms in the whale's victory against human predation.[9]

How do shipworms drill with long, fleshy bodies? Instead

of two prominent valves enclosing their soft parts, these clams have a pair of small, ridged calcareous shells toward their front ends that are hard enough to abrade woody tissues. Their drilling rotates the valves clockwise, counterclockwise, and back again, while also moving them up and down.[10] This action tears plant fibers grown over centuries by trees that were unceremoniously harvested, processed, and reduced into planks or pilings. Once safely ensconced in their new homes, these clams secrete a thin layer of white calcite to reinforce their borings and protect themselves while continuing to grow and erode. Two calcareous structures called pallets provide further protection at their rear ends, close to the outer wood surface.[11] Eventually, multiple generations of clams result in overlapping holes denoted by long, calcite-lined tunnels. For at least 2,500 years, collective actions by these wood-boring bivalves have undermined human ships' integrity, weakening the same parts meant to keep them afloat for long voyages.[12] Wooden sailing ships also greatly expanded the ranges of different species of shipworms, introducing new marauders to docks and other ships with each excursion.[13]

The Europeans and later the Americans fought back against these marine opportunists, but eventually lost. They tried using different types of wood, chemical treatments, frequent repairs, and more, but with little success.[14] By the late nineteenth century, once ships began moving under their own power from steam produced by coal (and thus made early contributions to future heightening of seas everywhere), shipbuilding also switched from wooden to metallic hulls.[15] Much later, petroleum was used in another way to develop fiberglass and other plastic applications in hulls. After that, ships were less prone to sinking from clams, and more likely to suffer damage from armed conflict, reefs, and (of course) icebergs.

As a result, most vessels floating in marine environments—from lifeboats to massive cargo ships—have hulls built to discourage or repel these pernicious bivalves, even as wooden docks, seawalls, and other coastal structures still undergo active degradation. Yet these clams and other marine bioeroders of driftwood live on regardless of what we do, still making homes in tree parts that leave land to sail the seas for however long their integrity permits. Moreover, clams are not the only marine animals that grind or eat wood. For example, crustaceans called gribbles are similarly responsible for coastal consumption of former trees' remains. Then there is wood that floats above deep-sea areas, eventually sinks into abyssal depths, and attracts new bioeroders once it rests on the sea bottom, including more clams and other crustaceans.

: : :

As long as trees have existed, their remains have gone out to sea. As we learned earlier, the first forests, composed of closely spaced plants that grew high enough to warrant our calling them trees, started in the Devonian at about 390 MYA.[16] These plants owed their vertical enhancement to the evolution of vascular tissues that lent enough turgidity to leave mosses and liverworts behind and to stand up for themselves. Those trees also developed extensive root systems and clustered in greater densities, helping hold soils in place and permanently altering the forms of rivers. Until these evolutionary innovations, most rivers were braided, choked with sediment and bearing many narrow intertwined channels.[17] But the retention of soils by tree roots put rivers on low-sediment diets, shaping them to flow in now-familiar single-channel meandering patterns, with cut banks, point bars, levees, and much more. Had it not been for this ecological-geological interaction, Mark Twain's *Life on the Mississippi* (1883), the musical *Show Boat* (1927), and other creative works inspired by meandering rivers surely would have been treated far differently.

Soon after the advent of forests, woody debris from these plants washed into rivers and flowed downhill. Forests closest to Devonian oceans contributed their wooden bits to coastal rivers that emptied into these seaways, adding massive pulses of organic matter those environments had never before encountered. This overload of organics in turn gave aerobic bacteria in oceanic environments too much to eat, which consumed too much oxygen and led to oxygen-starved oceans.[18] Meanwhile, forests and oceans also acted as carbon sinks, taking in and holding on to so much carbon that carbon dioxide levels lowered and caused global cooling, like snatching a blanket off a bed. As a result of this cooling, sea levels went down, which diminished marine environments in general while also spreading low-oxygen zones into those areas.[19] Because many invertebrates and fishes required oxygen to breathe, they died, oceanic ecosystems collapsed, and a mass extinction took out nearly 80 percent of all species both in the seas and on land.[20] Thanks a lot, Devonian forests.

Fast-forward to the Early Jurassic, about 200 million years later. By then, the earth had witnessed two more mass extinctions, the end-Permian (252 MYA) and the end-Triassic (201 MYA).[21] Many land plants survived both extinctions, and their survivors had enough variety to populate and grow new and different forests each time. Because trees are superb organisms (but ultimately looking out for themselves), they formed the foundation of many terrestrial ecosystems for insects and other invertebrates, as well as early dinosaurs, pterosaurs, mammals, and other continental animals.[22]

Following the end-Triassic mass extinction and the recovery of land plants in the Early Jurassic, trees once again flourished. Dead trees from coastal forests once again floated away from land, sailing the seas long before armadas, regattas, cruise ships, and billionaire/supervillain yachts. However, this renewed influx of wood to oceanic environments was not enough to overwhelm world oceans and decomposers, nor did it trigger a mass extinction. At least a few of these logs drifted on oceanic currents for years, and our evidence for these long-lasting Jurassic voyages are the fossilized logs themselves, adorned with enormous crinoids.

: : :

As mentioned earlier, crinoids are animals and echinoderms, which include sea stars, sea cucumbers, sea urchins, and more. Like sea cucumbers, crinoids have a botanically alluding nickname, "sea lilies." However, crinoids lived in oceans well before vascular plants (let alone flowering plants), having first evolved about 480 million years ago in the Ordovician.[23] Both fossil and modern crinoids superficially resemble flowering plants because some have branching "roots" at their base, which connect to thin stalks and thread-like branches along their lengths. Stalks are also crowned by "petals," which are arranged radially around central knobs reminiscent of receptacles in a typical flower with stamens and pistils. Look closer, though, and you will see that the "stem" is composed of a series of calcite rings stacked on top of one another, and the thin "branches" are projections called cirri. The "petals" at the top are more akin to arms, but made of calcite plates and bearing thin, comb-like bristles (pinnules) that lead to food grooves running along the lengths of the arms. Finally, the "receptacle" in the center of a crinoid "lily" is its calyx, composed of an intricate mosaic of calcite plates.[24] This is the home of a digestive system, including a mouth and anus, that processes and powers each otherworldly being. Crinoids are not just animals, but exquisite creatures that implore you to weep with joy that such complexity and splendor has existed for so long on this earth.

Admittedly, I have yet to see a living crinoid (and am quite cross about that), but I have seen plenty of dead ones and innumerable pieces of these departed echinoderms. Crinoids are (and were) suspension feeders, meaning they gather particles of organic matter suspended in seawater with their arms, filter it through their pinnules, and direct this collected food down their food grooves to their mouths.[25] Most crinoids were (and still are) sedentary, attaching to one place and staying there for the rest of their lives. Yet others swim, and swim beautifully as their pinnule-laden arms propel them through the water with mesmerizing motions.[26] Other mobile crinoids use their arms to crawl along ocean bottoms, and at least one crinoid that washed onto a Middle Jurassic shoreline in what is now Portugal actually crawled to its death, its body still at the end of its trail.[27]

For the first 300 million years or so of their existence, crinoids mostly lived on sea bottoms, settling on and attaching to soft or hard substrates, and staying put.[28] But starting in the Early Jurassic, some crinoids became nomadic, attaching onto floating tree trunks and allowing these wooden vessels to carry them through ocean waters for life. And we know this courtesy of spectacular crinoids directly connected to fossil tree trunks from Early Jurassic rocks of Germany.

I will never forget when I first learned about these Jurassic crinoids and their attachment to wood: not by reading about them in a book or on a website, but by seeing the real thing. During a fun-filled and geologically educational visit to Germany in 1995, a paleontologist friend of mine, Reinhold Leinfelder, took me on a day trip from Stuttgart (his home then) to Holzmaden and the Urwelt Museum Hauff. This museum holds a world-famous fossil exhibit put together by a family-owned company that mined an Early Jurassic (~180 MYA) shale in the area known as the Posidonia Shale (in German, *Posidonienschiefer*).[29] This fine-grained dark-gray rock, erroneously labeled as "slate," was used as a roofing and paving stone. But quarry workers under the employ of family-owner Bernhard Hauff kept encountering finely preserved fossils in this shale, often enough to warrant their recovery and preparation. Hauff and others eventually built the on-site museum to display these fossils, and it is now the largest privately owned museum in Germany.[30] Fossils from the Holzmaden quarry include ammonites, sharks, bony fishes, marine crocodiles, and plesiosaurs, but its most famous fossils are its ichthyosaurs. Some ichthyosaurs not only consist of complete skeletons, but also have filmy carbonized outlines of their original skin. Moreover, a few mother ichthyosaurs from the Posidonia Shale died just before the act of live birth, their offspring exiting tail first but forever stuck with their mommas.[31]

Yes, these body fossils impressed me, and I remember a younger version of myself shamelessly gawking with slack-jawed amazement at the paleontological bounty on display. But the fossil specimen I remember best for its grandeur and posing of a paleontological mystery was a slab of shale hosting a 14 m (45 ft) long fossil log with more than a hundred crinoids attached to it. The shale was displayed on a wall like a massive work of visual art, its detail inviting close scrutiny but its size also requiring backpedaling for perspective. Some crinoids (*Seirocrinus* subangularis) have stalks several meters long, with grand arms, finely defined pinnules, and other parts perfectly preserved, whereas the original log—once a tree that lived in an Early Jurassic forest—was but a flattened carbonized residue of its former self.

After getting over my initial speechless awe, I remember talking excitedly with Reinhold about this magnificent fossil assemblage, and he explained to me the then-reigning hypothesis about its origin. Appropriately enough, these suspension-feeding crinoids were themselves suspended, with their bases attached to the bottom surface of the log and the rest of their bodies hanging below in the water. Their arms and pinnules acted like nets, gathering suspended organics from the water as the log bobbed at or near the ocean surface. Based on the crinoids' sizes, the log must have drifted for years, gaining new passengers as it moved. At some point, though, the log soaked up enough water and gained enough crinoids that its buoyancy was no more, and it fell to the ocean bottom. That environment had so little oxygen that no animals disturbed the log and its companions. Sediments later buried the log, preserving it and its passengers until people uncovered it about 180 million years later.

This remarkable fossil composition of what was nicknamed the "Hauff specimen" was my first experience in learning about marine species (crinoids) adapting to floating terrestrial debris (wood) during the Mesozoic. In today's oceans, though, we take such relationships for granted. Remember barnacles? Those crustaceans often attach to driftwood and other floating objects today, as do some sea anemones, bryozoans, and other animals. Yet this mode of life was unusual in this early part of the Mesozoic, especially for such extreme examples like this.

What I just described seems like an open-and-shut case: trees washed out to sea, floated for a long time, crinoids settled on them, they all eventually sank to the sea bottom, were buried, and fossilized, but then found by people 180 million years later. But was that the real story? We scientists propose alternative hypotheses, or you could say different stories, but based on the same evidence. For instance, did the crinoids actually float with the logs near the water surface, or did they attach *after* the logs had already sunk to the sea bottom? Oh, those scientists. How dare they inject doubt into such a perfectly good and simple story!

Another explanation for the Hauff specimen likewise states the log came from land and did indeed float out to sea, and that crinoids attached to it while it was at sea: these conclusions are tough to deny. Where it veers into different territory is when we consider the amount of time the log was afloat and when the crinoids attached to it. The alternative posits that the log's porous tissues may have filled with water and sunk to the sea bottom immediately, and only then did crinoids attach to it. Proponents of the "sunken place" hypothesis also pointed out that the mass of so many robustly bodied crinoids would have weighed down the log, helping it sink more quickly. In this scenario, crinoid larvae floated in the water until they found a hard surface (the log) to settle down, feed, and grow, but at the bottom of the ocean.[32] From then on, their stalks projected up above the ocean bottom and they fed their way into becoming stately behemoths of crinoids, more like "sea sunflowers" than "sea lilies."

However, one fatal flaw in this alternative hypothesis was the fatal nature of the Posidonia seafloor. The perfectly preserved fossils of ichthyosaur mothers giving birth and other animals, as well as the high organic carbon content of the shale, implied that bottom-water conditions in this part of the ocean had very little to no oxygen.[33] No animals were scavenging or otherwise breaking down these future fossils before they were buried, because no animals would have survived even a short visit to this no-breathing zone. Likewise, because crinoids are animals and need oxygen to live, their larvae could not possibly have survived a trip into this terrible environment, let alone settle onto a log there and live long enough to grow and form one of the most spectacular of all known fossil-crinoid colonies.

Still, the exact nature of this international log of mystery remained unresolved until 2020, when paleontologist Aaron Hunter and his colleagues published a thorough study of it and its crinoid colony.[34] One of the tools they used to test the "floating raft" hypothesis was math. First, they mapped and counted where the crinoids had attached along the length of the log ("bases"), as well as locations of their "crowns," with the calyx and arms of most extending more than a meter away from the fossil log. They then looked at the spatial distribution of the bases and crowns and tested whether or not these distributions were evenly spread along the log, or if they showed preferred clustering. The researchers reasoned that if the log truly served as a crinoid cruise ship, its suspension-feeding passengers would have been best fed by congregating toward the rear (stern) rather than the front (bow). In such a position, they would have caused that end of the log to dip farther down into the water, delivering more organic goodies than if positioned closer to the surface. On the other hand, if the crinoids were not clustered and spread throughout the log, this random distribution would favor that the log was more of a bottom-floor residence for its biota.[35]

The researchers' spatial analysis clearly showed clustering toward one end of the log, which accordingly would have dipped lower in the water and helped the crinoids' suspension feeding as they hung down. Based on calculations of growth rates for the crinoids, as well as the amount of time needed for water to infiltrate pores in the wood and saturate it so that it sank to the ocean bottom, the scientists estimated the log carried them for 15–20 years. The uneven distribution of the crinoids also tells us which end of the log with fewer crinoids was closer to the surface as it drifted on a Jurassic sea.[36]

As remarkable as these fossils might be, the gentle reader may be wondering what the Hauff specimen has to do with bioerosion, and specifically the bioerosion of wood at sea. To answer that question, notice what I did *not* say was part of this fossil assemblage, but was mentioned earlier with regard to the sinking of the Spanish Armada: wood-boring bivalves. This omission was not because I forgot, but because the Early Jurassic had no wood-boring bivalves. In other words, this mind-boggling extreme of giant crinoids colonizing wood that floated at sea for 15–20 years would have been downright impossible in a sea with wood-boring clams: no wood, no giant crinoids. Thus a golden age of grand crinoid voyages ended soon after the Early Jurassic, an extinction assured by the diminution of their seafaring vessels.

How did the advent of wood-destroying bivalves happen? A lineage of clams that normally would have bored into rock or coral acquired some symbiotic friends that helped them break into and break down these foreign organic materials in their oceans.

: : :

One species of wood-boring clam, the "shipworm" *Teredo navili*, is often singled out and solely blamed for destroying human-made wooden objects, from ships to docks to doors in Venice. But this species is not alone in its bioerosion, as marine wood-boring clams are diverse, numerous, and widespread. The genus *Teredo* alone has about fifteen species, and other genera, each with multiple species, include *Bankia*, *Lyrodus*, *Martesia*, *Nausitora*, *Xylophaga*, and more.[37] Although wood-boring clams are often divided into two "families," Teredinidae and Xylophagaidae, those groups share a common ancestor and belong to the clade Pholadidae.[38] Pholididae also includes rock-boring and firmground-eroding clams, which in turn implies that wood-boring clams originated from lineages already capable of drilling into hard materials. The division of wood-boring clams into two groups is also generally based on where they live. For example, most clams in the Teredinidae, such as *Teredo*, live in wood in shallow-marine environments, such as bays and mangroves. In contrast, most clams in Xylophagaidae, such as *Xylophaga*, live in wood that somehow made its way into deep-marine environments.[39] However, one species of shipworm, *Kuphus polythalamius*, does not live in wood at all, but instead dwells in shallow-marine muds of the Philippines and elsewhere in the western Pacific.[40] This clam, which has tiny valves compared to its long, fleshy body and secretes a calcareous covering around itself, descended from wood-boring species. *Kuphus* is also the world's longest living bivalve, with some specimens reaching 1.5 m (5 ft) long.

As far as eating goes, wood-boring clams are suspension

feeders. Like rock-boring clams, they have two siphons for suck-ing in suspended organic matter (incurrent siphon) and one for ejecting wastes (excurrent siphon). These long siphons stick out of the shell and are positioned close to a wood surface at the top entrance of their boring, whereas the clam's shell is at the bot-tom.[41] Unlike wood-boring insects, clams do not directly chew wood, and instead drill through it. Nearly all wood-boring clams also gain their nutrition from wood, but not directly. Like any re-lationship that involves more than one entity, it's complicated. These clams host bacteria in their gills that chemically break down cellulose in the wood and convert it into organic com-pounds the clams can digest.[42] Clam digestive tracts include a stomach, intestine, and caecum, which is a pouch off the stom-ach that stores and digests wood.

Because wood-boring clams inherited their abilities from rock-boring clams, the motions and methods for drilling into woody tissues are similar. Their drilling methods, first described in the late 1960s, differ only slightly between species. For ex-ample, *Martesia striata* extends and retracts its muscular foot to move its shell down in its boring, while its muscles contract or relax in pairs to partly close and open valves, respectively.[43] This causes the shell to rock back and forth, but also rotate it clock-wise and counterclockwise. This combination of up-and-down, lateral, and rotational movements of the shell against wood scratches it and deepens the boring, with the clam continuing until its shell is well below the wood surface. In its drilling behav-ior, *Xylophaga dorsalis* moves its shell much like *M. striata*, but differs by alternating its adductor-muscle contractions instead of pairing them.[44] Both species follow what are described as "bor-ing cycles," which I will not describe in detail lest they live up to their name. Just know that cycles are as fast as once every 15 sec-onds for *M. striata* and about every 10 seconds for *X. dorsalis*.

Carvings produced by these clams are stubby to lengthy cylin-ders, but with a round cross-section that widens into a bulb at its bottom. Hence these traces are often described as "club shaped," but more like a bashing club rather than an exclusive one.[45] A more colorful way of describing these borings is as elongated

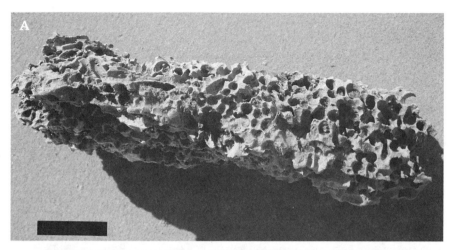

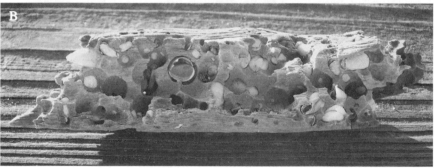

Bivalve-bored driftwood. A, Wood carved by wood-boring clams but lacking their shells, Ossabaw Island, Georgia; scale = 10 cm. B, Wood carved by clams (*Martesia striata*) with shells still in their former homes, Sapelo Island, Georgia.

"teardrops," or more royally, "Prince Rupert's drops."[46] The latter is in honor of Prince Rupert of the Rhine, a seventeenth-century warrior-scientist who showed how molten glass dropped into cool water formed solid glass with tadpole-like shapes and unusual physical properties. In finer-ground woody substrates, clam's shells can also leave striations on walls of borings, but in coarser material, original wood grains might also become apparent.

: : :

How much wood can a wood-boring bivalve bore? One clam by itself does not erode much wood, as they are often thin-bodied and limited by the amount of wood available. Yet these clams never seem to drill alone and are often densely populated, whether in individual chunks of driftwood, or in ship hulls, dock pilings, and peg legs. Given a surface with many holes in it made by these clams, volumes of missing wood can be measured, and given enough time and the right tools, scientists can calculate these volumes and rates of bioerosion.

In a 2015 study by marine biologist Diva Amon and her colleagues, they placed wood at three deep-sea sites ranging from 500 m (0.3 mi) to 4700+ m (3 mi) deep and later retrieved these samples to figure out volumes and rates of bioerosion.[47] From these experiments, the scientists calculated the range of volumes for individual borings made by different species of *Xylophaga*. They estimated wood erosion rates of as much as 22 percent per year, and that one hundred individuals of *Xylophaga* could erode 60 cm^3 (3.7 in^3) of wood per year.[48] These amounts may not sound like much, but add up quickly once millions of clams are involved over just a few years. For this study, Amon and others used computer tomography (CT) scans to render gorgeous 3D representations of the borings, many of which had the shells of their *Xylophaga* makers still in them.

In a 2021 study, marine biologist Irene Guarneri and her colleagues used a winning combination of X-rays and mathematics to calculate volumes of individual borings made by clams.[49] In this study, the researchers placed wood panels in shallow water at three places along the Italian shore, kept them immersed for three months, then pulled the panels out of the sea and put them into an X-ray machine. The resulting 2D images (radiographs) clearly showed shafts dug into the wood by clams, which the scientists converted into volumes by applying geometry. After some clever problem-solving, they successfully correlated shell diameters with volumes of their borings. This meant that when they saw shells in radiographs, they could also estimate volumes of bioeroded wood by simply measuring the sizes of the bioeroders' shells.

Given such tools and methods, scientists can reliably predict rates of bioerosion, amounts of bioerosion, and which clams bioerode more or less than other clams, whether in shallow- or deep-marine environments. All of this is not only fascinating, but also practical, bestowing a better understanding of how these animals affect our built environments, which annually inflict hundreds of millions of dollars' worth of damage.[50] Probably the most important facet of this understanding, though, relates to the long legacy of these clams.

Wood-boring bivalves seemingly have a single origin story from rock-boring ancestors, and one that coincided with their gaining bacterial helpers for digesting wood. Moreover, this evolutionary innovation was one of the most significant developments in the 500-million-year history of bivalves. The descent of modern wood-boring clams from one lineage of rock-boring clams is firmly supported by shared genetics, otherwise known simply as rRNA.[51] The rRNA evidence also suggests that symbiosis between cellulose-digesting bacteria and certain rock-boring clams happened at about the same time, enabling this great "wooden step" forward.[52] Such evolutionary codependency is analogous to how modern wood roaches and termites show their common ancestry via their shared symbiotic bacteria. A single origin of wood use in bivalves later led to their branching out into different marine habitats, with *Teredo* and its kin inhabiting wood near the ocean surface, and *Xylophaga* and its close relatives drilling into wood on abyssal plains.[53]

Marine wood-boring clams probably got their start sometime between the Middle and Late Jurassic between 165–60 MYA, which can be narrowed down thanks to fossil driftwood from just before and after then. For instance, giant floating crinoid colonies on wood at about 180 MYA imply no clams or other marine animals were boring into wood at that time. Similarly, fossil wood from Middle Jurassic (~167 MYA) marine deposits of Poland had plenty of fossil gastropods around them, but these mollusks were grazers, not wood-eroding.[54] This fossil wood is also striking for what was *not* there: borings or wood-boring bivalves. In contrast, Late Jurassic (about 160 MYA) fossil wood from Cuba contains borings that closely resemble those made by modern clams in driftwood.[55] Cretaceous examples are much more

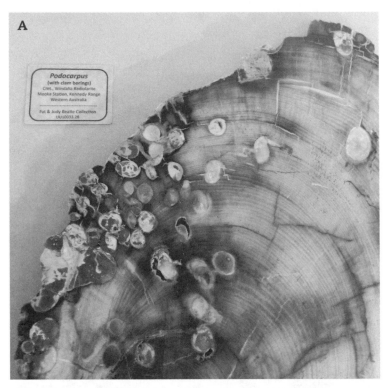

Fossil examples of wood bored by marine bivalves. *A*, Cross section of a fossil tree trunk (*Podocarpus*) with marine bivalve borings from the Windalia Radiolarite (Cretaceous, ~110 mya), Western Australia; specimen in the Museum of Natural History of Utah, Salt Lake City. *B*, Fossil log with sandstone-filled bivalve borings in the Tirikohau Formation (Miocene, ~20 mya), New Zealand.

common than those of the Late Jurassic, suggesting that wood-boring clams became more prolific and widespread throughout the rest of the Mesozoic.

Trace fossils attributed to wood-boring clams from the Late Jurassic to the near present were assigned the ichnogenus *Teredolites*, its name alluding to how modern *Teredo* clams make very similar traces.[56] *Teredolites* can be identified on the basis of two criteria. One is its shape as a stubby to lengthy cylindrical boring with a circular cross-section that widens at its bottom end and a rounded base, described earlier as "club shaped." A second trait, and an important one, is that the borings must be in fossil wood or cast by sediments that later filled the boring. I have seen many examples of the latter with *Teredolites*, in which the fossil driftwood that hosted the boring clams is gone, but its ghostly outline is defined by closely packed sand-filled borings, giving new meaning to the phrase "crowded clubs." Remarkably, most modern and fossil borings in the same chunk of wood manage to miss each other, as if these clams were respecting boundaries with their neighbors. Differently sized borings within the same assemblage can reflect growth stages of the clams, from baby boring bivalves to full-sized adults, similar to how beetle-larvae tunnels in wood do the same.[57] In a few lucky instances, the body fossil of the clam that made the trace fossil is also at the end of a *Teredolites* as a rare example of a tracemaker directly associated with its last behavior.[58]

After clams began breaking down wood in the Jurassic, they were eventually joined by two groups of wood-eroding crustaceans, gribbles in shallow-marine and squat lobsters in the deep sea. Although their wood-eroding methods differ from those of clams and are not quite as effective, these crustaceans nonetheless added their effects to marine wood cycles. The oldest known body fossils of squat lobsters are from the Middle Jurassic, or shortly before wood-boring clams got their start, which may or may not be a coincidence.[59] Their trace fossils are unknown and would be extremely difficult to distinguish from those of actual lobsters or other decapods, whether as tracks or pinches taken from fossil wood. Fortunately, wood-eating crustaceans have trace fossil representation, with ichnologists in a 2020 study recognizing gribble borings in fossil wood from the Early Cretaceous (about 110 MYA) of Alberta, Canada.[60] These trace fossils, assigned the ichnogenus name *Apectoichnus*, gave us further insights on how the bioerosion of marine wood involved more than just clams for more than a hundred million years.

However unlikely it might seem, this long relationship between terrestrial wood and marine wood-boring clams and other invertebrates not only happened, but also persisted, and likely will continue well into the future. As long as forests keep sending dead trees into the oceans, and humans keep using the remains of dead trees in and around ocean environments, these clams will keep on living and drilling.

: : :

Once paleontologists and geologists realized the abundance and extent of trace fossils made by wood-boring marine animals in the past, they began thinking about how to apply this knowledge to interpreting the history of world oceans. Although interrelationships between forests and oceans might seem counterintuitive, the preceding observations should provide hints of how geologists discern when sea level went up, went down, or stayed the same in the geologic record. And an integral part of interpreting these changes depends on recognizing trace fossils of wood-boring marine clams and other animals.

When wood from forests is deposited in shallow-water environments, it is not always as entire trees, or even just their roots or branches. Instead, most wood is broken down physically into tiny pieces that are easily carried in suspension and deposited much later and farther away from their original life places. Rivers could have carried this initial woody load from forests to the sea, with river currents and rocky bottoms breaking it down further along the way. Once in oceanic environments, waves and tides that washed out trees from coastal environments break them apart, as do clams and gribbles. I have sometimes seen accumulations of such comminuted plant debris flushed by rivers and carried by waves on Georgia beaches, evident at low tide as extensive dark-brown layers that look like wet coffee grounds.[61] In places where yet more woody material meets the sea, such as in tropical deltas, this organic debris forms thick layers that are buried under sediments and compressed. But if sea level rose and ocean waves eroded sediments above, these compacted layers of woody tissues were exposed on the seafloor. This is when ground wood turned into a woodground. Given such substrates, floating larvae of marine wood-boring clams descended and settled in, not knowing or caring whether the wood was fresh or old. As far as they were concerned, it was home, and it was food.

Thanks to ichnologists and geologists, we know such woodground surfaces with their characteristic borings have existed since at least the Late Jurassic, and that these surfaces related to sea level changes of the past. In 1984, three ichnologists—Richard Bromley, George Pemberton, and Ray Rahmani—published a paper that reported trace fossils of marine-clam borings (*Teredolites*) in the top of a coal bed from the Late Cretaceous (~70 MYA).[62] The borings were preserved as natural casts, filled with sand from above and rendered as beautifully defined structures. A few of the trace fossils also contained shells of their makers, which were related to wood-boring clams in shallow-marine environments. Based on this find, the authors then proposed a new perspective on fossil woodgrounds and their trace fossils, while also highlighting their significance as geological markers.

When an assemblage of trace fossils found together can be linked to a specific set of paleoenvironmental conditions like this, and the geologic record has multiple examples of such assemblages, ichnologists refer to it as an ichnofacies (where *ichno* = "trace" and *facies* = "appearance").[63] Ichnofacies have been a helpful tool for geologists who interpret environmental change through time, using trace fossil assemblages representing environments from the land to the deep sea. However, an important point Bromley and his colleagues made about these marine woodgrounds and their trace fossils is that they were totally unlike ichnofacies made in softgrounds, firmgrounds, or hardgrounds that most sedimentary geologists might be comfortable discussing. After all, these layers were not composed of minerals, but of organic material that used to be on land that later formed extensive surfaces on ocean bottoms. Hence these ichnologists proposed such distinctive assemblages of marine-woodground trace fossils as the *Teredolites* ichnofacies.

Since this paper was published in 1984, geologists have found many other examples of fossil woodgrounds that likewise fulfill the definition of the *Teredolites* ichnofacies.[64] But more importantly than just naming things, geology advanced as a science because of these fossilized woodgrounds. Time and again, geologists linked this ichnofacies with rising sea level, where they can point to a surface filled with trace fossils made by wood-boring marine clams and other animals, and say confidently that the sea rose and moved across that place as a transgression.[65] The modern transgression linked to climate change is likewise eroding down into wood-bearing shoreline deposits of the past, exposing these older layers so that modern wood-boring marine animals can settle in. In this sense, then, the past and present of wood-eroding animals are coming together, both conceptually and literally.

: : :

Evolution works in ways we don't always predict, which makes new discoveries of evolutionary innovations all the more exciting. One recent discovery that inspired sheer delight in ichnologists, ecologists, biologists, and other people who love new natural-history discoveries was a 2019 study reporting on a new species of freshwater clam in the Philippines that bored into rocks. The clams live in the Abatan River of Bohol Island in the south-central part of the Philippines, and although locals knew about them, they and their rock-boring ways were previously unknown to academic scientists. The group of scientists, led by marine biologist J. Reuben Shipway, assigned the species name *Lithoredo abatanica* to this clam, and they thoroughly studied its anatomy, life habits, genetics, biochemistry, and (best of all) borings.[66] The clams drilled into limestone bedrock exposed in parts of the Abatan River, creating enough holes to supply homes for other invertebrates living in the river. In this respect they were ecosystem engineers, carving out new niches in this river ecosystem.

Up until then, ichnologists, freshwater ecologists, and many other natural scientists would have readily agreed that drilling into rocks is a job best suited for marine rock-boring clams. Indeed, freshwater animals rarely bore into rock, represented by only a few types of insect larvae and a freshwater mussel (*Lignopholas fluminalis*) related to marine rock-boring clams.[67] Similarly, wood-boring bivalves are unknown in freshwater environments. Hence this discovery prompted all of us to revise our mental notes and think outside the drillhole.

The evolutionary twist in this marvelous story of scientific discovery was that these rock-boring freshwater clams were actually shipworms, and that they had descended from marine wood-boring species. Did they have small, ridged valves at their front ends, pallets at their rear ends, and a long fleshy body in between? Yes. Did their genetics show they were closely related to a marine genus of shipworm, *Teredora*? Yes. Did they have microbial symbionts in their gills? Also yes. Yet they lacked a caecum, which is a standard-issue internal organ for wood-digesting clams. Similarly, their guts contained no woody material, but instead were filled with minerals matching those in the limestone, showing they were eating the limestone.

Just what was the origin of this species? Like many rivers, the Ataban connects to the ocean. So somehow generations of shipworm larvae from a formerly seafaring species traveled up into the river system, and some larvae adapted to low-salinity water. Also, where lacking wood, those larvae found limestone, settled onto it, and drilled into it. Without a need for wood as food, there was also no need for a caecum, which probably first became a vestigial organ (like our appendix) and later vanished. These shipworms were also feeding off the limestone, perhaps by consuming green algae and cyanobacteria on the rock, with that diet supplemented by suspended organics.[68] As of this writing, the exact role of the gill-riding bacterial symbionts is still unknown, but *Lithoredo* had obviously traded in its cellulose-digesting pals for more rock-appropriate helpers. In short, natural selection eventually caused the descendants of marine wood-boring clams to declare, "No wood? No problem."

This behavioral shift from boring into rock to wood to rock again over the course of more than 150 million years thus completed a sort of evolutionary circle for these bioeroding bivalves. It was analogous to that of four-limbed aquatic vertebrates crawling up onto land, evolving for millions of years, and crawling back into the ocean to become aquatic once more. It is through this and other examples from the world of bioerosion that nature continues to surprise us.

: : :

Wood-boring clams of the deep sea share their dark realm with other animals that feed not on dead wood, but on vertebrate carcasses from above, meals delivered by the inevitability of both gravity and death. After the flesh is gone and the bones of these vertebrates are exposed, a special suite of consumers begins its work on the skeletal framework of these corpses, turning their solid essence into sustenance. Who are these animals that live off the deceased, and so far removed from the earth's sunlit surface?

Chapter 9

Bone Eaters of the Deep

After the largest animal in the history of the earth exhales its last breath and falls, it makes almost no sound. This seeming paradox happens not because no one is there to hear it, but because the meeting of blue-whale body with what lies beneath is separated by several thousand meters of ocean, a drop slowed by friction with the water and perhaps taking almost an hour. Its downward journey passes from sunlit to murky to black as its outline merges with the colder surroundings. Pressure squeezes the whale body in ways it could not have experienced in life, compressing gases and hastening the journey down. Once the bulk of this former life meets deep-ocean sediments, its collision is muted by the slowness of descent and softness of flesh meeting mud: less of a crash, and more of a thud.

For those animals unfortunate enough to be on the ocean bottom at the site of this collision, they are either crushed or pushed down into the mud by the 150 tons of dead weight above, perhaps unable to resurface. There, their buried bodies join the skeletal remains of millions of previous lives. Animals that can sense the approaching massive disturbance and can swim, crawl, or burrow do so, then return once the body has settled into its last place, with some of them ready to eat. Skin is consumed first, this outermost layer probed, peeled, and pierced by deep-sea fishes, such as sleeper sharks, hagfishes, and rattails. Most extractions are taken in small bits, whereas others are detached in chunks wider than a human. Bottom-dwelling crustaceans join these swimmers, including giant isopods, amphipods, and squat lobsters. Octopi move in and eat alongside these animals, as do thin-armed brittle stars. Just like on land, some diners take advantage of holes afforded by vertebrate anatomy and enter through orifices—ears, mouth, anus, and especially eyes—making their ways to the corpse's interior. Once entry is granted, these guests consume from within and diminish the once-smooth profile of this former whale, its tissues supplying energy to an enormous number and variety of animals for months, stretching into several years. Meanwhile, the area around and underneath the whale carcass changes chemically as the water and sediments become enriched in organics, a boost encouraged by a gentle snow of decayed tissues, feces from the scavengers, and a proliferation of microbes. Suspension-feeding animals colonize the sediment beneath and around the body, while microbial life thrives. Among the microbes are heterotrophic bacteria and chemosynthetic archeans superbly adapted to living in cold places without light.

The biologically mediated flensing by the scavengers eventually exposes bone. An enormous rib cage emerges from beneath skin, muscle, and blubber, ribs that protected lungs, a digestive tract, and the world's largest heart, once composed of about 180 kg (400 lb) of blood-pumping muscle. Vertebrae are at the top of an archway defined by these ribs, a cathedral honoring the dead with a celebratory feast. Patches of bare skull emerge, whitish areas that expand and eventually connect to better define what housed the brain of an animal that felt and cared for its own. Near its front, fine combs of baleen define its former mouth, a site of nourishment for the formerly living transferring its essence to new lives.

This unveiling of mineral biomass attracts a suite of new and radically different feeders. They apply acids, boring by the thousands into dense cortical bone and more porous cancellous bone below. These actions expose more bone underneath, much of it filled with lipid-rich marrow that the worms digest while also allowing entry for microbes, nematodes, and other polychaetes to live in and consume these parts. These are the *Osedax* worms, polychaetes with life cycles dependent on bone.

Over the course of several years, the once-sharp outlines of bones disappear, their former edges eroded and defined by a pinkish fuzz of densely populated *Osedax*. Penetration by the worms of inner bone encourages the growth of anaerobic bacteria that convert sulfates from the lipid-rich bones into sulfides, exuding a biogeochemical stew around the fall site. Mats of sulfide-loving microbes spread, with their organics bringing in grazers, such as limpets and other gastropods. If one could step outside a submersible and sniff the water, the rotten-egg smell of these sulfides would overwhelm just before the crushing pressure takes away all sensations.

Decades later, a new assemblage of encrusting and filter-feeding animals moves into and onto the remnants of the nearly unrecognizable skeleton, much of which has contributed to the life cycles of others. The filter feeders are there for the suspended organic matter now concentrated around this place, one of thousands like it in the abyss and in various states of succession, but connecting life-forms from anaerobes to sharks. This is when the hard bits of whale serve as attachment surfaces for clams and other sedentary animals, and the site becomes a reef of sorts, but one without corals, parrotfishes, or other colorful biota we normally associate with the reefs of the near-surface world.

And so the ecosystem engineer of the shallow sea continues an afterlife role as a different type of engineer for the deep sea. For more than 50 years before its death, it changed the world above, and for more than 50 years after its demise, it drives the diversity and evolutionary future of a community in a far different ecosystem, and one it never could have visited while alive. This is an ecological succession in which the most voluminous engineers from a luminous realm unwittingly conspire with some of the smallest of ecosystem engineers in one of the most inaccessible places on Earth, a place where the bone eaters of the deep reclaim the bodies of the gigantic.

Skeletal remains of a whale serving as the foundation for an ecological community with bacterial mats, bivalves, crabs, and bone-eating *Osedax* worms living at a depth of 1,674 m (5,492 ft). Whale carcass was placed there in 1998, hence this community took just 6 years to grow on, in, and around it. Photo by Craig Smith in January 2004, public domain via NOAA.

: : :

During the past 55 million years of cetacean lives and deaths, the cast of characters in deep-sea communities adjusted through time, with some relationships staying the same or shifting accordingly in response to climate and evolutionary changes. Fortunately, trace fossils on whale bones show who was eating them and when, including the traces of animals that might not otherwise leave behind any body parts.

Deep-ocean basins cover more than 60 percent of the earth's surface, and humans only recently developed the technological means to reach and explore these vast, dark, and cold expanses. As a result, we are very much innocent explorers, still observing what is down there and how those places relate to whales. The aftermath of intensive whaling during the eighteenth and nineteenth centuries also decimated whale populations in the twentieth century, but I actually mean that in the reverse (and much worse) sense of "decimated," as only about 10 percent of the original whale numbers survived.[1] Such cetacean scarcity means that the chance of humans encountering a dead whale on the deep-sea floor is akin to winning a lottery. Thus it is not so surprising that humans walked on the moon (1969) eight years before they saw a whale skeleton in the deep sea (1977); another 10 years passed before marine biologists assessed the ecological significance of what they called "whale falls."[2] Fortunately we are now in a renaissance of whale-fall research, whether it comes from deep-sea scientists surreptitiously encountering whale falls during research dives, or deliberately placing the remains of formerly beached whales on the ocean bottom as experiments.

From this intensive research, marine biologists sketched out four basic stages in the post-death history of a whale skeleton, from entire body to shadow of a former whale self:

1. An all-you-can-eat buffet for mobile scavengers, such as sharks, hagfishes, and decapods;
2. Colonization by enrichment opportunists, species that take advantage of organically enriched sediments around the whale carcass and exposed bones for food and homes (like *Osedax*);
3. A "rotten eggs" phase, with anaerobic bacteria converting sulfates in bone lipids to sulfides, making a hydrogen-sulfide halo that only sulfur-loving (sulphophilic) organisms can tolerate; and
4. Whale reefs, in which all organics from the skeleton are gone, leaving just a few eroded bones as hard substrates in the deep sea for encrusting suspension feeders, like deep-sea clams.[3]

Based on what marine scientists have observed so far, the first stage (scavenging) may take a few months to more than a year, the second stage (enrichment) could take almost five years, the third (sulphophilic) might be decades, and the fourth (reef) happens after all of the third stage runs out of food.[4] So far the fourth stage is only predicted and not yet observed. However, considering that marine biologists estimate thousands of whale falls are in ocean basins, and that stages of their ecological succession overlap and form a continuum, a fuller picture of their post-death histories will no doubt emerge in the near future. Regardless, these revelations about whale falls have revolutionized our understanding of the ecology of the deep ocean, while also showing how much of this ecology depends on essential organisms like the bone-boring *Osedax* worms.

Marine biologists who study whale falls also take care to distinguish between ecological processes in shallow whale falls versus deep whale falls. Shallow whale falls are on continental shelves with water depths that are typically less than 200 m (~650 ft). Because these environments receive sunlight, the foundations of their ecosystems are based more on phytoplankton, which photosynthesize. Animal life in the shallows is also vastly different, with many more scavengers and more encrusting invertebrates, such as sponges, bryozoans, bivalves, and corals. Added to this mix are aerobic bacteria, which hasten decomposition in warmer waters. Owing to a combination of such factors, whale carcasses in shallow environments tend to vanish more rapidly.[5] Meanwhile, in the deep sea, chemosynthetic organisms—such as microbes that derive energy from sulfides and sulfates—are at the base of these ecosystems, and decomposition is anaerobic. Colder temperatures slow bacterial decomposition, while also limiting animal reproduction rates, growth rates, and biodiversity. Although scavengers in the deep sea are numerous and varied, they may not be as abundant as in shallow shelf environments, either. These are reasons why ecosystems created by whale falls in the deep sea might last longer than the actual life of an adult whale, an ecological echo that persists and affects future generations of deep-sea life.

Another point about the hereafter of a whale is that just because it dies, it does not stop moving. For instance, a whale that normally lives above a shallow-marine shelf could have its body deposited in the deep sea. This scenario happens when a whale body still has air in its lungs and bacterial decay produces enough gases to keep it afloat so that it drifts with surface currents.[6] Paleontologists morbidly call this the "bloat and float" hypothesis, which they use to explain how bodies or body parts of land animals, such as dinosaurs, were buried in marine sediments.[7] Whale bodies may likewise sink, resurface, and then sail far from where they originally died and drop into deeper waters. Their final vertical journey only happens once enough surface scavengers—such as sharks and other fishes, as well as seabirds—have punctured the carcass and released its gases so it sinks.[8]

With these whale-corpse perspectives in mind, we can add time as a factor. Given time, whale falls and bone-eating worms likely contributed to rapid evolutionary rates in deep-sea animals and boosted the biodiversity of these formerly out-of-sight, out-of-mind environments. The harsh physical conditions of abyssal plains—cold temperatures, high pressures, little to no oxygen, and unreliable food sources—certainly take natural selection up a notch, in which traits for physiological efficiency were favored. Marine biologists who study *Osedax* have also wondered if different species of these worms may represent niche partitioning, in which they use slightly different feeding methods in bone to reduce direct competition for scarce resources.[9] Whereas *Xylophaga* and other marine wood-boring animals may represent "wooden steps" for increasing deep-sea biodiversity,[10] *Osedax* and its bone-centered communities were more like "bony steps" in the evolution of ecosystems that operate at extremes. The evolution of these bone-eating worms surely changed the flux of matter and energy in the deep sea, while also acting as linchpin species in the complex ecological succession of whale falls.

: : :

How does a soft-bodied worm like *Osedax* make its way into a whale skeleton and turn it into whale dust? Like many other circumstances in the history of life, natural selection helped make it happen. But like many other instances of life-forms that break down hard substrates, it also required multispecies assistance.

Osedax worms are polychaete worms, so they share a common ancestry with other polychaetes that bore into molluscan shells and limestone, as well as with other segmented (annelid) worms, such as earthworms and blood-sucking leeches. Species of *Osedax* and their closest kin belong to a group of polychaetes called Siboglinidae, which live almost exclusively in deep-sea environments, whether in sediments or attached to hard substrates.[11] A few of the preferred places for siboglinid worms are: methane seeps, where frozen methane is released out of the seafloor; hydrothermal vents near plate-tectonic margins; and organic-rich areas, such as where excessive plant or animal material was dropped, such as wood falls and whale falls.[12] So it is fair to say these polychaetes live in extreme environments with unique ecological factors at work, especially when compared to polychaetes that live in shallow-marine environments.

How long have we known about *Osedax*? Put it this way: if the genus name *Osedax* were a person, it would belong to Generation Z, as it was not named until 2004. The first people to see *Osedax* live and in color were Robert (Bob) Vrijenhoek and other marine scientists in 2002 while they were on a research vessel (the *Western Flyer*) in Monterey Bay, California, and operating a remotely operated vehicle (ROV), the *Tiburn*.[13] Like many great scientific discoveries, this one was serendipitous, as their goal was to prospect for deep-water clams by using sonar. When one of the sonar signals showed something big on the seafloor almost 2,900 m (9,500 ft) down, they sent the *Tiburn* to check it out. The scientists were then surprised by imagery of a gray whale (*Eschrichtius robustus*) skeleton, but with a red fuzzy margin. The "fuzz" turned out to be the feathery gills of thousands of *Osedax* embedded in the bones, a revelation that must have been both extraordinarily thrilling and eerie. The *Tiburn* recovered samples of these worms, which were quickly confirmed as relatives of other deep-sea polychaete worms (siboglinids) but otherwise were new to science. So in 2004, Gregory Rouse, Shana Goffredi, and Vrijenhoek published a paper naming the new genus *Osedax* (*os* = "bone," and *edax* = "eat"), with two species belonging to the genus.[14] As of this writing, these biologists and others have named more than thirty species of *Osedax* from the Pacific, Atlantic, and Indian Oceans, and the Mediterranean Sea, including what is inarguably the greatest species name of all time, *Osedax mucofloris*, which translates to "bone-eating snot flower."

Osedax has no mouth, yet it must eat bone. Indeed, the incredible weirdness of *Osedax* is perhaps best appreciated by noting what it lacks: a mouth, stomach, anus, and eyes. Because they are polychaete worms, they also do not have legs, but their cylindrical and segmented worm trunks are topped by plume-like gills and bottomed by wide root-like structures. Colors vary along their lengths, too, as their gills and trunks are normally pink to red, but their roots can be yellow, green, or orange. All species of *Osedax* secrete mucus, a trait that led to their endearing nickname as "snot worms," although their feeding on and in dead bones also earned them the moniker "zombie worms." As an *Osedax*'s fine gills extract dissolved oxygen in water, they send it to both the animal and aerobic bacteria living in its root tissues. An *Osedax*'s roots produce a carbonic acid that can dissolve apatite, the main mineral in bone.[15] This acid etches the outermost surface of bone so they can send out their roots into more porous bone beneath with its nourishing marrow. This is where fats (lipids) are digested with the aid of the bacteria, which actually live in a bone worm's cells. The worms, in turn, digest some of these cells. *Osedax* is not on a fat-only diet, though, and can vary its diet by having its bacteria digest collagen, a common protein in connective tissues.[16]

Female *Osedax* are the first to arrive at exposed bone as swimming larvae, where they make themselves at home by boring into it; nearly every *Osedax* that settles in bones is a female of their species. When exuding acid from their roots, these worms make a circular hole as an opening and a short vertical shaft and space underneath, into which they fit their trunk and roots (respectively), with gills waving above.[17] Because the trunks of most *Osedax* species are only 1–3 mm (0.2–0.6 in) wide, openings are only slightly wider.[18] Depending on the species, female *Osedax* range from gummy bear to gummy worm length, or 2–8 cm (0.8–3.1 in) long. Below the entry hole and shaft, roots dissolve bone and spread laterally, forming a gallery much wider than the entry hole.

As female worms mature, they collect tiny larval males by the hundreds and keep them as an all-male harem underneath a gelatinous tube around their trunks. And by "tiny," I am talking about orders-of-magnitude size differences, in which females are thousands of times bigger than males.[19] Given such minuteness, *Osedax* males' only purpose in life is to gain enough matter and energy to produce sperm for fertilizing eggs. Not surprisingly, they are also parasites, depending on hardworking females for protection. Nevertheless, biology thrives on exceptions, and one exceptional species, *O. priapus*, bucks the tiny-captive-male trend of its relatives. Males of *O. priapus* grow to full adult size, which is about a third of the females'.[20] The males also feed themselves by living on their own in bone. How do they mate if living in their own borings? They reach out and touch some worm by stretching their bodies to about ten times their normal length.[21] All of these attributes add up to male *O. priapus* easily achieving peak *Osedax* manliness and more than living up to their species name as worm-gods of fertility.

Considering how *Osedax* lifestyles differ from your average worm, their borings are accordingly distinctive. For one, they can only make homes and feed in bones or teeth, and not rocks, shells, or wood. This means *Osedax* borings are substrate-dependent traces. For another, their traces have radically different dimensions that reflect differences between trunk and root widths of the original tracemakers. Boring dimensions and forms also may vary as each individual worm grows up: younger worms make simpler borings, whereas older worms create more complex ones.[22] Finally, flexible behaviors by *Osedax* in response to different types of bone—such as compact cortical bone versus porous cancellous bone—also can affect the form of an *Osedax* boring.[23]

As mentioned before, entrances to *Osedax* borings (apertures) are small-diameter and circular in outline, with most connecting to short vertical shafts that expand into chambers below. How small, how short, and how expansive? Based on measurements of borings made by nine different species of *Osedax*, apertures ranged from 0.3 to 1.9 mm (0.01–0.07 in) wide, into which angel-hair pasta and capellini would fit, but spaghetti would not. Shafts were 0.9–6.5 mm (0.04–0.26 in) deep, or less than half the

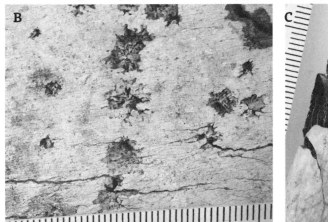

Whale bone and tooth with borings likely made by bone-eating worms (*Osedax*) from the Oligocene (28–30 mya), Ashley Formation, South Carolina. A, Mandible of whale *Micromysticetus* with many *Osedax*-like borings. B, Close-up of borings. C, *Osedax*-made boring in *Ankylorhiza* tooth (arrow), an example of a post-mortem cavity. Scale in photos in millimeters; all photos courtesy of Robert (Bobby) Boessenecker, and specimens are in the Mace Brown Museum of Natural History, College of Charleston, South Carolina.

length of a hole made by a standard pushpin. Chambers of the borings were likewise minuscule, varying from 2.2 to 98.5 mm³ (0.0004–0.02 US tsp) in volume, with the greatest volume less than one-fifth of a drop from a standard eyedropper.[24]

Although *Osedax* traces are individually quite small, their multiplied effects are massive. So once both biologists and paleontologists realized that species of *Osedax* are widespread, diverse, and extremely abundant, while also showing a ruthless dedication to consuming bones, they began talking about "the *Osedax* effect."[25] Although the *Osedax* effect might sound like a spatial or temporal anomaly that might arise from, say, the Oura megasurface, it actually more closely resembles a plot from the TV series *Breaking Bad*. In one of its episodes, two main characters try to cover up a murder by dissolving the victim's body with acid. (Spoiler alert: the acid dissolved more than the body.) Similarly, the effect of billions of bone-dissolving worms operating in oceans far and wide over geologically lengthy time spans probably erased countless whale skeletons from the seafloor before any of their parts were buried and fossilized. To wit, these teeny worms caused huge gaps in the fossil record of the world's largest animals.

Fortunately, whale fossils are still relatively abundant and some specimens are complete enough that paleontologists should have plenty of bones to study going well into the future. Yet knowing that many specimens and species of fossil whales didn't stand a chance of having their remains preserved is admittedly vexing. The *Osedax* effect thus points to a new and important facet of bioerosion that paleontologists hadn't previously considered. As soon as the ancestors of *Osedax* began their bone-dissolving ways, the fossil record of large marine vertebrates began to vanish, a "delete" button pressed on and off over perhaps the past hundred million years.

: : :

Before the evolution of *Osedax* and other bone eaters, there of course had to be bone. Despite the too-frequent headlining status of vertebrates in nature specials and documentaries, these animals are descended from relatively less adored marine invertebrates. In the Cambrian, at about 520 MYA, one lineage of marine invertebrates evolved into marine chordates, which were the predecessors to vertebrates.[26] Distinguishing traits of these vertebrate progenitors were gill slits, upwardly oriented (dorsal) notochords to stiffen their backs, and dorsal aortas in their circulatory systems. Some branches of those chordates then evolved the ability to take calcium and phosphorus out of the water to make apatite, which formed bone and teeth. This biomineralization in vertebrates nearly coincided with increased biomineralization in invertebrates, which produced calcite and aragonite.[27] We still don't quite understand whether this coincidence of skeletal evolution was an evolutionary response to changing ocean chemistry, predator-prey dynamics, or both. Regardless, it happened, and here we are.

Bones of marine animals before the Mesozoic show little evidence of invertebrates eating them, with most trace fossils on these bones suggesting predation or scavenging by other vertebrates. This is not so surprising, because only a few marine vertebrates of the Paleozoic were large enough to warrant epic post-death ecological communities. However, I expect paleontologists would be pleased to see evidence indicating that, say, giant Devonian fishes or other such vertebrates attracted varied and populous crowds at their wakes. Regardless, let's revisit the opening scenario of this chapter, but replace the blue whale with other animals that were also big, fleshy, and delicious to a wide range of deep-ocean scavengers. Although whales represent the zenith of mass in the evolution of vertebrates, a few marine vertebrates of the past rivaled or exceeded the sizes of most modern whales. These capacious contenders were marine reptiles of the Mesozoic, as well as a few fishes.

For the sake of simplicity, let's first look at Mesozoic marine reptiles and consider them as six groups, most of which have already been introduced: placodonts, ichthyosaurs, plesiosaurs, mosasaurs, crocodilians, and sea turtles. Of those groups, placodonts went extinct at the end of the Triassic, and ichthyosaurs were gone by the middle of the Cretaceous. When the end-Cretaceous mass extinction happened about 66 million years ago, it wiped out almost all large animals on land (dinosaurs) and in the oceans (mosasaurs, plesiosaurs). So of these six Mesozoic reptile groups, crocodilians and sea turtles are still with us, although only a few crocodilians venture into marine environments.[28] This big difference in big reptiles meant that Mesozoic seas offered much more scaly fodder for bottom-dwelling scavengers compared to the last 66 million years or so. Such a dissimilarity further implied that until the evolution of whales started about 55 million years ago, bottom-dwelling scavengers that specialized on reptile falls either went extinct or eked by, eating vertebrates other than reptiles or whales during those 10 million years or so.

Stout-bodied placodonts were among the first reptiles adapted to coastal environments, having evolved from land-dwelling ancestors in the Triassic soon after the end-Permian mass extinction.[29] As you may recall, placodonts took advantage of a post-extinction population boom in mollusks by developing shell-crunching teeth, which worked well for them until it did not. None of the placodonts achieved great size, though, with some only about 3 m (9.8 ft) long.[30] Placodonts shared Triassic seas with ichthyosaurs for a while, but the latter group survived the end-Triassic extinction and thrived throughout the Jurassic and part of the Cretaceous before their demise.[31] Even so, Triassic ichthyosaurs were not to be ignored, as some became enormous. For example, *Shastasaurus* from the Middle and Late Triassic of North America and China, as well as *Shonisaurus* from the Late Triassic of Nevada (US), were more than 20 m (66 ft) long,[32] nearly matching the length of a standard semi tractor-trailer.

Plesiosaurs, and particularly the short-necked pliosaurs, like-wise had a few hefty species, such as *Pliosaurus* from the Late Jurassic of northern Europe and *Kronosaurus* from the Early Cretaceous of Australia. Some species of *Pliosaurus* may have been as long as 15 m (49 ft), whereas *Kronosaurus* was 10–11 m long (33–36 ft).[33] Mosasaurs arrived a little later on the Mesozoic marine scene (about 100 million years ago), but by the end of their hey-day in the Cretaceous, some species, like *Mosasaurus hoffmanni*, were 15+ m (49+ ft) long.[34]

As for sea turtles, the largest was the Late Cretaceous *Archelon*, topping out at about 4.5 m (15 ft) long and 4 m (13 ft) wide, bigger than a compact car.[35] Mesozoic marine crocodilians varied greatly in their sizes, but a few, such as the Late Cretaceous *Deinosuchus* of North America, would have struck fear in the hearts of tyrannosaurs. Take the largest known modern crocodilian—the saltwater crocodile (*Crocodylus porosus*) of Australia—and then double its length, and you would have *Deinosuchus* (about 10–12 m/33–39 ft), its immense bulk justifying at least twice as many "crikey" utterances.[36]

Now, let's compare these Mesozoic marine reptiles to modern killer whales (*Orcinus orca*), or to use their less incriminating names, orcas. The biggest orcas today are about 8 m (26 ft) long,[37] which is only one-third the length of the Triassic ichthyosaurs *Shastasaurus* and *Shoniosaurus*, and half the length of the biggest plesiosaurs (*Pliosaurus*), mosasaurs (*Mosasaurus*), and crocodilians (*Deinosuchus*). The weightiest of orcas are about 6 metric tons (6.5 tons), and although figuring out weights of long-extinct marine reptiles from just their bones is problematic, we are still quite sure the largest ichthyosaurs, plesiosaurs, and mosasaurs were far heavier than these whales. Indeed, they must have exceeded or at least overlapped with the sizes of most modern baleen whale species.

Last and certainly not least, a few Mesozoic fishes also could have served as gigantic reservoirs of bone meals, some of which included the largest known bony fishes, such as *Leedsichthys* from the Middle to Late Jurassic (about 165–55 MYA). Specimens of this genus are estimated to have been 14–16 m (46–52 ft) long, matching sizes of the largest pliosaurs and mosasaurs.[38]

So to make a long marine-reptile and fish story just a little shorter, both shallow and deep seafloor communities during the Mesozoic had plenty of opportunities to consume flesh and bones of giant dead vertebrates dropping from above. Do we have any evidence that such dining in the dark happened then? Yes. But bioerosion of bone in marine environments was far different from today, especially before the Cretaceous. For instance, as of my writing this, no one has reported any trace fossils of marine invertebrates dining on placodont bones, or any other Triassic marine reptiles, for that matter. That all changed in the Jurassic.

∴ ∴ ∴

In the Late Jurassic (about 160 MYA), at least one "ichthyosaur fall" happened, and with a very different set of bone eroders than what we see with whales. In a 2014 study by paleontologist Sil-

via Danise and her colleagues, they described bones of the ichthyosaur *Ophthalomosaurus* in southern England with trace fossils of animals that both scavenged and grazed on these bones.[39]
Evidence for scavenging consisted of small, shallow, and short
grooves on some of the ribs, which the paleontologists interpreted as those of toothed fishes. They further reasoned these
cuts were from scavenging, not predation, because they were so
minor compared to the size of the ichthyosaurs.

Far more interesting, though, were the traces of grazing, also
on the ichthyosaur ribs. And just how did an animal graze on ichthyosaur bones? Or, a better question might be, *why* did an animal graze on ichthyosaur bones? The best way to tempt certain
invertebrates to scrape the surfaces of these bones would be to
cover them with organic goodness so delectable that they had no
choice but to rasp. Yes, I am of course talking about microbial
films and those lovers of microbial films on hard surfaces, sea
urchins. In this instance, bacteria and algae grew in thin layers
while the ichthyosaur bones were exposed on the seafloor. These
films attracted urchins, which put their Aristotle lanterns to
good use as they scraped organics off the bones, while also eroding bone with each scrape.[40] The five-pointed star-like grooves
formed by their activity left little doubt that an echinoderm
made them; spines of the echinoid *Rhabdocidaris* near the bones
further verified their probable maker.[41] Added to the larger trace
fossils were many microscopic tunnels in upper surfaces of the
bones attributed to bacteria, algae, and fungi.

Overall, these paleontologists proposed ecological stages in
this "ichthyosaur fall" similar to those of modern whales: (1) mobile scavengers (fishes) swam in to chomp on dead flesh; (2) opportunistic animals, such as gastropods and clams, took advantage of changing local conditions and settled in; (3) microbial
mats grew, followed by grazers mowing them down; and (4) the
site of the former body turned into a new reef community.[42] This
trace fossil assemblage of scavenging, scraping, and microboring
was the first of its kind reported for a Mesozoic marine-reptile
fall, and it helped establish both similarities and major differences between these and Cenozoic whale falls.

Both modern and Cretaceous bones provided more tantalizing insights into the evolution of big-dead-marine-animal falls during the Mesozoic. For instance, several researchers in 2008 noted fossil gastropods closely associated with Cretaceous plesiosaur bones, and then speculated that such snails might represent part of a plesiosaur deadfall community.[43] In 2010, Shannon Johnson and her colleagues made those speculations more real with their discovering two new species of modern deep-sea gastropods feeding on whale bones.[44] The modern gastropods represented a new genus, *Rubyspira*, which the biologists split into two species (*R. osteovora* and *R. goffredi*). The snails, found on and near bones of a gray whale at a depth of almost 2,900 m (9,500+ ft) in Monterey Submarine Canyon (California), had bone bits in their stomachs and feces, which admittedly was already suspicious. But then carbon- and nitrogen-isotope ratios from the snails also matched ratios measured from *Osedax* worms, further testifying to their bone-eating guilt.[45] However, the researchers concluded that one species (*R. goffredi*) ate directly from the bone, whereas the other species (*R. osteovora*) ate bone-chip leftovers in sediment around the whale carcass.

Cretaceous reptile bones later supplied the next clues about the evolution of bone-eating communities, which became a little more familiar to those who study fossil and modern whale deadfalls. In a 2015 study by Silvia Danise and Nicholas Higgs, they described *Osedax*-like borings in 110–100 MYA plesiosaur and sea turtle bones from southern England.[46] This trace fossil evidence helpfully extended the geologic range of these worms while synching with molecular phylogenies that implied their Cretaceous origin. Danise and Higgs also suggested that sea turtles, birds, and fishes might have sustained bone-eating worms during millions of post-Cretaceous lean years, and that these plesiosaur bones were in shallow-marine deposits. This suggests that ancestors of modern deep-sea *Osedax* might have originated in warmer, clear-lit waters. Only later did these worms divide ecologically into "shallow" and "deep" species, which may have been driven by the advent and spread of massive baleen whales that increasingly swam into and died in open-ocean environments.

Did any ammonite-munching mosasaurs also have their bones turned into food? All signs point to yes, demonstrated in two successive discoveries of altered mosasaur bones. In a 2019 study by paleontologists Marianella Talevi and Soledad Brezina, they described a mosasaur vertebra from Antarctica in 67 MYA rocks, just before mosasaurs went extinct.[47] Part of the vertebra was perforated with microborings, which the paleontologists concluded were made by bacteria and fungi, not worms or other animals. Other than being the first reported borings in mosasaur bones, the paleontologists also noted that microbial bioeroders were following former vascular channels in the bone. Moreover, only the outermost part of the vertebra was bioeroded, which probably meant this surface was exposed at the seafloor while the rest was buried in sediment below.

A 2020 study of mosasaur and turtle bones the same age as the Antarctic bones (67 MYA), but from the Netherlands, showed these were likewise bioeroded. The bones, examined by John Jagt and other Dutch researchers, belonged to two genera of mosasaurs (*Mosasaurus* and *Plioplatecarpus*) and the sea turtle *Allopleuron*.[48] The bones held gouges left by fish teeth and the distinctive five-pointed grooves of grazing sea urchins. Fish trace fossils were likely left by sharks stripping the last bits of flesh off the bones, whereas sea urchin borings must have been from echinoids scraping algae off bones, just as their Late Jurassic predecessors did with ichthyosaur bones in England.

: : :

Although these examples imply that worms acting very much like *Osedax* and other bone-eroding organisms were doing their work on Mesozoic reptile bones, the severe big-bone shortage caused by the end-Cretaceous extinction should have put a damper on such activities. So which animals' bones did they eat or otherwise erode before whales evolved and started dropping bodies? This question was better answered once paleontologists became more aware of the awesome bone-eating powers of modern *Osedax* and other marine bone eroders. Given this knowledge, they knew what to look for, and they more closely examined fossil bones of non-whale vertebrates for *Osedax*-like trace fossils.

But just before paleontologists started looking for worm-riddled fossil bones, marine biologists were already thinking outside of whales and considering other modern animals' bones as fodder for *Osedax*. In a 2008 study, William Jones, Shannon Johnson, and a few other researchers designed and conducted an experiment in which they enticed *Osedax* to bore into bovine bones.[49] All of the bones were femurs involuntarily donated by domestic cattle (*Bos taurus*) and cut longitudinally to expose inner bone surfaces. The researchers then secured the bones with cable ties onto six branches of a vertical PCV-pipe "tree" set in a concrete-filled plastic bucket, which should have either won a national art award as a commentary on the meat industry or been voted the worst Christmas tree ever. ROVs ferried these "bone trees" to four locations off the coast of Monterey (California) varying in depths from 385 m (1,263 ft) to 2,893 m (9,491 ft). Follow-up visits showed that six of the eight known species of *Osedax* in this offshore area colonized the bones within two months at one site, although the shallowest site took almost a year. Reproductive activities were already on their agendas, too, as some female worms were producing eggs and attracting their usual harems of minute males.

Soon after this study, in 2011, a few marine biologists joined Greg Rouse in a fishy test of *Osedax* eating preferences. In this experiment, they put shark cartilage, bony-fish parts, and a cattle femur in wire cages and placed them about 1,000 m (~3,300 ft) deep west of Monterey.[50] The cages were sited close to a blue-whale carcass to ease recruiting of local *Osedax* larvae and left for five months to give the larvae plenty of time to find the samples. Sure enough, three species of *Osedax* found the cow bone and fish bones, although the shark cartilage was completely destroyed, but probably not by *Osedax*. Like the previous experiment, females with fertilized eggs and males were also present, showing they moved in, ate bone, and got busy within just five months. In short, these two experiments neatly demonstrated that non-whale mammal and fish bones worked just fine for at least a few *Osedax* species, showing they can switch from surf to turf and back again, while also finding time to make baby snot worms.

With an ichnological gauntlet thrown by marine biologists in 2009 to find *Osedax*-made trace fossils, paleontologists soon rose to the challenge. In a 2011 study, Steffen Kiel and his colleagues reported *Osedax*-like borings in diving bird bones from Oligocene (~23 MYA) rocks of Washington (US), the first fossil evidence of *Osedax* eating non-mammal bones.[51] In 2013, the same set of researchers documented more *Osedax* borings in Oligocene whale bones and teeth as well as in fish bones, also from Washington.[52] In between these discoveries, in 2012, Nicholas Higgs and a team of researchers found *Osedax* trace fossils in a single beaked-whale bone from the Pliocene (a mere 3 MYA) of Italy.[53] Using micro-CT scanning methods similar to those applied by Diva Amon in her study of deep-sea wood borings, the scientists created subsurface 3D images of the traces, which enabled them to link the forms of modern *Osedax* traces to these trace fossils. As a result, these researchers felt confident enough about their identity to assign them the ichnogenus name *Osspecus* (where *os* = "bone" and *specus* = "cavern").

Given all of these previous studies of modern and fossil *Osedax* borings, and their presence in a variety of animals' bones past and present, it only made sense for marine biologists to up the bone-boring ante. In this case, they designed a deep-sea deadbody project that would better connect with paleontologists. And what better animal to use than that iconic inspirer of Mesozoic nostalgia, the American alligator (*Alligator mississippiensis*)?

In 2019, Craig McClain and several other marine biologists took the bodies of three adult alligators ranging from 1.7 to 2 m (5.6–6.5 ft) long and had ROVs place them on the ocean bottom about 2,000 m (6,500+ ft) deep off the US Gulf Coast, south of the Mississippi River Delta.[54] They anchored the carcasses with weights to keep them on the seafloor and marked positions with floats so ROVs could find them later. Less than a day after the first alligator body was deployed on February 14, 2019, it became a romantic Valentine's Day dinner for a bevy of giant isopods

(*Bathynomus giganteus*), which immediately commenced chowing down on the body. Within two days, the isopods had exposed the alligator's ribs and jousted with one another for favorite feeding spots on and in the carcass. The second alligator body was sunk on February 20, 2019, and when revisited 53 days later, all of its soft tissues were gone, with intense scavenging having left behind nothing but bones. Its bones had also attracted a healthy population of maturing female *Osedax* that bored in and settled to feed; these *Osedax* were the first known from the Gulf Coast and a new species.[55] As for the third alligator body, placed on April 15, 2019, no one knows what happened to it, because when the biologists checked its site eight days later (April 23), it was gone. Sometime in those eight days, a large unidentified ocean scavenger apparently decided to do takeout instead of dining in. The weight, restraints, and marker float were the only signs of the carcass's former presence, with these less appetizing items slightly more than 8 m (27 ft) away from the alligator's original resting spot. McClain and his colleagues accordingly speculated that the most likely culprit was a large shark adapted for deep-ocean feeding.

Overall, these results showed not only that alligators are good food for deep-ocean scavengers, but also that their bones can host *Osedax*. The latter point especially bodes well for paleontologists and ichnologists, who might now expand their search images to look for these distinctive borings in crocodilian bones from marine deposits of the Mesozoic and Cenozoic Eras.

: : :

The happy burst of research over the past 20 years or so on modern whale falls, as well as trace fossils of *Osedax* or *Osedax*-like animals in fossil bones, shed light on a 100-million-year history of marine bones as hard substrates that took on "second lives." With the evolution of bone-eating worms, skeletons became homes, sources of food, and nurseries for these opportunistic animals. In this sense, the bones of dead whales, fishes, land mammals, birds, and reptiles in shallow- and deep-sea environments are like woodfalls in those same places with their specialized wood-eating communities, or snags in forests with their wood-chewing insects and woodpeckers. But on a broader scale, these worms are yet another example of how bioeroders can become ecosystem engineers at the center of nutrient cycling, even influencing evolution.[56]

Hence the bioerosion of marine animals' bones by the ancestors of modern *Osedax* may have been "jump-started" by the bones of big Mesozoic reptiles and fishes, but then was kept alive by the bones of fishes, turtles, crocodilians, and marine birds. These events were necessary steps before finally reaching a "golden age of snot worms" made possible by the evolution of whales in the Cenozoic, with the largest animals in the history of the earth enabling the continued existence of these tiny bone eaters of the deep.

Chapter 10

More Bones to Pick

When my future wife Ruth first moved in, she brought with her a cow skull, a memento spotted on a roadside during a trip to New Mexico. With much ceremony, we hung it on a brick wall in the courtyard outside our home. Considering our mutual love of natural history and its curios, the skull's ghastliness merged with hominess and was a welcome sight whenever we looked out the sliding-glass door onto our patio.

Unfortunately, our patio decoration wasn't meant to stay, as our Georgia town hosts small, furry, bone-destroying beings that descend from the trees and eat skulls. One day I was alerted of these beasts' malevolent presence by a persistent and rhythmic scratching on our patio. Intrigued, I glanced through the sliding-glass door and saw the bioeroding culprit in action. It was a gray squirrel (*Sciurus carolinensis*), perched atop the skull and gnawing relentlessly. Its sounds linked with actions as I watched its head move busily up and down, its incisors sliding against the bony surfaces. The squirrel likely made a short sojourn to our courtyard from a nearby oak tree, where its boughs held several nests, each housing a squirrel family. But rather than shoo it away, I watched and continued to study it and other squirrels whenever they were on the skull, whenever I could. After all, this was natural history happening on our patio, and a chance to have science come to me.

This squirrel-initiated attrition began in 2008 and ended by

Gray squirrel-gnawed cow skull, before and after. A, Entire skull hanging on a brick wall, with adult feline (Misha, RIP) and Mardi Gras beads for scale; photo taken on March 9, 2008. B, Remnant of same skull, its bulk worn down by bio-eroding gray squirrels; photo taken on December 21, 2013. C, Close-up of tooth traces on bone, matching the width and form of gray squirrel incisors.

2014, when the skull had been reduced to a mere fragment and the nail could no longer hold this leftover. Considering the average lifespan of a gray squirrel is less than 10 years,[1] but more like 5–7 years in a place with many hawks and cars, generations of local squirrels likely noshed on this skull. Regardless, its gradual disappearance was a wonderfully inadvertent experiment in preservation, one in which I wish I had been more of a scientist by measuring and documenting its diminishment. But sometimes home is an escape from such responsibilities.

Squirrels are rodents, and all rodents have prominent incisors they are quite good at using. One fine example in North America is beavers (*Castor canadensis*), which have teeth augmented by iron, helping them fell trees and ultimately reshape ecosystems.[2] However, tree-dwelling squirrels made temporarily homeless by beavers do not necessarily swallow bark and cambium when they chew. Yes, they bite into acorns and other nuts, but they also snip through thin branches to harvest clusters of nuts or to gather branches for their nests.[3] A horrifying aspect of rodent tooth growth, though, is that if they do not wear down their teeth regularly, they keep growing.[4] If left alone, their incisors would protrude from their mouths and curl underneath their chins, which would eventually hamper eating anything, nuts or otherwise. An added incentive for squirrels and other rodents to scrape bones is for their own bones, in that the mineral matter (apatite) supplies much-needed calcium in their diet.[5] In non-urban environments, bones are probably more available for squirrels, mice, and other rodents that need to gnaw. In contrast, squirrels in urban areas have far fewer opportunities to add calcium to their diet. So our neighborhood squirrels must have felt very lucky indeed to find such a cow-supplied bonanza of apatite to supplement their appetite.

Given this inadvertent ichnological lesson on my patio, I became curious about squirrels' and other rodents' gnawing bones and sought more information. It turns out that although many rodents scrape their teeth along bones, they may have different motivations for doing so. For instance, brown rats (*Rattus norvegicus*) prefer to consume fresher porous (cancellous) bones with fats still in them, extracting calories to go with their minerals.[6] On the other hand, gray squirrels go for older, dry bones devoid of fats or other organics, and they chew on compact (cortical) bone. Because these differences in bone feeding relate to postmortem ages of bones, knowing how to distinguish rat tooth traces from those of squirrels can be useful for aging them.

Although mammals get much attention today for their bone-eating abilities, some insects—such as some beetles, termites, and more—are bone eaters too. Non-mammal vertebrates also pierce or break bones, whether from attacking or feeding. Just how long have insects and vertebrates gnawed, crushed, or otherwise modified bone? This question is best answered by watching modern animals at work on newly dead bones, as well as studying body fossils (bones) and trace fossils (borings in bones), telling us who was eating skeletal bits and when, and most importantly, why. Yes, dead bones can tell tales, but so can their borings.

: : :

Insects that bore into bone are relatively few today, but may have been more common in the past, especially considering how Mesozoic dinosaurs left plenty of bones, some massive. The links between bone-eroding insects of the Mesozoic and those affecting bones today are clear, as trace fossils reflecting such behaviors are quite similar to modern traces. So to start our inquiry, a review of modern insects known for modifying bones seems appropriate. Among these six-legged skeletal depreciators are beetles, termites, ants, and a few others.

The most famous of bone-eroding insects are dermestid beetles. These beetles—belonging to the clade Dermistidae—have a range of common names, such as carpet beetles, carrion beetles, hide beetles, larder beetles, leather beetles, and skin beetles.[7] Considering that three of those names refer to the outermost coverings of mammal bodies, and one refers to dead bodies in general, one might get the (correct) impression that dermestid beetles are scavengers. They include nearly a hundred species of *Dermestes*, which are small (less than 10 mm/0.4 in long), dark-colored beetles that are extremely good at stripping away dry fleshy parts adhered to dead vertebrate bodies.[8] They accomplish this feat in tandem, as great numbers of adult beetles apply their mandibles to grasp and eat skin, muscle, tendons, and other soft tissues. Some species of *Dermestes*, such as *D. maculatus*, are so prized for their skeletonizing abilities that scientists at museums and universities have used them to "clean" bones for collections and displays, and still do, sparing interns of at least one fewer odious task.[9]

Although adult beetles mostly focus on eating soft tissues, their enthusiastic consumption of such morsels next to bone surfaces can leave paired grooves in the bone from their mandibles.[10] Yet much bone damage inflicted by dermestids is done as part of their life cycle. After hatching from eggs, larvae prepare for adulthood by also eating dried soft tissues on the bones, adding their feeding scratches to those of preceding adults. In several weeks, when larvae are ready to pupate, they carve out flask-shaped pupation chambers with "necks" about as wide as the original larvae and spherical to oval chambers recessed into bones.[11] However, because beetle larvae make pupation chambers in and under dried soft tissues and those tissues later decay, the chambers' full forms may not be preserved in fossil bones. Other borings made by dermestid larvae include: low-relief (less than a millimeter) straight to curving grooves with U- or V-shaped cross-sections; pits; rings; or rosettes, all of which may be closely spaced or overlap on bone surfaces.[12]

Considering the common association of termites with demolishing wood, these insects are perhaps unexpected bone-breaking allies to dermestid beetles. Nevertheless, some termites are documented for their scratching, pitting, and otherwise diminishing bone.[13] Like dermestid beetles, termites bite into bone with their mandibles, making star-shaped pits and long, parallel scratches, or they smooth bone surfaces via abrasion. Some researchers have even reported termite trails left on bones in tropical environments.[14] But why do termites erode bone, especially if wood or other sources of cellulose are available? In a 2020 study, archaeologist Lucinda Backwell and her colleagues concluded that such termites—which sometimes build their colonies around buried bone—harvest fungi growing on fresh bones and consume minerals on fossil bones.[15] The former behavior is analogous to how chitons, gastropods, and echinoids leave scratches in rocks when scraping algae off these substrates. A few bone-eroding termites also eat keratin, the main protein that forms hooves, claws, horns, or hair, all of which can be on top of bone and hence make those underlying parts more susceptible to termite-inflicted damage.[16]

Ants are relatively new additions to the list of bone-boring insects, but considering their ubiquity and adaptability, this is not so surprising. However, one of the challenges faced by entomologists and paleontologists alike is distinguishing ant traces from termite traces in bones. Both clades of eusocial insects are represented by thousands of species, all with divisions of labor connected to castes—such as workers, soldiers, and queens—that result in a range of varied and complex behaviors within those species. So we can all give scientists a break when they confess to a little ichnological uncertainty about termite and ant traces.

Still, recent insights offered by ant traces in modern human bones have helped clarify a few differences between ant traces and those of termites. In a 2018 study by anthropologist Matthew Go, he reported an ant colony living in a human skeleton that also modified its bones.[17] This accidental discovery happened when he and his research assistants began examining a skeleton temporarily interred in a rice sack at a cemetery in Manila (the Philippines). In what must have been a rather unsettling moment, ants burst out of both the sack and the bones. Also, some ants that emerged from the skeleton were workers carrying larvae, showing that a breeding colony was inside the bones. Go then used this unexpected opportunity to do science that contributed to taphonomy (the study of fossilization), ichnology, and paleontology. For one, he identified the bioeroding ants as a species of *Nylanderia*, nicknamed "crazy ants,"[18] but not previously known as crazy enough to bore into bone.

The earliest trace fossil evidence of insects rasping, scraping, or drilling into bones is from the Middle Triassic (about 237 MYA), from strata in Rio Grande do Sul of Brazil.[19] Borings are apparent as circular to oval holes, cylindrical tubes, shallow channels, and ellipse-shaped chambers in limb bones of *Dino-*

dontosaurus, a large, tusked herbivorous synapsid that lived during the Middle Triassic in South America. The trace fossils— described by Voltaire Neto and a team of Brazilian and South African paleontologists—show that their insect makers were probably carrion eaters that mined the bones while they were still relatively fresh, and may have pupated in them, too. More importantly, these trace fossils established that bone-chewing insects evolved in less than 15 million years after the end-Permian mass extinction.

Between the Middle Triassic and the rest of the Mesozoic, bone-eating insects apparently became more common and diverse, as their trace fossils show up in bones of dinosaurs and other land vertebrates from those times. Paleontologists also grew more adept at noticing and diagnosing insect trace fossils in dinosaur bones, especially as they heeded observations from entomologists who study bone-eating insects. After the end-Cretaceous mass extinction, bone-eating insects clearly survived with a few land vertebrates, as their trace fossils show up on reptile, bird, and mammal bones of the Paleogene Period.[20] Hence these bioeroding insects persisted, and some even left their sign on the remains of our human ancestors,[21] an intersection of insects with us via bones as places for them to eat, grow, and reproduce.

: : :

Dinosaur National Monument, located just north of Vernal, Utah, and managed by the US National Park Service, is not just a monument, but also a shrine, honoring some of the most iconic of the Late Jurassic (~150 MYA) dinosaurs of the American West.[22] Among these luminaries are: the herbivorous, long-necked, and weighty dinosaurs *Apatosaurus*, *Camarasaurus*, and *Diplodocus* (all sauropods); the carnivorous and formidable *Allosaurus* (a theropod); and the original all-American stegosaur, *Stegosaurus*, with its small plant-ingesting head, platy back, and spiky "thagomizer" tail.[23] Sightseers visiting the Monument are welcome to drive into it and gaze upon its arid and variegated landscapes, all while imagining dinosaur lives represented by bones hidden from view below those surfaces.

But there is no need to limit imaginations to geomancy when reality is so well displayed in the Dinosaur National Monument Quarry Exhibit Hall. Inside this building is the showcase feature of the Monument, an outcrop of the Late Jurassic Morrison Formation containing a stunning bas-relief sculpture. The outcrop, its sandstone and mudstone beds tilted upward by tectonic forces, displays about 1,500 exquisitely preserved dinosaur bones, ranging from tailbones to femurs and minuscule to magnificent. After American paleontologist Earl Douglass discovered the bone bed in 1909, its outward face became a combined natural and human artwork of workers who diligently performed their own bio-erosion to partially expose the bones for study and public viewing.[24] Two walkways in the Exhibit Hall, one at ground level and another above, allow visitors to stare open-mouthed and emit appreciative monosyllables while strolling along the soccer-field length of the outcrop.

Yet within those dinosaur bones at Dinosaur National Monument and in the Morrison Formation outside the exhibit hall are perhaps millions of other fossils, clues that reveal themselves to scrutinizing paleontologists who get up close and personal with dinosaur skeletons. These are the trace fossils made by insects on and in dinosaur bones soon after their former owners died, and just before their burial.

In studies from 2008 and 2020, paleontologists documented a remarkable variety of insect borings in dinosaur bones from the Morrison Formation. In the 2008 study, paleontologist Brooks Britt and his colleagues described insect borings in a *Camarasaurus* skeleton from Medicine Bow, Wyoming.[25] The trace fossils consisted of varied grooves, pits, tunnels, and furrows. Grooves were paired millimeter-long gouges evidently made by paired mandibles, also evident in the shallow to deeper pits. These insects did not just scratch the surface, though, but also probed long and deep, making meandering tunnels as they apparently consumed lipid-rich porous bone below.[26] Further evidence of bone-eating activities was reflected by what came out insect ends. Abundant millimeter-wide bits of broken bone filling some tunnels and other borings were likely feces, skeletal roughage that qualified as traces within traces. The researchers concluded these borings were nearly identical to those of modern dermestid beetles, and that most borings were likely made by bone-hungry Jurassic beetle larvae.[27]

The 2020 study of Morrison Formation dinosaur bones bearing borings comes from a place near Fruita, Colorado, called the Mygatt-Moore Quarry, one of the richest dinosaur bone beds of the Morrison Formation in western Colorado. Paleontologists and geologists who have worked at this site for the past 30+ years agree that the sediments and thousands of bones there were deposited in a pond that only occasionally received water.[28] Oddly, most bones represent just two dinosaur genera, the sauropod *Apatosaurus* and theropod *Allosaurus*. After decades of working at the site and extracting its paleontological bounty, geoscientists also noticed that many of Mygatt-Moore's bones carried vestiges of bone-eroding animals. Some trace fossils were grooves imparted by bladed teeth linked to *Allosaurus* and other theropods, which scavenged on bodies of both *Apatosaurus* and their own kind.[29] Yet other traces in the bones were much smaller, with most less than a millimeter deep or millimeters wide.

These minute traces painted a bigger picture of the original environments that preserved and eventually buried these dinosaur bones, as well as implying an odor. Paleontologist Julie McHugh and her colleagues found more than 2,000 small trace fossils on nearly 900 bones, the ichnologically richest Jurassic bone deposit documented anywhere in the world.[30] They interpreted trace fossils on about 16 percent of all examined bones as mostly those of insects. The traces consisted of pits, rosettes, scrapes, furrows, and other such violations of bone boundaries that the paleontologists credited to dermestid beetles. However, some borings they also attributed to at least two unknown insects and possibly from gastropods preserved in the bone bed, which may have chipped bone while grazing on organics.[31]

Artistic recreation of a Late Jurassic (~150 mya) scene at what is now the Mygatt-Moore dinosaur quarry in western Colorado, with dermestid beetles and canni-balistic *Allosaurus* imparting traces on bones; artwork by Brian Engh and in McHugh et al., "Decomposition of dinosaurian remains" and Drumheller et al., "High frequencies of theropod bite marks."

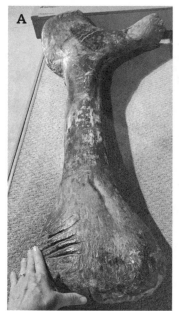

Tooth traces in an *Apatosaurus* (sauropod) bone from the Morrison Formation (Late Jurassic), attributed to the carnivorous theropod *Allosaurus*. A, *Apatosaurus* ischium with tooth traces on lower left. B, Close-up of striated furrows formed by serrated teeth pulling across the bone. Specimen displayed in the Dinosaur Journey Museum, Fruita, Colorado.

Thousands of dinosaur bones from the Late Triassic through the Late Cretaceous, and from around the world, contain evidence of insects actively chewing on dinosaur bones. So although paleontologists may say that the evidence of dinosaurs eating insects is scanty, we also can be absolutely sure that insects ate dinosaurs.

: : :

A few leaps through geologic time and space bring us to the La Brea Tar Pits and the George C. Page Museum, located in Los Angeles, California. Although *"la brea"* translates from Spanish as "the tar," and hence the name La Brea Tar Pits is repetitively redundant, we can forgive this if expressed out of enthusiasm. The pits consist of a number of closely spaced pools of black

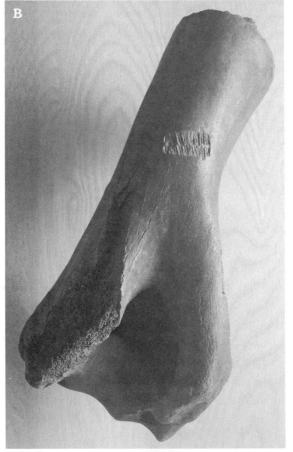

Fossil insect and rodent traces in bone. A, Insect borings (pits and rosettes) in dinosaur bone from Morrison Formation; specimen displayed at Dinosaur National Monument, Utah. B, Rodent tooth traces in mammal bone from the La Brea Tar Pits (Late Pleistocene), specimen displayed at George C. Page Museum, Los Angeles.

hydrocarbons (more properly called asphalt) that accumulated as near-surface bubbling crude starting in the Late Pleistocene about 50,000 years ago.[32] These pits of asphalt ensnared biota ranging from pollen grains to megafauna that then sank into its viscous depths, sealing off bodies from oxygen and aerobic bacteria. This in turn delayed decay and preserved bones, leaves, insects, and much more.[33] Such qualities eventually made this area the best Late Pleistocene fossil deposit in the world, with more than a million bones recovered, prepared, and cataloged since the early twentieth century.

Whose bones are interred in these petroleum pits of woe? One of the tar pits outside the museum provides an answer: a life-sized fiberglass recreation of a mammoth-family tragedy, with a mother mammoth stuck in the tar and bellowing a final farewell to baby and father mammoths just above. So yes, remains of the Columbian mammoth (*Mammuthus columbi*) are at La Brea, as well as those of another North American elephant, the American mastodon (*Mammut americanum*). The pits also contain many other large herbivorous mammals, such as giant ground sloths, bison, camels, horses, deer, antelopes, and tapirs.[34] Strangely, carnivores are overrepresented at La Brea, composing about 90 percent of the vertebrates, which is far more abundant than in their original ecosystems.[35] The most impressive of the former meat-eaters are the saber-toothed cats (*Smilodon fatalis*) and short-faced bears (*Arctodus simus*), but they also include an American lion (*Panthera atrox*), giant jaguar (*Panthera onca*), dire wolves (*Canis dirus*), and others. Why so many predators? One hypothesis is that whenever a big, hairy package of fresh flesh wandered into a tar pit and got stuck, it enticed great numbers and varieties of hungry patrons that also got stuck, which in turn brought in even more carnivores: trap, die, repeat.[36]

Tooth traces on bones show us exactly which carnivores ate which herbivores at La Brea, but insect trace fossils on bones show us much more than just identities of bone donors and bioeroders. In a 2013 study of insect-nibbled bones from there, paleontologist Anna Holden and her coauthors noted that bones with insect trace fossils are apparently rare at La Brea, but more than make up for their scarcity with useful information.[37] First, insects seemed to avoid carnivore bones. Second, they went mostly for camel, bison, and horse foot bones, rather than big limb bones, vertebrae, or skull parts. The most surprising find, though, was that the trace fossils came from two different types of beetles. Dermestids were there, leaving their distinctive excavations as they have done since the Mesozoic, but so were tenebrionid beetles, which were previously not known to eat bones.

Tenebrionids are more popularly called darkling beetles. This nickname alludes to their Latin root *tenebrio*, meaning "seeker of darkness," such as inside flour bags in cabinets, which is where most people have encountered them. Represented by nearly 20,000 species, tenebrionids scavenge on plant and animal debris, but were only suspected of bone feeding.[38] So along with Holden and her colleagues' descriptions of the fossil beetle borings, they did experiments with live dermestids (*Dermestes maculatus*) and tenebrionids (*Tenebrio* and *Eleodes*) feeding on modern chicken and pig bones. From these experiments, Holden and her colleagues demonstrated that tenebrionids made bigger and more extensive borings, a direct result of having mandibles four times larger than those of the dermestids.[39] Plenty of dermestid and tenebrionid body fossils in the same tar deposits also could be reasonably identified as the penetrating perpetrators.

Despite the impressive results of this research, I nevertheless should mention that insect trace fossils from the La Brea deposits present an ideal situation for the researchers, in that they also had beautifully preserved insect body fossils they could point to and say, "That one did it." In contrast, in nearly every other instance—in fossil bones from the Triassic to the near present—we do not have any of the actual insects. Thus the greatest importance of insect trace fossils in fossil bones is that they are almost always the only evidence we have of insects interacting with vertebrates.

: : :

Whenever I think back on my wife's cow skull and its gradual consumption by squirrels, I am also reminded to wonder when mammals first started wearing down bone. In contrast to bone-eroding insects that started in the Middle Triassic, the oldest known tooth traces attributed to mammals are in dinosaur bones from the Late Jurassic (~160 MYA) in China.[40] Despite the almost 80-million-year gap between these trace fossils, we know that both insects and mammals were modifying bones by the Late Jurassic, and probably did so all through the Cretaceous. Nevertheless, mammal trace fossils in Mesozoic dinosaur bones are still rare, with most coming from the Late Cretaceous.[41] So this means either mammals rarely scraped dinosaur bones for nearly 100 million years, or (more likely) we just haven't yet recognized their trace fossils.

A big setback for bone-eroding insects and mammals was at the end of the Cretaceous (about 66 MYA), as a meteorite impact and its ecological consequences erased all big-boned land animals in a geological instant. Because modern dermestid beetles, termites, and other bone-eroding insects are here now and chewing away, we assume at least a few of their ancestors survived the impact. But we do not know if the ancestors of bone-eroding mammals survived, or whether this behavior evolved again independently in mammals, eventually leading to situations where urban squirrels got to have more cow bone.

What we do know, though, is that bone consumption by modern herbivorous mammals, which is *not* directly related to carnivory, is actually more common than one might suppose. For instance, rats and porcupines also work their incisors on bones, and for the same reasons as squirrels, which is to add calcium to their diets while also reducing their ever-growing teeth.[42] But if you were surprised that squirrels eat bone, then you might be downright shocked to know that deer and some of their hoofed relatives also partake in skeletal snacks. Herbivorous hoofed mammals (ungulates) that chew bones include at least seven species of deer, bighorn sheep, sable antelopes, gemsbock, kudu, wildebeest, and domesticated animals, such as sheep, goats, and pigs.[43] In perhaps the creepiest example, a white-tailed deer (*Odocoelius virginianus*) was recently documented chewing on human bones.[44]

Among the more intriguing of bone-eroding hoofed herbivores are those seemingly all-vegan browsers of trees and the
tallest mammals in the world, giraffes. Giraffes have a good skeletal fossil record, one showing the evolution of their famously
long necks from what were originally short-necked ancestors.[45]
Given their impressive heights (4–5.5 m/15–18 ft), giraffes can
browse a variety of high-growing vegetation, helped by agile
tongues and flexible upper lips that can grab and pull off leaves
from tree branches. Such anatomical attributes compensate for
an absence of front teeth on their upper jaws. However, giraffes
still have incisors and canines on their lower jaws, as well as robust premolars and molars on both jaws.[46] These teeth do their
everyday duty by grinding leaves and twigs, while also being well
adapted for abrading bones. Just to be clear, though, giraffes do
not actually break up and swallow entire bones. Instead, they
only scrape their teeth along bone surfaces to get just what they
need: more like taking a calcium supplement rather than having
a full-bone meal.[47]

Traces left by giraffes and other hoofed mammals that gnaw
on bone are distinctive and often described as "forked zigzags."
These gouges split from one into two (that's the "fork") and sets
of those forks can change direction along a bone surface (that's
the "zigzag"). Because most teeth involved in gnawing have flat
surfaces, trace cross-sections are more often U-shaped than
V-shaped. In contrast, carnivorous mammals typically leave
straight-down punctures or V-shaped sections as they pull their
pointy teeth across bones.[48]

Knowing the differences between nibbles left by insects and
herbivorous mammals on bones versus those of carnivorous
vertebrates is a handy skill for paleontologists to have. For one,
they can tell whether a fossil bone was used for: eating flesh and
bone, while also producing new generations (insects); scraping
bone to shorten teeth and ingest calcium (rodents); or just to get
calcium (ungulates). Paleoanthropologists also benefit from this
knowledge by more accurately discerning traces made by non-
carnivores versus those of carnivores, including humans or human relatives that used stone tools to cut bones.[49]

Which brings us to the traces of vertebrate-eating carnivores past and present, and especially to those animals not only supremely adapted for eating meat, but also for crushing bones. And given such evidence, we know minimally how long this bone crushing, puncturing, scoring, and other destruction has been happening, and how.

: : :

Petrified Forest National Park in Arizona is often nicknamed "Triassic Park," and with good reason. This is partly because it is one of the best places in the world to learn about continental ecosystems from the Late Triassic (237–201 MYA), but also because it *is* a park, managed by the US National Park Service.[50] As mentioned earlier, its name comes from its extraordinarily abundant and large, colorfully mineral-filled conifer-tree trunks lying on otherwise sparsely vegetated badland exposures of the Chinle Formation.

But Petrified Forest National Park also holds much more evidence of this slice of Earth history than "just" trees. For most paleontologists who worked regularly in the park, its body fossils are the stars. Among these are the remains of many vascular plants other than the conifers, as well as freshwater horseshoe crabs, crustaceans, insects, and molluscans. Vertebrates were also represented by bony fishes, giant amphibians called metoposaurs, lizards, and a clade of reptiles called archosaurs.[51] Archosaurs included armored and spiky aetosaurs, multi-ton herbivorous synapsids, as well as phytosaurs and rauisuchids, both of which looked like huge crocodiles. Dinosaurs were there too, but most were minor and not nearly as intimidating as their Late Triassic cohorts. In the late 1990s I repeatedly visited the park to study its trace fossils, which include insect borings in its trees (mentioned previously), burrows of crayfish-like animals, vertebrate tracks, and coprolites.[52] To study the coprolites, I enlisted the help of an undergraduate research student, Allison Wahl. Other than noting ground-up bone in the coprolites, Wahl's most important find was tiny burrows in them, which she interpreted as the work of feces-eating insect larvae.[53]

Although coprolites from the Petrified Forest show that insects interacted with vertebrate feces then, the coprolites' bone-
infused content also demonstrated that at least a few vertebrates
in these environments ate other vertebrates. Primary suspects include metoposaurs, rauisuchids, and phytosaurs, all undoubted
carnivores. Metoposaurs and phytosaurs were water dwellers that were likely confined to river channels and their banks,
whereas longer-legged rauisuchids probably hunted on land.[54]
Although we did not determine whether the coprolites fell on
land or in the all-natural toilets of streams and lakes, knowing
this would have been another factor in helping identify who did
do the doo-doo.

Fortunately, newer research since my too-brief times in the
Petrified Forest and on the Chinle Formation revealed much
more about who ate who during the Late Triassic. In 2014, paleontologist Stephanie Drumheller and her colleagues described
two large rauisuchid femurs from the Chinle Formation in New
Mexico.[55] Both femurs had multiple sets of D-shaped holes that
were wide at their tops and tapered to points at their bottoms,
penetrating bone at varying depths. These holes were assuredly
not from insects, but from large carnivores. Moreover, the bigger of the two femurs—which belonged to an 8–9 m (26–29 m)
long rauisuchid—told of an animal that dared to attack this
huge predator, and a hole with an embedded tooth revealed that
at least one bite came from a phytosaur.[56] Even better, the bone
around this tooth and four other nearby holes showed signs of
healing, implying that the rauisuchid was very much alive and
escaped the phytosaur attack. Drumheller and her colleagues
then reconstructed an entire animal from a single tooth to estimate the size of the assailant. Based on a simple principle that
the bigger the tooth, the bigger the phytosaur, the researchers
concluded it was 5–6 m (16–20 ft) long: so not as big as the rauisuchid, but still big enough to chomp it.[57]

How did a phytosaur bite the rauisuchid in the first place,
considering they were apex predators in supposedly separate
realms, one in the water (phytosaur) and one on land (rauisuchid)? Perhaps the rauisuchid walked down to a riverbank for

Row of tooth traces (arrow) by a large phytosaur in bone from the Chinle Formation (Late Triassic), New Mexico; specimen displayed at the New Mexico Museum of Natural History, Albuquerque.

a drink and got careless. Or the phytosaur, like some modern
crocodilians, may have taken a brief stroll on land. Regardless,
the phytosaur bite penetrated skin and muscle deeply enough to
reach the surface of the rauisuchid's femur. Either the force of
the phytosaur bite or the escape of the rauisuchid then wrenched
a tooth out of the phytosaur's jaw, leaving it as a reminder of this
epic battle of Triassic titans.

And with that, welcome to the inner world of pierced and
broken bones, skeletal traumas that live on as trace fossils in remains of the dead.

: : :

Very soon after bones evolved, other bone-bearing animals
evolved ways to break them. The evidence for such breaks var-
ies in accordance with what I like to call the "holy trinity of ichnology": substrate, anatomy, and behavior (amen). For traces in
bones, the substrates are made of the mineral apatite, and their
owners were either live or dead. If dead, then we must ask how

long they were dead. Also, a single bone can have different densities, with compact cortical bone on the outside and porous cancellous bone underneath. As for anatomy, vertebrate traces in bones are most often related to teeth or other bony parts in jaws, but also can be inflicted by horns or other weaponry. Given the vast array of teeth possessed by vertebrates over the past 300+ million years, but the uniqueness of some, paleontologists can often connect a tooth trace with the tooth (and hence animal) that made it. Finally, behavior refers to how a bite or other types of blows also reflect intent, or how a healed bone shows both a failed attack and a successful escape.

The earliest examples of the behavioral shift of vertebrates from making bones to breaking bones are from the Devonian, sometimes called "The Age of Fishes." The Devonian was a time of explosive diversification of swimming vertebrates after the Silurian Period, and placoderms were especially prominent among these fishes. Placoderms (where *placo* = "plate" and *derm* = "skin") seemingly competed with one another to see who could build the strongest body armor as protection against predation, with interlocking bony plates covering upper, lower, and lateral surfaces. This change from soft- to hard-bodied was a vertebrate example of the "marine predation revolution" that likewise affected shelled invertebrates during the early part of the Paleozoic.[58]

The threat of vertebrates preying on vertebrates came about early in the Devonian, and escalated dramatically with the evolution of jaws. The earliest fishes to swim in the world's oceans lacked this anatomy, which meant their food gathering was more passive. For instance, modern jawless fishes, such as lampreys, have parasitic lifestyles, latching on to and feeding on the flesh and blood of living fishes.[59] In contrast, biting requires a lever system similar to that used by shell-crushing crabs, and body parts—such as teeth or beaks—that can crack, crush, or otherwise penetrate hardened defenses. This adaptation changed vertebrates for the rest of their evolutionary history, enabling many lineages to kill and consume a variety of sternly packaged foods.

All jawed vertebrates are placed under the clade Gnathostomata (where *gnathos* = "jaw" and *stomata* = "mouth").[60] When viewed through the lens of evolutionary heritage, a stunning range of vertebrates is included in this clade. Think of all modern sharks, rays, bony fishes, amphibians, reptiles, birds, and mammals, comprising more than 60,000 species: all are gnathostomes. Now, extend that membership to extinct vertebrates—from placoderms to mastodons—that diverged from a common ancestor more than 400 million years ago. Jaws probably originated from two or three of the forward-most gill arches in jawless fishes, a hypothesis partly supported by subtraction: modern jawless fishes have nine arches, whereas jawed fishes have seven.[61] Over thousands of generations, these arches became hinges with opposing parts, an upper jaw (maxilla) and lower jaw (mandible), each reinforced by connective tissues, powered by musculature, and accompanied by shearing edges, such as bony blades or teeth. The earliest undoubted body fossils of jawed fishes are from the Silurian (~430 MYA), but older and tantalizing pieces of probable gnathostomes are also in Middle Ordovician rocks (~465 MYA).[62]

The lethal effects of jaws are apparent from early in the history of gnathostomes. Hard-bodied invertebrates were doubtless on the menu, as shell-crushing fishes arose and munched on shelled molluscans and brachiopods. Smaller placoderms even bit into crinoids, as shown by traces on Middle Devonian (~390 MYA) crinoids of Poland.[63] These small nips led to big changes, as crinoids evolved defensive strategies against fish attacks by developing thicker plates, withdrawing their arms, or displaying colors warning of toxicity.[64] Such adaptations eventually led to magnificent floating colonies of crinoids attached to logs during the Jurassic, until wood-boring clams put a stop to such planktonic grandeur.

Of course, placoderms did not restrict themselves to the spineless and soon went after one another. In a 2009 study conducted by paleontologist Oleg Lebedev and others, they found wounds in a variety of Devonian fishes from throughout eastern Europe.[65] Healed wounds showed these animals were alive when bitten and that they survived long enough to recover from a bite. In contrast, those fossils bearing traces with no signs of healing were either bitten as part of a predatory act or scavenged soon after dying. What these paleontologists found was that other Devonian fishes with jaws assaulted Devonian placoderms, and often. Injuries on top and lateral plates revealed that many attacks came from above and on the sides, rather than from behind and below. Shapes of tooth impressions and bites pointed toward shark-like fishes called acanthodians, shark-imitating placoderms called arthrodires, and early sharks.[66] Lobe-finned fishes (sarcopterygians) were additional biting suspects. In at least a few specimens, sizes of bite traces relative to their prey show that their placoderm predators took on fishes that were too big, trying to bite off more than they could chew. These and other reasons could be why predators' attacks sometimes failed. Yet we also must recall survivorship bias, and consider that these fossilized placoderm bits are from fishes that postponed death, whereas the remains of the killed and eaten may not have been preserved.

: : :

Rare is the paleontologist who will not admit that the zenith of Devonian placoderms is represented by one armored fish to rule them all, *Dunkleosteus*. The genus *Dunkleosteus* had ten species spread over the last 20 million years of the Late Devonian, contributing to a reign of oceanic terror ceased by the end-Devonian mass extinction (~358 MYA).[67] Its most famous species was *Dunkleosteus terrelli*, a marine predator unlike any then or since. Picture a fish about 9 m (29 ft) long, weighing several metric tons, and covered not by scales, but by body armor, and armor so extensive that it had circles of bone within its eyes. These swimming tanks completed that analogy by augmenting formidable defenses with dreadfully effective offensive weaponry. They did not have teeth, but instead possessed bladed bony plates that worked like massive multipronged pruning shears. Their maws opened so quickly that the rapid change in outside pressure sucked in whatever hapless prey was in front of it, where it would be sliced in half with one bite, bony armor notwithstanding.

Although paleontologists have known about *Dunkleosteus* body fossils since the late nineteenth century and found remains of its various species in North America, Europe, and Africa, few have tried to learn more about its behavior. Granted, anyone who gazes upon a complete skull of *Dunkleosteus* will shout "Predator!" out of sheer survival instinct, but details of its flesh-eating remained murky. Fortunately, paleontologists in the past few decades have studied it more closely and confirmed that these fishes were guilty of dreadfully awful biting during the Devonian.

Speculations about *Dunkleosteus* and its predatory prowess were rendered less mysterious when research published in 2007 by paleontologists Philip Anderson and Mark Westneat showed that its bites depended on what they termed "four-bar linkage."[68] This refers to linkages between joints and levers, with four joints used for opening and closing jaws; linkage bars were the connections between joints, and jawbones were moved by sets of muscles. Their results—based on computer simulations of inferred jaw motions—were astonishing. In the words of Anderson and Westneat, *Dunkleosteus* "had one of the most powerful bites in vertebrate history."[69] Simulations showed this giant placoderm could open its mouth in as little as 20 milliseconds, or far less time than it takes to say *"Dunkleosteus."* It then closed its mouth in 30–40 more milliseconds, meaning that suction was immediately followed by the shearing action of the bony plates.[70] And what shearing it was. Bite forces for a 6 m (20 ft) long *Dunkleosteus* were calculated at 4,400 N (~990 lbf, or pound-force) at the tip of the jaws and 5,300 N (~1,200 lbf) toward the back of the jaws.[71] Considering this deadly combo of rapid suction and instant slicing, we can safely surmise any prey that could fit in a *Dunkleosteus* mouth would not stay whole or survive for more than a second.

Computer simulations aside, curmudgeonly paleontologists might reasonably ask for actual evidence from the fossil record that *Dunkleosteus* could chomp harder than any other known Paleozoic animal. Fortunately, the fossil record obliged. In a 2016 study by paleontologist Lee Hall and his coauthors, they reported multiple bite traces on Late Devonian placoderms' bony plates.[72] Trace fossils of these attacks were deep furrows, fractures, and bent bone on the sides and tops of several placoderm heads, none of which show signs of healing. Moreover, wounds were inflicted on probable weak points in the placoderms' body armor, such as just below the eyes, near the gills, or where head plates met the rest of the body. The horrific effects of these impacts, as well as the sizes and shapes of the wounds, laid blame on one fish and its fearsome bladed mouth, *Dunkleosteus*.

You might wonder what placoderms could possibly withstand more than one bite from a *Dunkleosteus*, and still have anything left to preserve in the fossil record. Why, *Dunkleosteus*, of course.[73] What motivated these giant placoderms to bite one another? Cannibalism is one possibility, similar to how naticid gastropods have relatives over for dinner. But another (and more likely) reason for *Dunkleosteus* fights is competition, in which giant rivals fought over resources in their Devonian seas. One can imagine that such battles bordered on the legendary, but were probably short-lived, making them more suited for TikTok videos than feature films.

Sadly for those who love enormous marine predators of the past, *Dunkleosteus* and all other placoderms vanished from the tapestry of life with the end-Devonian mass extinction. Yet one of the *Dunkleosteus* pieces examined by Hall and his coauthors held clues of jawed fishes that succeeded them and eventually produced their own impressive consumers of flesh and blood. The bony plate held a series of thin scratches that were likely left by a small fish scavenging on the remains, a modest preview of bone-breaking ascendency represented by one of the most successful clades of marine vertebrates in Earth history: sharks.

: : :

During the Devonian "age of fishes," the first sharks swam alongside the more exotically dubbed placoderms, acanthodians, and sarcopterygians. As we learned earlier with rays and their shellcrushing ways, these fishes share a common ancestor with sharks and belong to the same clade (Elasmobranchii). Sharks today consist of more than 500 species with a worldwide distribution, and their sizes vary from the dwarf lanternshark (*Etmopterus perryi*), which is shorter than most adult shoes (about 15 cm/6 in), to the largest, the whale shark (*Rhincodon typus*).[74] This filter-feeding shark also doubles as the largest extant fish, with some perhaps more than 15 m (49 ft) long, exceeding the length of an average sailboat (about 10 m/33 ft). If sharks could tell people to get off their ocean, the one most likely to do so would be the Greenland shark (*Somniosus microcephalus*), as it is the longest-lived vertebrate, with ages perhaps as long as 400 years.[75] Depending on the species, sharks can live in sunlit shallow waters or exceed depths of several thousand meters.

Unlike their bony-fish companions, cartilage is the main structural support in shark jaws. However, most sharks compensate for a lack of mineral-laden struts with an impressive variety of pointed, bladed, serrated, and otherwise lethal teeth. Unlike our species and many other vertebrates, sharks continually replace lost teeth, meaning that one shark could have contributed hundreds of teeth to the fossil record during its lifetime. Along with such body-part donations, sharks have also imparted countless scrapes, punctures, and scores on vertebrate bones. Based on their trace fossils, these fishes have bitten into and through bones for more than 350 million years, ranging from the Late Devonian through today.

Considering that sharks survived four mass extinctions, a list of shark bite recipients would compose one of the world's longest and most varied menus. Among the fossils with shark bite traces or shark teeth in their bones are, in alphabetical order: birds, bony fishes, cetaceans (both toothed and baleen), crocodilians, dinosaurs, mosasaurs, pinnipeds (seals, walruses, and their kin), plesiosaurs, pterosaurs, and sea turtles.[76] One of the reasons why sharks have eaten such a range of vertebrates is because the ocean often has contained both local and imported animal protein. For the latter, if marine sharks ever tired of eating seafood, they could also depend on occasional influxes of land animals' bodies washed out to sea. Paleontologists have even documented shark-bitten coprolites, in which the sentiment "Eat feces!" was taken literally.[77]

Who is the best biter among modern sharks? That title clearly belongs to the world's biggest predatory fish, the great white shark (*Carcharodon carcharias*). Fully grown female great white sharks can be 6 m (20 ft) long and weigh almost two metric tons (4,400 lb), with males not far behind in both categories.[78] Although these sharks are unfairly maligned as voracious killers that can never pass up an opportunity to get them some sweet, delicious human flesh, attacks on people are extremely rare. Great white sharks are much more likely to work their jaws on marine mammals (seals and sea lions), rather than pasty New England–bred prey. Still, their tastes are wide ranging, as they also consume other fishes (including other sharks), seabirds, sea otters, cetaceans, and sea turtles.[79]

Turtle-killing bites represent enormous bite forces, but despite our long fascination and fear of great white sharks, we mostly see just the results of their bone-slicing strength. A 2008 study by paleontologist Stephen Wroe and many coauthors helped better quantify what happens when this shark bites.[80] For their study they used mathematical and computer modeling of a great white shark's anatomy. Part of the scientists' careful approach to modeling was also based on how shark jaws are supported by cartilage, an anatomical difference affecting the distribution of stresses and strains across their jaws. Their results showed that a modestly sized 2.5 m (8.2 ft) long specimen could have bitten with about 1,600 N (~360 lbf) of force at the front of its mouth, but almost double that (3,100 N/~700 lbf) with bites toward the rear.[81] When scaled up to a much larger great white shark, such as one 6.4 m (21 ft) long and weighing 3,324 kg (~7,328 lb), the front-jaw and rear-jaw bite forces would have been an impressive 9,320 N (~2,100 lbf) and 18,216 N (~4,100 lbf). These forces, coupled with huge, bladed, and serrated teeth, mean that prey animals' bones can be easily sheared or crushed by great white sharks, as they have been, and as they shall be. All such insights should suggest the need to acquire a bigger boat.

Nonetheless, we cannot talk about bite forces of just great white sharks without mentioning the greatest biter of all. This was *Otodus (Carcharocles) megalodon*, often called "*Megalodon*" or "The Meg" by its close friends and the film industry. Although very much extinct, this gigantic shark and other *Otodus* species lived in the world's oceans during the Miocene and Pliocene,

from a little more than 20 MYA until a mere 3.5 million MYA.[82]
O. megalodon was originally classified as a relative of great white
sharks, but actually was more closely related to modern mack-
erel sharks.[83] Its enormous, pointy, and finely serrated teeth represent most of its body fossil record; the largest teeth easily cover
most adult humans' palms, measuring 16–19 cm (6.5–7.5 in) long.
These teeth also testify to why "Megalodon" was so named, as it
means "big tooth." How large was O. megalodon? It depends on
which paleontologist you ask. Using a number of techniques to
extrapolate body size based on its teeth, most estimates place it
at 15–20 m (49–66 ft) long and weighing more than 50 metric
tons (55 tons),[84] comparable to the most massive of Mesozoic marine reptiles. Why so big? Paleontologists propose that O. megalodon's great size was an adaptation for hunting whales, which also
trended toward larger sizes then.

O. megalodon also bit big. In their same 2008 study of great
white shark bite forces, Stephen Wroe and his colleagues calculated forces for an assumed 50+ metric tons (55+ tons) O. megalodon, and 93,127 N (20,935 lbf) and 182,201 N (40,950 lbf) for its
front and back bite forces, respectively.[85] In a 2021 study by paleontologists Antonio Ballell and Humberto Ferrón, they modeled
stresses for teeth of a similarly sized O. megalodon and came up
with a 49,000 N (11,015 lbf) puncture force for its front teeth and
about 96,000 N (21,581 lbf) for its back teeth.[86]

So imagine being a prey animal floating or swimming non-
chalantly near the water surface when a 50-ton shark speedily
approaches from below. Even without biting, a collision with this
shark would have killed anything smaller than it. Unfortunately
for paleontologists, though, a shark large enough to swallow all
or bite through its prey was less likely to leave much evidence of
these fatal encounters. Still, a few fossil bones recorded whale
interactions with O. megalodon or a closely related species. For
example, a whale vertebra from the Pliocene of Venezuela had
an embedded O. megalodon tooth in it, which is rather damn-
ing.[87] Traces of another O. megalodon attack consist of gouges in
three Miocene whale tail vertebrae from coastal Maryland.[88] One
of the vertebrae has scores from adjacent teeth that match the

tooth row of *O. megalodon*, giving that shark reason to ask, "Do I have someone stuck in my teeth?" Because these wounds were inflicted on tailbones, this bite was probably intended to disable the whale. Perhaps the most dramatic of all inferred *O. megalodon* trace fossils, though, are three scores on a whale tooth, and not just any whale, but a predatory sperm whale.[89] The tooth is from either Miocene or Pliocene (7–3 MYA) sediments of North Carolina, and probably belonged to a 4 m (13 ft) long whale. Considering that a toothed whale's head would have been less desirable for scavenging, the scores—caused by two successive bites—were more likely from an attack aiming to dispatch the whale. This and other evidence show that *O. megalodon* used its unprecedented bite forces to kill and eat whales.

: : :

After the end-Triassic extinction, a clade of comparatively smallish reptiles, Crocodylomorpha, became one of the most successful of all tetrapod clades, spreading into continental and aquatic environments during the Jurassic and Cretaceous. At the end of the Cretaceous (66 MYA), they succeeded further by surviving the same mass extinction that erased their distant archosaur relatives, the dinosaurs and pterosaurs. Crocodylomorphs today are represented today by twenty-four species of crocodilians (crocodiles and alligators) that mostly live in temperate to tropical freshwater environments.[90] Yet in the past, they ranged from iguana-sized plant-eating land dwellers to monstrous coastal carnivores, the latter represented by the 11+ m (36+ ft) long *Deinosuchus* of the Late Cretaceous (about 80–75 MYA).[91]

Crocodilian tooth trace fossils in bones are almost as old as their clade, but most recognizable traces from the Mesozoic are Late Cretaceous (80–66 MYA). Some of these trace fossils are invaluable for shedding light on ecological relationships between crocodilians and dinosaurs that we might not otherwise know. For example, crocodilian tooth traces and a wayward tooth left on juvenile dinosaur bones from the Late Cretaceous Kaiparowits Formation of Utah tell us that smaller crocodilians ate them.[92] This evidence implies that differently sized crocodilians might not have competed for the same foodstuffs. Tooth trace fossils on a Late Cretaceous dinosaur bone from marine sediments in New Jersey also show that coastal crocodilians fed on bloat-and-float dinosaur remains washed out to sea.[93]

Fortunately, modern crocodilians give us living examples of saurian biters that can be compared to those of their long-gone relatives. This means scientists can directly measure crocodilian bite forces and examine extant animals' bones punctured or shattered by their teeth and jaws. Bite-force experiments on alligators and crocodiles—many done by paleontologist Greg Erickson and his colleagues over the past 25 years—involve instruments they've nicknamed "bite bars," which directly measure forces imparted by a crocodilian bite. In 2012, Erickson and others published a paper that compiled bite-force measurements taken from all living species of crocodilians.[94] Their results were remarkable, showing that modern crocodilians are, kilogram for kilogram, the strongest biters we have today (sorry, sharks). Also, body size and bite forces correlate in modern crocodilians. American alligators (*Alligator mississippiensis*) have bite forces ranging from 2,400 to 9,500 N (540–2,135 lbf), which is impressive, but they are bested by saltwater or "saltie" crocodiles (*Crocodylus porosus*) of Australia, with recorded bites as much as 16,400 N (3,687 lbf).[95]

Knowing these data, researchers developed statistical predictions they applied to extinct crocodilians to recreate bite forces from the past. So, how about *Deinosuchus*? Based on a presumed mass of an adult *Deinosuchus* (about 5 metric tons/5.5 tons) and following trends of modern crocodilian bite forces, its extrapolated bites were more than 102,000 N (22,930 lbf), overlapping those of *O. megalodon*.[96] This also means that saying "*Tyrannosaurus* was the apex predator during the Late Cretaceous" should always be followed by the qualifier "on land."

:   :   :

Still, we love to talk about *Tyrannosaurus*, which somehow has its name or abbreviation (*T. rex*) shoehorned into virtually every paleontology news article, like so: "Trilobites died out almost 200 million years before the first *T. rex* roamed the earth." Nevertheless, I will begrudgingly admit that *Tyrannosaurus* deserves admiration for its bone-crunching ability, which was probably the most powerful of all land animals' and penetrated other dinosaurs' bones with ease.

But just how did paleontologists test and calculate a tyrannosaur bite without using live lawyers? They tried reproducing trace fossils observed in dinosaur bones. The trace fossils, described in 1996 by paleontologists Greg Erickson and Kenneth Olson, were punctures showing exactly where *Tyrannosaurus* teeth penetrated the hipbones of a *Triceratops*.[97] In another article, Erickson and other researchers reported their attempts to mimic these traces and measure the bite forces needed for such penetration.[98] First, they learned that *Tyrannosaurus* was responsible for the punctures by molding putty to the holes, which neatly made casts of the *Tyrannosaurus* teeth. (Paleontologists sometimes refer to *T. rex* dental implements as "lethal bananas," as the serrated teeth of a full-grown adult were about the length and width of this breakfast berry.) The experimenters later made metallic models of tyrannosaur teeth and mounted them into a mega-bite machine that correlated forces with depths penetrated by these false teeth. Their results indicated that 6,410 to 13,400 N (1,441–3,012 lbf) of force made traces in cattle bones similar to those in *Triceratops*.[99] This study and many others since support the fair-minded notion that tyrannosaurs bit hard and deep, which worked well for handling both live and squirming prey, as well as for cutting through skin, muscles, and bones of dead-dinosaur bodies.

More evidence that tyrannosaurs included bone in their diets is embodied in their fossil wastes. The first presumed tyrannosaur coprolite was found in the Late Cretaceous Frenchman Formation of Saskatchewan (Canada) in the 1990s. This deposit of former feces, with an estimated volume of more than 2 L (about 2 qt), was composed of tiny pieces of dinosaur bones cemented by apatite. When paleo-poopologist Karen Chin and several other scientists reported on it in a 1998 paper, they pronounced it a "king-sized coprolite," one that could only have been left on the throne of the tyrant king, *Tyrannosaurus rex*.[100] Its ground-up bone, though, also showed that *Tyrannosaurus* fractured bones by chewing, rather than by gulping big chunks. In 2017, paleontologists Paul Gignac and Greg Erickson supported Chin and her colleagues' conclusions with a study of *T. rex* jaw movements, bite forces, and tooth pressures.[101] They convincingly demonstrated that this dinosaur's eating methods were unusual in at least two ways. For one, it had massive bite forces—recalculated as 35,000 N (~7,870 lbf)—and tooth pressures that shattered bones on impact. For another, it chewed its food repeatedly and in the same places, continuing to splinter these bones and thus converting bites into bits. These adaptations led the researchers to declare *T. rex* as guilty of "extreme osteophagy."

Given the horrendous bite forces of *Tyrannosaurus* and undisputable evidence of its bone chomping, one might ponder how another dinosaur could have survived a close encounter of the wrong kind with this theropod. Yet we know that some dinosaurs were bitten by a tyrannosaur and lived. How to see such evidence for yourself? It requires a trip to the Denver Museum of Nature and Science in Denver, Colorado. In one room is a specimen of a dinosaur that has captivated me each time upon viewing, probably because it also reminds me of my career in academia. The skeleton is of an *Edmontosaurus annectens*, a large herbivorous hadrosaur (nicknamed a "duck-billed" dinosaur) that lived in the western part of North America during the Late Cretaceous (70–66 MYA). And this one was a survivor. It is displayed with all four feet on the ground, its head up and looking to its left, its back straight and parallel to the ground in a pose suggesting alertness. As your gaze moves from left to right, starting with the head and moving back along its torso, you might note the pleasing symmetry of its skeleton, an expressive beauty that makes a few damaged tailbones all the more striking. According to one paleontologist, Kenneth Carpenter, this is where a tyrannosaur clamped down its jaws on the tail of a live and very likely terrified *Edmontosaurus*.[102] But this is not a mark of death, but one of escape and subsequent healing, in which the hadrosaur lived long enough afterward for its bones to repair.

During the century-long study of *Tyrannosaurus* and its kin

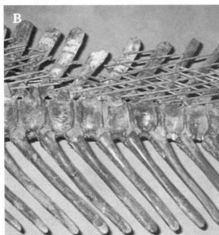

Late Cretaceous hadrosaur (*Edmontosaurus annectens*) with healed bite trace in its tail vertebrae, displayed at Denver Museum of Nature and Science, Denver, Colorado. A, Entire specimen with bite trace indicated (arrow); *Triceratops* skull for scale. B, Close-up of injured tail vertebrae.

as presumed top predators, a controversy arose when Hollywood consultant Jack Horner successfully trolled the paleontological community with a different take. In his 1993 book *The Complete T. Rex*, Horner proposed that *Tyrannosaurus* was more a scavenger than a predator.[103] Since then, plenty of tyrannosaur trace fossils have indicated otherwise, including the Denver specimen of *Edmontosaurus* and other healed tyrannosaur bites, which shows at least one healed tooth trace with a *Tyrannosaurus* tooth still there.[104] Would a tyrannosaur pass up a happy meal offered by an already-dead *Edmontosaurus* or *Triceratops*? Of course not, just as most modern carnivores easily switch between predation and scavenging for their daily flesh.

Considering that acquiring food, fighting over territory, and seeking mates are a normal part of life for many animals, one might justifiably ask if tyrannosaurs were ever tempted to bite one another. In an intriguing 2021 study by paleontologist Caleb Brown and his colleagues, they answered that question with a scientifically qualified "Yes, but . . ." In their research they examined more than 500 facial bones from about 200 specimens of different tyrannosaur species and documented that more than half of the bones had healed lesions.[105] Surprisingly, a closer look revealed all lesions were from bites inflicted by other tyrannosaurs, and that the bite traces and healed wounds were concentrated on the upper and lower jawbones. This implied that tyrannosaurs were non-fatally fighting other tyrannosaurs and making future scars by biting one another's faces. However, such bite traces were missing from juvenile tyrannosaurs and only found in half-grown to full-sized adults. Tooth spacing of the bites further showed they came from same-sized individuals.

Such oddities led Brown and his coauthors to propose that these trace fossils represented behaviors of sexually mature tyrannosaurs competing with one another. Although they could not determine the sexes of the biters and the bitten, the researchers quite reasonably speculated that male tyrannosaurs were competing for mates. Scars in facial bones of other theropods from the Late Triassic through Late Cretaceous likewise support that those animals competed, too.[106] Nevertheless, such scars are *not* in facial bones of theropods more closely related to birds, which were likely feathered. An absence of bite traces in those dinosaurs suggests that once feathers became more common and used for showy displays, sexual competitions in theropods shifted from face-biting smackdowns to face-saving dance-offs.

: : :

While large crocodylomorphs and dinosaurs dined on sizable vertebrates in their respective Mesozoic ecosystems, mammals stayed mostly out of their way, living in burrows and otherwise avoiding snack-attack status. However, turnabout was fair play, demonstrated by fossils of mammals that ate at least parts of dinosaurs. For instance, as we learned earlier, Late Jurassic tooth trace fossils show that small mammals nibbled on dinosaur bones. But we also know that an opossum-sized mammal, *Repenomamus*, from the Early Cretaceous (~125 MYA) of China, consumed not just a few scrapes of a dinosaur bone, but a dinosaur.[107] Granted, it was a baby dinosaur and that of an herbivore (*Psittacosaurus*), but its skeleton inside a *Repenomamus* rib cage offers a token sense of Mesozoic retribution for generations of mammals either swallowed whole or squashed underfoot.

Following the mass extinction at the end of the Cretaceous 66 million years ago, mammals filled vacant ecological niches and otherwise diversified quickly, occupying nearly all environments within the next 10 million years or so.[108] Mammalian carnivores made especially impressive advances during that time, with some developing awesome bone-crunching abilities that persist today. Among present-day bone eaters with impressively large ancestors in the recent geologic past are felids (lions, tigers), ursids (bears), and others (oh, my). However, the mammals best known for breaking other mammals' bones are modern hyenas, such as the spotted hyena (*Crocuta crocuta*) and striped hyena (*Hyaena hyaena*) of Africa. With powerful jaws and robust teeth, hyenas have also long fascinated paleontologists, taphonomists, and zoologists because of their ability to scatter and fragment the skeletal remains of other mammals, including those of our human ancestors.[109]

Yet the surprise mammalian champion of skeletal smashers was not on land, but in the sea, and was a whale: *Basilosaurus*. Because lineages of toothed whales evolved quickly from ponysized to more whale-appropriate proportions early in the Cenozoic, part of their predatory competence is attributed to greater skull size and jaw musculature.[110] Despite its erroneous "*saurus*" suffix, *Basilosaurus* was not a reptile, and lived during the latter part of the Eocene (40–34 MYA), with one species (*B. cetoides*) in the southern US and another (*B. isis*) in northern Africa and the

Middle East.[111] *Basilosaurus* was probably the biggest animal on Earth during its slice of time, with adults reaching 18–20 m (59–66 ft) long, larger than most modern sperm whales. *Basilosaurus* was also a carnivore, which was bad news for sharks, bony fishes, and other toothed whales wherever it swam. A close look at *Basilosaurus* teeth further reveals damage caused by biting into hard objects. However, scratches and other minor signs of wear on their teeth suggest *Basilosaurus* also augmented its diet with shelled molluscans and crustaceans.[112] Regardless, *Basilosaurus* teeth tell us that they chewed their food, rather than just biting and swallowing.

In a 2015 study published by paleontologist Eric Snively and coauthors, they calculated bite forces for *Basilosaurus* at a maximum value of 20,000 N (~4,500 lbf).[113] This number far exceeds bite forces calculated for any land mammals extant or extinct. Indeed, this bite put *Basilosaurus* in a chomping category more akin to giant marine reptiles of the Mesozoic. I should also note that *B. isis* was the smaller of the two *Basilosaurus* species, implying that *B. cetoides* may have bitten slightly more ferociously.

: : :

Given the long history of insects and vertebrates as bone consumers, and the impressive bone-breaking start of early mammals, other mammals of the Cenozoic also evolved to break down hard objects, including rocks, shells, and wood. Among these mammals were some of the largest land animals of all time, but they also included a species that changed an entire planet with its bioerosion, rendering a lasting effect on the surface and biota of the earth.

Chapter 11

The Biggest and Most Boring of Animals

When an elephant needs salt, it will remove mountains to get it. And on the flanks of Mount Elgon—a dormant volcano straddling the border between Kenya and Uganda[1]—that is exactly what generations of elephants did, carving ballroom-sized caves in its side as lasting evidence of their saline cravings. As far as rocks go, the volcanic sediments gouged by their tusks were relatively soft, but solid enough to form the foundation of the mountain.

At some point in the past 10,000 years or so, a few African elephants (*Loxodonta africana*) discovered these mineral deposits, began breaking up the rock with their tusks, and used their trunks to ingest chunks of the salt-laden rock. Knowledge of these deposits, such as where to find them and how to mine them, was apparently passed down from matriarch to matriarch, and hence from generation to generation. Over time, the elephants wore down paths leading to and from the mountain with their creation of Kitum Cave and more than thirty other caves on its eastern (Kenyan) side, some as deep as 150 m (about 500 ft) and 60 m (~200 ft) wide.[2] Cave walls today are crisscrossed by tusk-wide and tusk-deep linear scratches from up, down, and sideways movements against the rock. This weakening of cave walls led to occasional ceiling collapses, which in turn heightened the floor and caused the elephants and the caves to move upward. Elephant droppings in and near the cave confirmed their rock-eating habits, adorned with chunks that had passed through as mineral supplements. All of these clues and actual eyewitnessing of elephants breaking apart and eating the bedrock led to the inevitable conclusion that these elephants, which can be as tall as 4 m (13.1 ft) and weigh as much as 6 metric tons (7 tons), are also the largest living bioeroders of rock.[3]

Such significant alterations of rocks by land mammals that are motivated by a need for salt or other minerals are either rare or underappreciated. Geologists Charles Lundquist and William Varnedoe posited the latter, that some big holes in rocks simply labeled as "caves" might actually be the result of rock-eating behaviors by large herbivores. In a 2005 article, they termed such geologic features as "salt-ingestion caves," with Kitum Cave and similar elephant-made caves in Kenya as prime examples.[4] In their definition, a salt-ingestion cave needs to fulfill two criteria. First, it must qualify as a cave, which is a cavity in rock large enough for an adult human to enter. Second, it must be a cavity made by vertebrates consuming rock for its salts. Other modern examples of caves formed by mammals licking or otherwise breaking down rocks include caves from the Altai Mountains of central Asia, British Columbia (Canada), Cambodia, Sarawak (Malaysia), and Mississippi (US).[5] As for the cave in Mississippi (pithily dubbed "Rock House Cave"), people discerned its mammalian origins in the mid-nineteenth century. Since then, scientists think white-tailed deer (*Odocoileus virginianus*) and bison (*Bison bison*)—animals native to that area—were responsible for its initial formation, with its volume later enhanced by cattle.

Do trace fossils of salt-ingestion caves exist as mega-borings carved by megafauna? Lundquist and Varnedoe suggested at least one example from Chile, la Cueva del Milodón ("Milodon Cave"). This and other nearby caves may have been made (or at least enlarged) by Pleistocene giant ground sloths, such as *Mylodon darwinii*, named after well-known barnacle appreciator Charles Darwin. This cave-creation credit is assigned to *Mylodon* because its bones, patches of its hairy skin, and *Mylodon*-sized coprolites are in these caves.[6] Other possible mammal-made caves are on Cyprus, a place that once had dwarf hippopotamuses (*Hippo-*

potamus minor) and dwarf elephants (*Palaeoloxodon cypriotes*). Coastal caves on Cyprus hold many bones of both species, which lived on the island until their extinction just 9,000–10,000 years ago.[7] Although labeled as "dwarf" elephants and hippos because they were significantly smaller than their mainland counterparts, both species probably weighed as much as 200 kg (440 lb), which is still hefty. Regardless, how did these animals "shrink," especially when compared to the average Pleistocene-Holocene megafauna? Because Cyprus has always been a Mediterranean island separate from the rest of Asia, ancestors of the dwarf elephants and hippos must have swum there, and in enough numbers to breed. Later, the limited resources of their island habitats led to natural selection that favored the survival and reproduction of smaller-sized individuals, eventually differing noticeably from their larger ancestors.[8] In a 2008 study by paleontologists Eleftherios Hadjisterkotis and David Reese, they inferred that these elephants and hippos might have expanded preexisting caves while seeking water, shelter, and minerals, including salt.[9]

Other caves attributed to Pleistocene megafauna include tunnels in Argentina and Brazil that were once regarded as erosional features formed by groundwater, but later recognized as the work of animals. Argentine and Brazilian paleontologists identified these tunnels—some more than 4 m (13 ft) wide, 2 m (6.6 ft) tall, and more than 100 m (330 ft) long—as former homes of giant ground sloths and giant armadillos that lived in South America during the Pleistocene.[10] Like artists who used distinctive brushes and brushstrokes, deep and long scratches on tunnel walls matched the huge-clawed hands of sloths and armadillos, revealing their creators. However, as impressive as these structures might be (and they are), they are burrows, not borings. The reason for this pedantic distinction is because most of the tunnels in Argentina were made in unconsolidated sediments. Also, the tunnels in Brazil cut through deeply weathered bedrock that was soft enough that giant ground sloths and armadillos could dig into and move through it.[11] So although ichnologists can rightfully point to these as the largest known nonhuman burrows, they can also give credit to salt-ingestion caves as the largest known borings made by nonhuman animals.

Given this knowledge of the biggest bioeroding animals, which animals are the grand champions of bioeroding, breaking rock, shells, wood, and bones on a scale that ensures their traces will surely outlast their species, and thousands more species they take with them? For a representative example of this species, look into a mirror. Or if a reflective surface is not available, look at a current photo of *Tȟuŋkášila Šákpe* (= "Six Grandfathers"), the Lakota Sioux name for a dramatic outcrop in what is now called South Dakota.[12] There, in granite that cooled from deeply seated magma about 1.7 billion years ago,[13] uplifted and eroded by weather and biota alike since, is a far more rapidly made bioerosion structure depicting the faces of four people from the past few hundred years. Not coincidentally, the collective lifespans of these four people also represent a time of unprecedented and accelerated bioerosion, especially from mining, drilling, and deforestation. The mining included that for coal and drilling for oil and gas, with their burning contributing to atmospheric carbon dioxide levels that increased 33 percent since the birth of the oldest of the four people, and 25 percent in just the 80 years since their carved visages were completed.[14] So while gazing at either your features or those of these recent historical figures, take the time to muse on how you are viewing a transitional species, but one that through its bioerosion changed Earth's history in a shockingly short amount of time.

: : :

Elephants and humans have lived together for hundreds of thousands of years, especially in their mutual place of origin, Africa. The oldest known fossil elephants are from the Paleocene (~58 MYA) of Morocco; they evolved and spread throughout African landscapes for more than 50 million years before a certain lineage of primates descended from trees and joined them on the surface.[15] The near-world distribution of elephants was aided by plate tectonics, which closed a great seaway separating Africa from Europe and Asia, thus granting them passage to those landmasses.

Once in these novel environments, elephants evolved more quickly, with great size and social cohesiveness among the many traits selected and inherited by future generations. And thanks to trace fossils, we know that elephants were already living together as extended families by about 7 million years ago. A spectacular fossil elephant tracksite in Abu Dhabi shows that thirteen elephants of varying sizes—presumably composed of at least one adult matriarch and younger members—traveled together, their trackways crossed by that of a larger lone bull elephant.[16] Much later (in elephant time), mammoths and other elephants wandered across the Bering Strait between what is now Russia and Alaska. From the Arctic region they moved south through the Americas, passed the equator, and reached the Atlantic and Pacific coasts of both continents.[17] This remarkable journey of elephants into all continents other than Australia and Antarctica was one of the greatest feats achieved by any lineage of large mammals, deserving our awe and respect. We can also be sure that all species of elephants altered their environments while on their journeys, whether as salt-ingestion caves like those in Africa, Asia, and North America, or in other ways.

Late in elephant history in the Americas, some mammoths and other megafauna initiated a different form of bioerosion, one that might have escaped our notice if not for modern animals doing the same. The elephantine evidence is in rocks along the coast of California and more specifically in Sonoma County, famed for its growing and fermentation of grapes into popular beverages. These outcrops and boulders of coastal Sonoma have been nicknamed the "Mammoth Rocks," an appellation that succinctly but dramatically summarizes their proposed origin.[18]

However, this wearing of rock was far subtler than excavating caves in mountainsides for ingesting salt and was prompted by a different biological need. Large, hairy, and warm-blooded are all attractive qualities for parasites that can easily find and attach to mammals with such traits. Those animals with excessive parasites, such as lice and ticks, had to find ways to alleviate the

"Mammoth Rocks" in Sonoma State Park, California, metamorphic rocks that became ancient scratching posts worn smooth by Pleistocene megafauna; human for scale (right). Photo by Thewellman, Wikimedia Commons.

itching and other discomforts caused by their inadvertent little passengers. So they rubbed, but being elephants, they rubbed grandly. Similar to the development of caves, many generations of mammoths, mastodons, ground sloths, camels, and other big mammals once native to North America probably contributed to polishing these rocks.[19] Although lower polished areas might have come from bison and smaller mammals suffering the same discomfort—including more recently introduced domesticated cattle—the highest areas that are about 4 m (13.1 ft) off the ground limit who rubbed where and when.

Given such needs and their great sizes, mammoths, mastodons, and other big mammals sought outcrops of durable rocks along the California coastline for itch relief. The rocks, consisting of chert and blueschist, are originally of sedimentary and metamorphic origin (respectively), formed in the deep sea but then tangled and mangled by subsequent events into a geological mess called a mélange.[20] These rocks were once part of a submarine trench formed by colliding plates during the Jurassic and Cretaceous, while near-surface seas hosted ichthyosaurs, plesiosaurs, and mosasaurs. Oceanic sediments in these areas were squeezed and transformed by heat and pressure into rocks. Later, like snow plowed into cohesive sheets, they rode up and on top of one another and were shoved onto landward environments.[21] Much later, during the Pleistocene at around 120,000 years ago, some of the rocks were eroded by waves near the current coast and formed sea stacks. These actions happened when sea level was 5–6 m (16.4–19.7 ft) higher than today, but the rocks are above the present shore because of tectonic uplift. This is not surprising, considering the San Andreas Fault is just offshore from these rocks, a transform fault in which the Pacific and North American tectonic plates move past one another.[22] So like the rocks of Mount Elgon, these former sea stacks are another outcome of crustal turmoil that affected the lives of elephants and of elephants modifying the products of that unrest.

Other than the height of this unusual polish, what other evidence supports an animal origin for the erosion of these rocks? Their position relative to the sea is one, as the sides of these former sea stacks facing the ocean and offshore winds are not polished, whereas the leeward sides are. This textural contrast implies that large animals sought protection from the near-constant buffeting of coastal winds when moving their bodies against the rocks. The polished surfaces are also closely associated with points and edges, which would have served as spots for maximum relief from itchiness. Moreover, a closer look at the rock surfaces with scanning electron microscopes reveals many long but minute parallel scratches, signature traces of where furry bodies imbued with fine-grained quartz-rich sediments moved back and forth across them.[23] Mammal hairs are too soft on their own to erode the rock, but sand and silt clinging to the hairs can, much like how soft-soled shoes bearing grit can wear down stone steps of older buildings over time. Linear scratches of these surface textures contrast with non-polished surfaces, which bear circular scratches imparted by wave-powered pebbles and cobbles in these former sea stacks.[24] Yet another point supporting the origin of these altered rocks is modern examples of parasite-alleviating surfaces in Africa and Asia, where elephants take regular mud baths and rub against trees or rocks. Zoos also install rubbing posts for the comfort of their elephants and rhinoceroses, which produce similar textures. Hence ichnologists and paleontologists can reasonably propose these traces as analogues for probable trace fossils reflecting the same behaviors.

The final pieces of evidence supporting the origin of these "Mammoth Rocks" are those of mammoths and other mammal body fossils verifying that oversized rubbers were indeed in the area. For instance, bones of a Columbian mammoth (*Mammuthus columbi*) were located just a few kilometers south of these rocks, and more Pleistocene megafauna remains come from other parts of Sonoma County and adjacent Marin County, including mastodons (*Mammut americanum*), giant camels (*Camelops hesternus*), giant ground sloths (*Glossotherium harlani*), and two extinct species of bison (*Bison latrifons* and *B. antiquus*).[25] All of these animals were potential polishers of the stones, each according to their different heights but with all but the tallest overlapping their traces; modern cattle and horses later added their rubbings to the ancient ones below. These polished stones are thus fine examples of bioerosion trace fossils that were also group projects, with recent domestic mammals belatedly contributing well after their due date.

How does fine-grained sediment get onto mammal bodies? Wind-blown clays, silt, and sand can adhere and collect on hair over time, especially if a mammal rarely bathes in water. But a far more direct and efficient means for getting gritty is to bathe in dust or mud. As for dust bathing, I have seen 2 m (6.6 ft) wide depressions of bare, dry soil in otherwise grassy areas of Yellowstone National Park that mark where modern bison (*Bison bison*) living there habitually rolled on their sides and backs.[26] The dust, composed of dry clays and silt, helps dislodge or suffocate pesky parasites. However, mud baths can work even better, as parasites are enveloped by mud and removed by rubbing. Modern elephants do exactly this, seeking out muddy pools as sources of all-natural sunscreen, cooling, and parasite treatments. In videos recording these behaviors, clearly gleeful elephants sit down in and roll around and cover their bodies in mud.[27] As one might imagine, such vigorous activities by a number of weighty mammals can expand these muddy pools while also redistributing sediments from one place to wherever the elephants roam afterward.

Did mammoths and other Pleistocene fauna take mud baths, and then wear down local rocks with that mud and their motions? Very likely so, and a subtle and large (0.5 hectare/1.2 acre) depression between the main Mammoth Rocks on the Sonoma coast hints at the spot where such joyful bathing would have happened. Wetland plants currently live in the depression, their growth encouraged by a groundwater seep there, but its position and size also suggest a previous history as a mud wallow.[28] Hence one can imagine a Pleistocene vista along this coast about 20,000 years ago, with landmarks like the Mammoth Rocks as familiar points, but in an ecosystem inhabited by the massive animals that shaped them.

∶ ∶ ∶

As for the ancestors of modern humans in Africa who lived alongside elephants, they were much smaller but similarly modified their ecosystems, especially through their accidental or intentional use of fire. The bioerosion of these early humans and their immediate ancestors—such as australopithecines or other species of *Homo*—mostly modified rocks, shells, wood, and bones for tools. By about 3 million years ago, early human relatives began making stone tools by choosing chert and similar silica-rich rocks that, when struck by a similar rock, broke so these developed finer edges.[29] This technique, called percussion flaking, produced the first blades, which were used to scrape mammal skin from flesh and cut flesh from bones. These actions also left distinctive traces on bones, called bone-surface modifications, or BSM for short.[30] Other tools included axes and hammers used on their own but later affixed to carved wooden handles.

Although these changes in hard substrates required only modest bioerosion of the materials themselves, the tools themselves later contributed to major changes in terrestrial ecosystems. For instance, tools made of altered rock and wooden resources were used as weapons that enabled these relatively slight and slow-moving mammals to hunt and kill much larger and faster mammals, which in some instances may have hastened their extinctions.[31] Eventually humans followed large mammals (including elephants) out of Africa, and throughout Eurasia they constructed new tools out of stiff materials along the way, leaving evidence via reshaped rocks, shells, wood, and bones.

As for bones, a sort of bioerosion repurposing happened in which hard parts of the largest bioeroders were modified into tools by relatives of the most bioeroding species. In 400,000-year-old deposits near Rome, Italy, archaeologists uncovered nearly a hundred tools made from the main shafts (epiphyses) of elephant limb bones.[32] The tools—some of which were pointed and bladed—were interpreted as mostly scrapers and smoothers, the latter used to press against and smooth dry mammal hides. Bone tools dating to 1.6–1.3 MYA in Ethiopia give a minimum time for when human relatives (hominins) began using vertebrate remains as part of their technology, augmenting rocks and wood as raw materials.[33] Later, coastal cultures added marine-molluscan shells to hard stuff that was reworked into useful items needed for survival. All such durable artifacts were carried or created in place by humans wherever they went in Africa, Asia, Europe, and Australia, and over several hundred thousand years.

When did people in the Americas start bioeroding rocks, shells, wood, and bone? Recently discovered human tracks in New Mexico show they were in North America minimally 23,000 years ago, whereas bone-surface modifications from a site in Uruguay suggest a human presence there about 30,000 years ago.[34] Regardless of when humans arrived, these traces and other evidence tell us that elephants preceding them in their trip from Asia had to deal with them yet again. Back in Europe, Neanderthal tracks in the same places and horizons as elephant tracks in southern Spain also suggest relationships between these people and those behemoths on shared landscapes co-shaped by their presence.[35]

∴ ∴ ∴

Zoologists used to regard tool use in nonhuman animals as an aberration or abomination, insisting that tools were what separated us from "lesser" creatures. Fortunately this anthropocentric nonsense has since been sacked, and the list of animals that employ objects for accomplishing the tasks of everyday life continues to grow as acceptance of their use spreads. For example, earlier we learned about: fishes slamming clams on rocks; gulls dropping molluscans on hard-packed surfaces to break their shells; and sea otters picking out rocks for hammering clams on their chests, or putting clams on rocky surfaces used as anvils. In all of these instances, tool use is aimed to rapidly break hard materials while also using other hard materials to aid in that fracturing, or bioerosion by proxy.

Nonhuman primates, and specifically a few species of monkeys and apes, have similarly adopted hard-tool behaviors. In some monkeys their tool use resembles that of otters by using rocks as hammers and additional rocks as anvils. But instead of breaking clams, these primates open nutritious nuts. Although the lasting effects of this behavior are not nearly as impressive as, say, caves scraped out of volcanic mountains, monkeys did have an impact with their impacts by wearing down rock surfaces. Research of this tool use led to more surprises, such as how this bioerosion was not merely the result of "find nut, smash nut" behaviors. Rather, it reflected a more deliberate and sophisticated decision-making process applied to the use of both tools and nuts, ultimately telling us much about monkey cultures.

In several studies of bearded capuchin monkeys (*Sapajus libidinosus*) in Brazil, primatologist Dorothy Fragaszy and other scientists confirmed the monkeys were themselves scientists and engineers.[36] For one, the capuchins picked stones they could pick up with both hands, but that were also hard enough to survive pounding against their favorite nuts, usually from piassava palms (*Orbignya* sp.), as well as nuts of other palms. Second, the monkeys chose sizable, durable, and stationary rocks—whether as boulders or outcrops—that did not move when struck with another hard rock and a nut between them. Third and most importantly, the capuchins learned optimal positions for placing nuts on anvil surfaces so these did not roll off those surfaces, whether before or after striking them.

To test these observations more precisely, Fragaszy and her

Young black-striped capuchin monkey (*Sapajus libidinosus*) putting a seed on an anvil rock and readying a hammer rock to open it; Sierra da Capivara, Brazil. Photo by Tiago Falótico, Wikimedia Commons.

colleagues marked lines (meridians) along the nuts showing where they ceased rolling on flat surfaces. Then they gave the marked nuts to the monkeys, which promptly got cracking. The scientists observed that the capuchins almost always placed the nuts on anvil surfaces with meridians at less than 30° from vertical, which kept them in place during and after smashing.[37] In what must have been the most amusing and self-deprecating phase of the research, the scientists made monkeys out of volunteers by blindfolding them and having them try to replicate the capuchins' methods. Deprived of sight, the humans had to orient nuts by sensing their stability on anvil surfaces. Sadly, in what was surely a missed opportunity for video outtakes, the volunteers did not then blindly smash nuts with rocks. Still, they did put the nuts in the same positions as the monkeys, affirming how it could be done.[38] More human testing of nut stability likewise corroborated that the monkeys were indeed finding the best positions for breaking open these nutritious treats.

Although the nuts themselves did not contribute to the lowering of anvil surfaces, the repeated collision of capuchin-wielded rock hammers with anvils did indeed erode anvil surfaces, forming saucer-shaped depressions of varying widths on these rocks. These depressions in turn help keep nuts from rolling away with striking, which means skilled capuchins probably seek out these long-used anvils.[39] Also, hammer rocks kept nearby have been reused and passed down over time, which probably chipped away their original volumes. In contrast, most anvils stay in the same place, implying that an anvil site might have experienced tens of thousands of blows from thousands of stones, over hundreds or thousands of years. Anvil rocks also had their minerals loosened and swept away by wind and rain over time, meaning they hold records of their histories through what is not there. In 2019, Tiago Falótico and his colleagues verified long-term capuchin bioerosion when they documented 3,000-year-old monkey-made stone tools in Brazil.[40]

The erosion of rocks over time thus reflects generations of capuchins adjusting their nut-cracking methods with different and changing anvil surfaces, which required flexible thinking and regular shifts in tactics. Perhaps what was most significant about this research, though, is that it shows the importance of hard tools that could be passed down from one generation to the next, along with the methods for using them.[41] Once selected and used, hammers had new wielders taught by previous generations of adults, and older anvils. Occasionally new hammers were constructed as needed, but anvils were picked for their stability and reliability, staying exactly where they were found.

This dual form of bioerosion—the breaking of compact woody tissues by rocks and the wearing down of rocks used as tools—suggests how the earliest primate bioerosion may have started. Moreover, it tells us the importance of culture and the passing down of well-tested techniques and knowledge to future generations, which in our species eventually led to animal-caused bioerosion of unprecedented amounts.

: : :

Whenever I converse with people for the first (or second) time, and I tell them I am a paleontologist, some well-meaning folks

then conflate me with archaeologists, who study human artifacts, or paleoanthropologists, who study the fossil record and
evolution of humans. So my usual response is to utter an Ace
Ventura line, "I don't do humans," which helpfully leads me to a
short explanation of how I'm also not Ross Geller, Alan Grant, or
(I wish) Ellie Sattler. I also explain that although dinosaur bones
are also not so much my thing, I like dinosaur tracks and other
traces enough to have written a rather thick book about them.[42]
And if they're still with me? Good, because then we can move
on to discuss something really thrilling, such as moon snail
drillholes.

Yet there's a way we can bridge the divides between archaeology, paleoanthropology, and paleontology, and that is
through ichnology. Back in the day—which is to say, the Pliocene
(~3–4 MYA)—the ancestors of modern humans left many traces
of their behaviors, such as footprints and stone tools, and eventually traces imparted by stone tools on bones.[43] As mentioned
before, the earliest stone tools made by hominins were simple
implements made by striking rocks with other rocks, a method-
ology that quickly revealed which rocks were softer or harder
than others. This implies the earliest stone-using hominins—
like the capuchin monkeys of Brazil—grasped the main concepts
of the Mohs hardness scale and applied it to rock identification
long before introductory-geology students did the same in college classrooms. Hominins applied other forms of bioerosion as
well, by modifying bones, wood, and shells. In a sense, australopithecines and the earliest representatives of *Homo* were the
most skilled and varied of bioeroders in primate history before
modern humans. This knowledge made a difference in their subsequent evolutionary histories, particularly in a species that became the most boring of them all.

When did humans become bioeroders in a major way, significantly altering rocks and other hard materials? The discovery that certain minerals or rocks could be used for tools or
other resources was only a start in this respect. Such realiza-
tions then led to the birth of mining, whether by simply picking
up and carrying rocks out of riverbeds, or digging into soils for

more of the "right" materials. Artistic expression also required finding appropriately hard substances for chipping into rock itself to make petroglyphs, which often depicted animals hunted with sculpted spear or arrow points. Or, softer tools were used to make pictographs on rock walls, but this artwork was made possible by pigments derived by crushing minerals. For example, red ocher, which came from the mineral hematite in the Ngwenya mine of Eswatini in southern Africa, was extracted more than 40,000 years ago, making this site the oldest known human mine.[44]

The discovery of metallic ore deposits and subsequent treatments for extracting their metals happened at least 8,000 years ago, with the start of the Copper Age.[45] Smelting of metallic ores initially used the renewable fuel of wood, but greater demand for fuel meant cutting down more trees. Then, with the discovery that a compact organic-rich black rock (coal) could be burned as fuel for steam power, it too was mined. By the mid-nineteenth century the fuel potential of petroleum deposits and natural gas beckoned,[46] which led to the development of drills that used the metals derived from ore deposits and were powered by coal, which led to more bioerosion. And with that, the human-bioerosion genie was out of the bottle, as much more intensive mining, drilling, and other processes tore apart once-solid rock. Human bioerosion on a massive scale is thus ultimately responsible for many of the challenges that we are facing as a society, civilization, and species, along with the fates of many other species.

: : :

While growing up in Indiana during the 1960s, I longed to dive deeply into an ocean, any ocean. At least part of this geographically inappropriate hankering was prompted by watching Jacques Cousteau scuba-diving or stuffing himself and other people into tiny submersibles that went to great depths, leading the way to similarly thrilling escapes from earthly matters. Fortunately for me, my parents paid attention to their strange kid who longed for watery environments. Despite a limited income, they invested in swimming lessons for me at a local YMCA rather than, say, dental insurance. So I swam, and eventually became less than terrible at it, which led to new dreams involving water. That was when I heard about people who swam long distances, including across the English Channel, or *La Manche* ("the Sleeve") if you're French. However, it might have been after looking at a map and seeing the distance between England and France (about 33 km/22 mi) that I decided to take up long-distance running instead.

Of course, since the invention and reinvention of boats—wooden or otherwise—people had no need to swim from England to France, or vice versa. But how about driving? Ferries at first accommodated this automotive dream by carrying cars back and forth across the Channel, probably with at least a few people (namely, Americans) who sat in their cars behind the steering wheel for the entire journey and imagined it as a road trip. More enlightened folks who supported rail travel likewise pined for trains smoothly taking them from one landmass to another. So eventually commerce beckoned, shoved, and pushed into the very bedrock underlying the Channel, evolving into what became the third longest tunnel in human history, known as either the Channel Tunnel, the Chunnel by many American tourists, or the "tunnel under the Sleeve," by you guessed it.

Given that wooden sailing ships were relatively slow, moved on water, and could be easily sunk by wood-boring clams, people imagined ways they could travel from France to England or back again while still using ground transportation: or rather, underground transportation. French and English scientists and engineers throughout the early nineteenth century addressed this desire by proposing they dig a tunnel underneath the channel.[47] Because these plans were drawn well before automobiles, railroads were imagined as the main means of transport. Enthusiasm for this tunnel even resulted in exploratory bioerosion in 1882, with the English and French each using coal-rock-powered rock-boring machines to produce combined tunnel lengths of 3.5 km/2.2 mi.[48] Still, this effort only represented about 10 percent of the channel width, and the dream of boring gigantically remained deferred until the late twentieth century. Once the geological information, engineering strategies, and far more contentious political and economic concerns were finalized, construction began in 1988, using TBMs (tunnel boring machines) that started from either end and met in the middle.[49] Six years later, the Channel Tunnel officially opened, a feat of undersea linkage that far exceeded nineteenth-century imaginations. Indeed, when combined with high-speed rail connections on either side of the water, the tunnel made it possible for someone to travel by train from London to Paris in a bit more than two hours.

The Channel Tunnel may have been a product of human imagination and technology, but it was made possible by geology. For one, the sedimentary rock hosting the gigantic boring was well suited for the task. Given a channel's fill of water above the tunnel and the ever-present factor of gravity, overlying and underlying strata needed to limit fracturing and otherwise prevent water leakage into the tunnel. The rock also had to be sturdy enough to prevent the tunnel from collapsing under the weight of strata and water above, but soft enough to allow a TBM to chew through it. Based on seismic surveys of the underlying strata and drillholes that brought up samples of the rock, geologists and engineers decided that one of the Cretaceous formations, the West Melbury Chalk Marl, would do nicely for their grand bioerosion project.[50] The lower part of this formation was also underlain by an impermeable Cretaceous formation, the Glauconitic Marl.

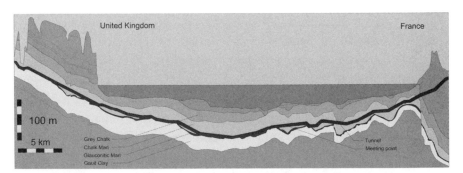

Cross-sectional view of subsurface geology associated with the tunnel under the English Channel ("Chunnel"), one of the largest human-made bioerosion structures. Illustration by Commander Keane, Wikimedia Commons.

Marls are fine-grained rocks that consist of clay- and silt-sized sediments, but with mixtures of calcium carbonate and silicate minerals. Once detected and mapped, the contact between the two formations was used as a guide for workers operating the TBMs.[51]

Incidentally, among the calcium carbonate sediments in the Cretaceous marl were the mineralized remains of single-celled marine algae and protozoans, many of which contained borings from cyanobacteria, algae, and fungi. Cretaceous oysters and other molluscan shells also contained these traces, as well as those of bioeroding sponges. This meant that when the TBMs tore through the marl, one of the largest life-caused borings affected some of the smallest, a chipping away of solid substances separated by more than 70 million years.

: : :

A common theme of this book is that bioerosion in its various forms, from microbes to elephants, has affected the earth's oceans, atmospheres, and land surfaces for nearly as long as life has existed. The evidence for this vast history is in traces, and sometimes traces in body parts, collectively reflecting the myriad methods used by life to break down rocks, shells, wood, and bones. A companion theme is that bioerosion will continue to affect our ocean and land surfaces in many of the same ways as in the past. The effects of bioerosion will be imparted on rocks, shells, wood, and bones, whether these come from: rock-diminishing cyanobacteria, fungi, and lichens; rock-rasping chitons and echinoids; coral-munching parrotfishes; shell-drilling snails and shell-crushing crabs; wood-boring beetles, clams, and woodpeckers; bone-eating worms; bone-crushing sharks and hyenas; or mountain-eroding megafauna.

Added to these past, present, and future traces of bioerosion are those of rocks, shells, wood, and bones altered by human relatives in the past few million years, a degrading of hard stuff that accelerated dramatically in the past 10,000 years. Moreover, the explosive growth of technology that accompanied the Industrial Revolution in just a few hundred years also grinds away at what was once solid. These human-made mines and drillholes in rocks, crushed shells, cut trees, butchered bones, and other changes in durable materials will surely become part of the geologic record, whether our descendants are there to witness it or not. Also, the indirect effects of surface and shaft mines, drilling, and fracking—the deep fracturing of strata to more easily extract hydrocarbons—as well as deforestation, animal husbandry, and the like, will continue to affect the lives of nonhuman bioeroders.

Finally, with increased global temperatures predicted for the remainder of the twenty-first century, we should expect significant sea level rises with the addition of more water from continental ice sheets.[52] The marine cyanobacteria, algae, fungi, and invertebrates that survive the rapid change in ocean temperatures and acidity will then begin eroding the hard substrates of our civilization at the margins of the sea. Where limestone edifices now stand, their surfaces will be pockmarked by this evidence of life enduring in the durable, carrying on a biological process that made it past all other global calamities for more than 3 billion years. So for as long as the earth provides rocks, shells, trees, or bones, life will erode these, returning their essence for the next round of renewal, becoming solid again, however temporarily.

Acknowledgments

The overarching theme for this book first came to me in 2018, but the scaffolding and background research for it took place during a 2019 summer residency at the Hambidge Center for Creativity in the Arts and Sciences near Rabun Gap, Georgia. So the first of my many thanks are to the Hambidge Center for granting me a peaceful respite for creative musings that helped develop main themes and details alike.

After repeated rejections of the book proposal from potential publishers (their loss), my literary agent, Laura Wood (FinePrint Literary), found the perfect home for this book with University of Chicago Press. Senior Editor Joseph (Joe) Calamia at Chicago helped me with its gestation and birthing, nearly all of which happened as we all went through the tumultuous times of 2020–2022. Thank you, Laura and Joe, for making my initially unconventional idea of a book become real.

Others to thank at the University of Chicago Press as the manuscript went from rearranged electrons on a computer screen to something different include its editorial and production staff. Thanks in particular to Anjali Anand and Sara Bakerman, for their preliminary handling of the main text and figures, and Tamara Ghattas, who oversaw editing of the manuscript as it went into production. Freelance editor Marianne Tatom expertly copyedited the manuscript and saved me from numerous literary embarrassments , and Theresa Wolner expertly indexed it.. I am also grateful to the external reviewers of my first draft, John Long and an anonymous reader, whose comments, suggestions, and corrections helped make my manuscript shift from a rough draft to a smoother one. Of course, any mistakes of text or fact that readers may spot in this book (usually on the first page) are entirely my fault.

I struggled with finding a proper epigram to start this book, but I was inspired by one that blossomed from the wisdom of the thirteenth-century Persian poet Jalāl al-Dīn Muḥammad Rūmī, known by most as "Rumi." Thus I am grateful to my neighbor and friend Reza Sameni for giving me a proper interpretation of Rumi's original quotation and a deeper understanding of its context.

Ichnology is the science of life traces, and I am an ichnologist who knows many other ichnologists. Most of them (like me) are also paleontologists, and some are paleontologists or biologists who enjoy working with ichnologists. In what is surely an abbreviated list, these people are (in alphabetical order): Bobby and Sarah Boessenecker, Richard Bromley, Luis Buatois, Carlos Neto de Carvalho, Michelle Casey, Karen Chin, Al Curran, Stephanie Drumheller, Tony Ekdale, Jorge Genise, Patrick Getty, Jordi Gilbert, Murray Gingras, Roland Goldring, Murray Gregory, Ashley and Lee Hall, Stephen Henderson, ReBecca Hunt-Foster, Patricia Kelley, Reinhold Leinfelder, Martin Lockley, Gabriela Mángano, Radek Mikuláš, Renata Guimarães Netto, George Pemberton, Cecilia Pirrone, Thomas Rich, Andrew Rindsberg, Joana Rodrigues, Ana Santos, Dolf Seilacher, Katy Smith, Alfred Uchman, Trevor Valle, Patricia Vickers-Rich, Christy Visaggi, Sally Walker, and Andreas Wetzel. A few of these scientists I have known for more than thirty years, and some have passed, with Patrick the most recent. Fortunately they left lasting traces of their thoughts, helping us to continue learning from them.

Sincere gratitude goes to more than thirty years of undergraduate students for educating me at my academic home of Emory University in Atlanta, Georgia. These students gave me both the incentive and the experience to convey sometimes complex scientific concepts into simpler (but never simplistic) ways for audiences other than academic colleagues. I wish them all well on their personal excursions and want them to know that I treasure their cultivating me into the scientist, teacher, and writer I am today.

My habitual book writing has been affirmed and supported by my wife Ruth, but this book was far more challenging for us than its predecessors. My writing it in earnest started in late 2019, just before the COVID-19 pandemic spread throughout the world in early 2020. Somehow I wrote words, sentences, and paragraphs about natural history in between the pandemic's horrific effects, as well as economic turmoil, protests for human rights, a contentious election, an insurrection, climate-change-fueled disasters, war crimes, and more. Amid such bleakness and with heightened empathy for those who suffered, my writing could have easily and understandably ceased under the weight of emotional exhaustion. But I have Ruth to thank for her patience and encouragement during such a difficult time for us all. Thank you, Ruth, for your strength, for your love.

Finally, there are the tracemakers, which, despite our efforts to alter the planet to better accommodate us and us alone, will surely include species that outlast us, and ones that break down rock, shells, wood, and bones, including those of our own design and bodies. Thank you for teaching us about your lives and your evolutionary histories, giving us much-needed perspective as we share this exceedingly thin slice of Earth history, together.

Notes

Chapter One

1. L. M. Chiappe and M. Qingjin, *Birds of Stone: Chinese Avian Fossils* (Baltimore: Johns Hopkins University Press, 2016); B. L. Stinchcomb, *Mesozoic Fossils II: The Cretaceous Period* (Atglen, PA: Schiffer Publishing, 2015).

2. M. J. Everhart, *Oceans of Kansas: A Natural History of the Western Interior Sea* (Bloomington: Indiana University Press, 2017).

3. B. T. Huber et al., "The rise and fall of the Cretaceous hot greenhouse climate," *Global and Planetary Change* 167 (2018): 1–23; W. W. Hay, "Toward understanding Cretaceous climate—an updated review," *Science China Earth Sciences* 60 (2017): 5–19.

4. C. N. de Carvalho (ed.), *Icnologia de Portugal e Transfronteriça* [*Ichnology of Portugal and Cross-Border*]. *Comunicações Geológicas* 103 (2016).

5. A. Santos et al., "Two remarkable examples of Portuguese Neogene bioeroded rocky shores: new data and synthesis," *Comunicações Geológicas* 103 (2016): 121–30.

6. A. Conrad Neumann first coined the term "bioerosion" in a 1966 research article reporting on modern sponge bioerosion: A. C. Neumann, "Observations on coastal erosion in Bermuda and measurements of the sponge, *Cliona lampa*," *Limnology and Oceanography* 11 (1966): 92–108.

7. A. J. Martin, *Life Traces of the Georgia Coast: Revealing the Unseen Lives of Plants and Animals* (Bloomington: Indiana University Press, 2013); A. J. Martin, *Dinosaurs Without Bones: Dinosaur Lives Revealed by Their Trace Fossils* (New York: Pegasus Books, 2014); A. J. Martin, *The Evolution Underground: Burrows, Bunkers, and the Marvelous Subterranean World Beneath Our Feet* (New York: Pegasus Books, 2017); A. J. Martin, *Tracking the Golden Isles: The Natural and Human Histories of the Georgia Coast*. (Athens: University of Georgia Press, 2020).

8. P. Brannen, *The Ends of the World: Volcanic Apocalypses, Lethal Oceans,*

and Our Quest to Understand Earth's Past Mass Extinctions (New York: Ecco, 2017); R. Black, *The Last Days of the Dinosaurs: An Asteroid, Extinction, and the Beginning of Our World* (New York: St. Martin's Press, 2022).

9. D. Erwin and J. Valentine, *The Cambrian Explosion: The Construction of Animal Biodiversity* (Greenwood Village, CO: Roberts and Company, 2013). Although now dated in its scientific details, I also recommend reading the following book, a paleontological classic: S. J. Gould, *Wonderful Life: The Burgess Shale and the Nature of History* (New York: W. W. Norton, 1990).

10. B. Deline et al., "Evolution and development at the origin of a phylum," *Current Biology* 30 (2020): 1672–79; A. Kroh, "Phylogeny and classification of echinoids," *Developments in Aquaculture and Fisheries Science* 43 (2020): 1–17.

Chapter Two

1. S. Hamada and H. D. Slade, "Biology, immunology, and cariogenicity of *Streptococcus mutans*," *Microbiological Review* 44 (1980): 331–84; S. Mounika and J. M. Nithya, "Association of *Streptococcus mutans* and *Streptococcus sanguis* in act of dental caries," *Journal of Pharmaceutical Sciences and Research* 7 (2015): 764–66; A. L. Bisno et al., "Diagnosis of strep throat in adults: are clinical criteria really good enough?," *Clinical Infectious Diseases* 35 (2002): 126–29.

2. P. D. Marsh, "Dental plaque as a biofilm: the significance of pH in health and caries," *Compendium of Continuing Education in Dentistry* 30 (2009): 76–78.

3. B. Wopenka and J. D. Pasteris, "A mineralogical perspective on the apatite in bone," *Materials Science and Engineering, C* 25 (2005): 131–43.

4. G. Falini et al., "Control of aragonite or calcite polymorphism by mollusk shell macromolecules," *Science* 271 (1996): 67–69; S. B. Pruss et al., "Calcium isotope evidence that the earliest metazoan biomineralizers formed aragonite shells," *Geology* 46 (2018): 763–66.

5. W. W. Gerberich et al., "Toward demystifying the Mohs hardness scale," *Journal of the American Chemical Society* 98 (2015): 2681–88.

6. S. G. Dashper and E. C. Reynolds, "Lactic acid excretion by *Streptococcus mutans*," *Microbiology* 142 (1996): 33–39. Also, I brought up the movie *Alien* because it is also among my favorite science-fiction films, which is at least partly related to the protagonist's (the alien's) use of chemical bioerosion against the antagonists (the humans), and as an important part of the plot. The corrosiveness of the alien's blood limits choices for the humans inexplicably trying to destroy these poor, misunderstood creatures that just need to eat and make baby aliens.

7. C. F. Demoulin et al., "Cyanobacteria evolution: insight from the fossil record," *Free Radical Biology & Medicine* 140 (2019): 206–23.

8. A. C. Allwood et al., "Stromatolite reef from the Early Archaean era of Australia," *Nature* 441 (2006): 714–18.

9. B. E. Schirrmeister, "Cyanobacteria and the Great Oxidation Event: evidence from genes and fossils," *Palaeontology* 58 (2015): 769–85; A. P. Gumsley, "Timing and tempo of the Great Oxidation Event," *Proceedings of the National Academy of Sciences* 114 (2017): 1811–16.

10. A. Oren and S. Ventura, "The current status of cyanobacterial nomenclature under the 'prokaryotic' and the 'botanical' code,'" *Antonie Van Leeuwenhoek* 110 (2017): 1257–69.

11. Demoulin et al., "Cyanobacteria evolution."

12. L. Sagan, "On the origin of mitosing cells," *Journal of Theoretical Biology* 14 (1967): 225–74. Margulis was married to famed astronomer and science communicator Carl Sagan then and took on his surname, but later published under her original last name.

13. M. W. Gray, "Lynn Margulis and the endosymbiont hypothesis: 50 years later," *Molecular Biology of the Cell* 28 (2017): 1285–87.

14. A. Bodył et al., "Organelle evolution: *Paulinella* breaks a paradigm," *Current Biology* 22 (2012): R304–6.

15. A. Pantazidou et al., "Euendolithic shell-boring cyanobacteria and chlorophytes from the saline lagoon Ahivadolimni on Milos Island, Greece," *European Journal of Phycology* 41 (2006): 189–200.

16. B. S. Guida and F. Garcia-Pichel, "Extreme cellular adaptations and cell differentiation required by a cyanobacterium for carbonate excavation," *Proceedings of the National Academy of Sciences* 113 (2016): 5712–17.

17. Guida and Garcia-Pichel, "Extreme cellular adaptations."

18. I. Glaub et al., "The role of modern and fossil cyanobacterial borings in bioerosion and bathymetry," *Ichnos* 8 (2001): 185–95; I. Glaub et al., "Microborings and microbial endoliths: geological implications," in *Trace Fossils: Concepts, Problems, Prospects*, ed. W. Miller III (Amsterdam: Elsevier, 2007): 368–81; I. Glaub and K. Vogel. "The stratigraphic record of microborings," *Fossils & Strata* 51 (2004): 126–35.

19. S. Golubic et al., "Diversity of marine cyanobacteria," in *Marine Cyanobacteria*, ed. L. Charpy and A. W. D. Larkum (Monaco, *Bulletin de l'Institut Oceanographique* Special Issue 19 [1999]): 53–76.

20. S.-J. Balog, "Boring thallophytes in some Permian and Triassic reefs: bathymetry and bioerosion," in *Global and Regional Controls on Biogenic Sedimentation. I. Reef Evolution*, ed. J. Reitner et al. (*Göttinger Arbeiten Geologic Paläontologie* 2 [1996]): c305–9.

21. K. Vogel et al., "Experimental studies on microbial bioerosion at Lee Stocking Island (Bahamas) and One Tree Island (Great Barrier Reef, Australia): Implications for paleobathymetric reconstructions," *Lethaia* 33 (2000): 190–94.

22. J. B. Anderson et al., "Clonal evolution and genome stability in a 2500-year-old fungal individual," *Proceedings of the Royal Society of London, B* 285 (2018): 20182233.

23. G. A. Hoysted et al., "A mycorrhizal revolution," *Current Opinion in Plant Biology* 44 (2018): 1–6.

24. R. Gacesa et al., "Rising levels of atmospheric oxygen and evolution of Nrf2," *Scientific Reports* 6 (2016): 27740.

25. S. Bengston, "Fungus-like mycelial fossils in 2.4-billion-year-old vesicular basalt," *Nature Ecology & Evolution* 1 (2017): 0141.

26. T. Gan et al., "Cryptic terrestrial fungus-like fossils of the early Ediacaran Period," *Nature Communications* 12 (2021): 641. Older (1.0-billion-year-old) possible fungi fossils are in rocks in Canada, but these also might have been marine: C. C. Loron et al., "Early fungi from the Proterozoic Era in Arctic Canada," *Nature* 570 (2019): 232–35.

27. R. Honegger et al., "Fertile *Prototaxites taiti*: a basal ascomycete with inoperculate, polysporous asci lacking croziers," *Philosophical Transactions of the Royal Society, B* 373 (2018): 20170146.

28. A. Cherchi et al., "Bioerosion by microbial euendoliths in benthic foraminifera from heavy metal-polluted coastal environments of Portovesme (south-western Sardinia, Italy)," *Biogeosciences* 9 (2012): 4607–20; T. B. de Oliveira et al., "Thermophilic fungi in the new age of fungal taxonomy," *Extremophiles* 19 (2015): 31–37.

29. S. Golubic et al., "Endolithic fungi in marine ecosystems," *Trends in Microbiology* 13 (2005): 229–35.

30. Golubic et al., "Endolithic fungi."

31. Golubic et al., "Endolithic fungi."

32. Glaub et al., "Microborings and microbial endoliths"; N. Meyer et al., "Ichnodiversity and bathymetric range of microbioerosion traces in polar barnacles of Svalbard," *Polar Research* 39 (2020): 3766.

33. T. H. Nash, *Lichen Biology* (Cambridge: Cambridge University Press, 2008).

34. Nash, *Lichen Biology*.

35. R. A. Armstrong, "The use of the lichen genus *Rhizocarponin* in lichneometric dating with special reference to Holocene glacial events," in *Advances in Environmental Research*, ed. J. A. Daniels (Hauppauge, NY: Nova Science Publishers, 2015), 23–50.

36. Nash, *Lichen Biology*.

37. J. Chen et al., "Weathering of rocks induced by lichen colonization: a review," *Catena* 39 (2000): 121–46.

38. J. Chen et al., "Weathering of rocks."

39. J. Chen et al., "Weathering of rocks."

40. J. Chen et al., "Weathering of rocks."

41. J. Chen et al., "Weathering of rocks."

42. S. M. Morrison et al., "The paleomineralogy of the Hadean Eon revisited," *Life* 8 (2018): 64.

43. M. Brown et al., "Plate tectonics and the Archean Earth," *Annual Review of Earth and Planetary Sciences* 48 (2020): 291–320.

44. D. C. Catling and K. J. Zahnle, "The Archean atmosphere," *Science Advances* 6 (2020): eaax1420.

45. Brown et al., "Plate tectonics and the Archean Earth."

46. T. M. McCollum, "Miller-Urey and beyond: what have we learned about prebiotic organic synthesis reactions in the past 60 years?," *Annual Review of Earth and Planetary Sciences* 41 (2013): 207–29; M. S. Dodd et al., "Evidence for early life in Earth's oldest hydrothermal vent precipitates," *Nature* 543 (2017): 60–64.

47. T. M. Lenton and S. J. Daines, "Matworld—the biogeochemical effects of early life on land," *New Phytologist* 215 (2016): 531–37.

48. Gumsley, "Timing and tempo of the Great Oxidation Event."

49. J. P. Pu et al., "Dodging snowballs: geochronology of the Gaskiers glaciation and the first appearance of the Ediacaran biota," *Geology* 44 (2016): 955–58.

50. F. H. Gleason et al., "The roles of endolithic fungi in bioerosion and disease in marine ecosystems. I. General concepts," *Mycology* 8 (2017): 205–15; F. H. Gleason et al., "The roles of endolithic fungi in bioerosion and disease in marine ecosystems. II. Potential facultatively parasitic anamorphic ascomycetes can cause disease in corals and molluscs," *Mycology* 8 (2017): 216–27.

Chapter Three

1. R. Levi-Setti, *The Trilobite Book: A Visual Journey* (Chicago: University of Chicago Press, 2014).

2. F. J. A. Slieker, *Chitons of the World: An Illustrated Synopsis of Recent Polyplacophora* (Ancona: L'Informatore Piceno, 2000).

3. A. Kinder, "*Cryptochiton stelleri*: giant Pacific chiton," *Animal Diversity Web* (Online), accessed October 28, 2021, at https://animaldiversity.org /accounts/Cryptochiton_stelleri/.

4. I. Irisarri et al., "A mitogenomic phylogeny of chitons (Mollusca: Polyplacophora)," *BMC Ecology and Evolution* 20 (2020): 22.

5. Slieker, *Chitons of the World*.

6. D. Joester and L. R. Brooker, "The chiton radula: a model system for versatile use of iron oxides," in D. Faivre, *Iron Oxides: From Nature to Applications*, ed. D. Faivre (Hoboken, NJ: Wiley, 2016), 177–206.

7. N. Sealey, *Bahamian Landscape: Introduction to the Geology and Physical Geography of the Bahamas* (3rd ed.) (New York: Macmillan Publishing, 2006).

8. D. T. Gerace, *Life Quest: Building the Gerace Research Centre, San Salvador, Bahamas* (Port Charlotte, FL: Book-Broker Publishers, 2011).

9. H. A. Curran and B. White, "Ichnology of Holocene carbonate eolianites on San Salvador Island, Bahamas: diversity and significance," in *Proceedings of the 9th Symposium on the Geology of the Bahamas and other Carbonate Regions*, ed. H. A. Curran and J. E. Mylroie (San Salvador, Bahamas: Gerace Research Centre, 1999), 22–35; A. J. Martin, *Trace Fossils of San Salvador* (San Salvador Island, Bahamas: Gerace Research Centre, 2006).

10. Slieker, *Chitons of the World*.

11. B. M. Lawson, *Shelling San Sal: An Illustrated Guide to Common Shells*

of San Salvador Island, Bahamas (San Salvador Island, Bahamas: Gerace Research Centre, 1993).

12. T. F. Donn and M. R. Boardman, "Bioerosion of rocky carbonate coastlines on Andros Island, Bahamas," *Journal of Coastal Research* 4 (1988): 381–94.

13. M. A. Fedonkin et al., *The Rise of Animals: Evolution and Diversification of the Kingdom Animalia* (Baltimore: Johns Hopkins University Press, 2007).

14. M. A. Fedonkin et al., "New data on *Kimberella*, the Vendian mollusc-like organism (White Sea region, Russia): palaeoecological and evolutionary implications," in *The Rise and Fall of the Ediacaran Biota*, ed. P. Vickers-Rich and P. Komarower (London: *Geological Society of London, Special Publications* 286 [2007]), 157–79.

15. Fedonkin et al. 2007, "New data on *Kimberella*."

16. M. A. Fedonkin and B. M. Waggoner, "The Late Precambrian fossil *Kimberella* is a mollusc-like bilaterian organism," *Nature* 388 (1997): 868–71.

17. J. G. Gehling et al., "Scratch traces of large Ediacara bilaterian animals," *Journal of Paleontology* 88 (2015): 284–98; R. P. Lopes and J. C. Pereira, "Molluskan grazing traces (ichnogenus *Radulichnus* Voigt, 1977) on a Pleistocene bivalve from southern Brazil, with the proposal of a new ichnospecies with *Radulichnus tranversus* connected to chitons (vs. *R. inopinatus*)," *Ichnos* 26 (2019): 141–57.

18. M. Gingras et al., "Possible evolution of mobile animals in association with microbial mats," *Nature Geoscience* 4 (2011): 372–75.

19. A. J. Martin, *The Evolution Underground*, and specifically chapter 7, "Playing Hide and Seek for Keeps" (pp. 183–213), which provides an overview of how the "agronomic revolution" hypothesis developed, with citations of primary literature on that topic.

20. H. Hua et al., "Borings in *Cloudina* shells: complex predator-prey dynamics in the terminal Neoproterozoic," *Palaios* 18 (2003): 454–59.

21. S. Reynolds and J. Johnson, *Exploring Geology* (4th ed.) (New York: McGraw-Hill Education, 2016).

22. G. Nichols, *Sedimentology and Stratigraphy* (2nd ed.) (Oxford: Wiley-Blackwell, 2013).

23. J. A. MacEachern et al., "The ichnofacies paradigm: a fifty-year perspective," in *Trace Fossils: Concepts, Problems, Prospects*, ed. W. M. Miller III (Amsterdam: Elsevier, 2007), 52–77; J. A. MacEachern et al., "Uses of trace fossils in genetic stratigraphy," in *Trace Fossils: Concepts, Problems, Prospects*, ed. W. M. Miller III (Amsterdam: Elsevier, 2007), 110–34.

24. M. A. Wilson, "Macroborings and the evolution of bioerosion," in *Trace Fossils: Concepts, Problems, Prospects*, ed. W. M. Miller III (Amsterdam: Elsevier, 2007), 356–67.

25. Wilson, "Macroborings and the evolution of bioerosion."

26. D. E. G. Briggs. "The Cambrian explosion," *Current Biology* 25 (2015): R864–68.

27. T. Servais and D. A. T. Harper, "The Great Ordovician Biodiversification Event (GOBE): definition, concept and duration," *Lethaia* 51 (2018): 151–64.

28. M. A. Wilson and T. J. Palmer, "Patterns and processes in the Ordovician Bioerosion Revolution," *Ichnos* 13 (2006): 109–12.

29. C. Van Der Wal and S. Y. W. Ho, "Molecular clock," *Encyclopedia of Bioinformatics and Computational Biology* 2 (2019): 719–26.

30. M. Dohrmann and G. Wörheide, "Dating early animal evolution using phylogenomic data," *Scientific Reports* 7 (2017): 3599.

31. G. D. Love et al., "Fossil steroids record the appearance of Demospongiae during the Cryogenian Period," *Nature* 457 (2008): 718–21; D. A. Gold et al., "Sterol and genomic analyses validate the sponge biomarker hypothesis," *Proceedings of the National Academy of Sciences* 113 (2016): 2684–89.

32. I. Bobrovskiy et al., "Algal origin of sponge sterane biomarkers negates the oldest evidence for animals in the rock record," *Nature Ecology & Evolution* 5 (2021): 165–68.

33. E. C. Turner, "Possible poriferan body fossils in early Neoproterozoic microbial reefs," *Nature* 596 (2021): 87–91.

34. S. S. Asadzadeh et al., "Hydrodynamics of sponge pumps and evolution of the sponge body plan," *eLife* 9 (2020): e61012.

35. Asadzadeh et al., "Hydrodynamics of sponge pumps."

36. Asadzadeh et al., "Hydrodynamics of sponge pumps."

37. Asadzadeh et al., "Hydrodynamics of sponge pumps."

38. K. Rützler, "The role of burrowing sponges in bioerosion," *Oecologia* 19 (1975): 203–16; S. A. Pomponi, "Cytological mechanisms of calcium carbonate excavation by boring sponges," *International Review of Cytology* 65 (1980): 301–19; R. G. Bromley and A. D'Alessandro, "Ichnological study of shallow marine endolithic sponges from the Italian coast," *Rivista Italiana di Paleontologia e Stratigrafia* 95 (1989): 279–340.

39. D. M. de Bakker et al., "Quantification of chemical and mechanical bioerosion rates of six Caribbean excavating sponge species found on the coral reefs of Curaçao," *PLOS ONE* 13 (2018): e0197824.

40. German paleontologist and geologist Heinrich Georg Bronn (1800–1862) originally named *Entobia*, but he did not know what produced these borings: H. G. Bronn, "*Lethaea geognostica, 2: Das Kreide und Molassen-Gebirge*" (1837–1838): 545–1350. Sponges were accepted as the makers of these distinctive borings by the mid-twentieth century: C. L. Schönberg, "A history of sponge erosion: from past myths and hypotheses to recent approaches," in *Current Developments in Bioerosion*, ed. M. Wisshak and L. Tapanila (Berlin: Springer), 165–202.

41. Wilson, "Macroborings and the evolution of bioerosion."

42. R. W. Rouse and F. Pleijel, *Polychaetes* (Oxford: Oxford University Press, 2001).

43. Rouse and Pleijel, *Polychaetes*.

44. J. Lachat and D. Haag-Wackernagel, "Novel mobbing strategies of a fish population against a sessile annelid predator," *Scientific Reports* 6 (2016): 33187.

45. L. H. Spencer et al., "The risks of shell-boring polychaetes to shell-fish aquaculture in Washington, USA: A mini-review to inform mitigation actions," *Aquaculture Research* 52 (2020): 438–55.

46. S. M. Haigler, "Boring mechanism of *Polydora websteri* inhabiting *Crassostrea virginica*," *American Zoologist* 9 (1969): 821; R. G. Bromley and A. D'Alessandro, "Bioerosion in the Pleistocene of southern Italy: ichnogenera *Caulostrepsis* and *Maeandropolydora*," *Rivista Italiana di Paleontologia e Stratigrafia* 89 (1983): 283–309; M. Çinar and E. Dagli, "Bioeroding (boring) polychaete species (Annelida: Polychaeta) from the Aegean Sea (eastern Mediterranean)," *Journal of the Marine Biological Association of the United Kingdom* 101 (2021): 309–18.

47. Bromley and D'Alessandro, "Bioerosion in the Pleistocene of southern Italy."

48. Bromley and D'Alessandro, "Bioerosion in the Pleistocene of southern Italy."

49. Wilson, "Macroborings and the evolution of bioerosion."

50. Wilson, "Macroborings and the evolution of bioerosion"; M. El-Hedeny et al., "Bioerosion and encrustation: evidences from the Middle–Upper Jurassic of central Saudi Arabia," *Journal of African Earth Sciences* 134 (2017): 466–75.

51. F. K. McKinney and J. B. C. Jackson, *Bryozoan Evolution* (Chicago: University of Chicago Press, 1989).

52. Z. Zhang et al., "Fossil evidence unveils an early Cambrian origin for Bryozoa," *Nature* 599 (2021): 251–55.

53. McKinney and Jackson, *Bryozoan Evolution*.

54. Wilson, "Macroborings and the evolution of bioerosion."

55. P. D. Taylor et al., "*Finichnus*, a new name for the ichnogenus *Leptichnus* Taylor, Wilson and Bromley 1999, preoccupied by *Leptichnus* Simroth, 1896 (Mollusca, Gastropoda)," *Palaeontology* 56 (2013): 456.

56. A. J. Southward (ed.), *Barnacle Biology* (Boca Raton, FL: CRC Press, 2018).

57. A. M. Power et al., "Mechanisms of adhesion in adult barnacles," in *Biological Adhesive Systems*, ed. J. von Byern and I. Grunwald (Vienna: Springer, 2010), 153–68.

58. Southward, *Barnacle Biology*.

59. C. R. Darwin, *Living Cirripedia, A Monograph on the Sub-class Cirripedia, with Figures of All the Species. The Lepadidæ; or, Pedunculated Cirripedes* (London: Ray Society, vol. 1, 1851–1852); C. R. Darwin, *Living Cirripedia, The Balanidæ, (or Sessile Sirripedes); the Verrucidæ* (London: Ray Society, vol. 2, 1854); and for an overview of how Darwin's long study of barnacles contributed to his development of evolutionary theory: R. Stout, *Darwin and the Barnacle* (New York: W. W. Norton, 2004).

60. Southward, *Barnacle Biology*; B. K. K. Chan et al., "The evolutionary diversity of barnacles, with an updated classification of fossil and living forms," *Zoological Journal of the Linnean Society* 193 (2021): 789–846.

61. Chan et al., "The evolutionary diversity of barnacles."

62. Wilson, "Macroborings and the evolution of bioerosion."

63. P. D. Taylor and M. A. Wilson, "Palaeoecology and evolution of marine hard substrate communities," *Earth-Science Reviews* 62 (2003): 1–103; Wilson, "Macroborings and the evolution of bioerosion."

64. "*Lithophaga* Röding, 1798," World Register of Marine Species (WoRMS), MolluscaBase editors (2021), MolluscaBase, *Lithophaga*, accessed October 29, 2021, World Register of Marine Species, https://www .marinespecies.org/aphia.php?p=taxdetails&id=138220.

65. B. Witherington and D. Witherington, *Living Beaches of Georgia and the Carolinas: A Beachcombers Guide* (Sarasota, FL: Pineapple Press, 2011).

66. L. S. Fang and P. Shen, "A living mechanical file: the burrowing mechanism of the coral-boring bivalve *Lithophaga nigra*," *Marine Biology* 97 (1988): 349–54.

67. K. Kleeman, "Biocorrosion by bivalves," *Marine Ecology* 17 (1996): 145–58.

68. Kleeman, "Biocorrosion by bivalves."

69. J. B. Benner et al., "Macroborings (*Gastrochaenolites*) in Lower Ordovician hardgrounds of Utah: sedimentologic, paleoecologic, and evolutionary implications," *Palaios* 19 (2004): 543–50; Wilson and Palmer, "Patterns and processes in the Ordovician Bioerosion Revolution"; Wilson, "Macroborings and the evolution of bioerosion."

70. L. Tapanila et al., "Bivalve borings in phosphatic coprolites and bone, Cretaceous-Paleogene, Northeastern Mali," *Palaios* 19 (2004): 565–73; C. I. Serrano-Brañas et al., "*Gastrochaenolites* Leymerie in dinosaur bones from the Upper Cretaceous of Coahuila, north-central Mexico: Taphonomic implications for isolated bone fragments," *Cretaceous Research* 92 (2018): 18–25.

71. K. Sato and R. G. Jenkins, "Mobile home for pholadoid boring bivalves: first example from a Late Cretaceous sea turtle in Hokkaido Japan," *Palaios* 35 (2020): 228–36.

72. W. I. Ausich and G. D. Webster (eds.), *Echinoderm Paleobiology* (Bloomington: Indiana University Press, 2008).

73. Ausich and Webster, *Echinoderm Paleobiology*. Of course, in humans this orifice can metaphorically reverse during adulthood and ascend to high political office.

74. Ausich and Webster, *Echinoderm Paleobiology*.

75. Ausich and Webster, *Echinoderm Paleobiology*.

76. E. Voultsiadou and C. Chintiroglou, "Aristotle's lantern in echinoderms: an ancient riddle," *Cahiers de Biologie Marine* 49 (2008): 299–302.

77. T. S. Klinger and J. M. Lawrence, "The hardness of the teeth of five species of echinoids (Echinodermata)," *Journal of Natural History* 19 (1984):

917–20; M. Reich and A. B. Smith, "Origins and biomechanical evolution of teeth in echinoids and their relatives," *Palaeontology* 52 (2009): 1149–68; S. R. Stock, "Sea urchins have teeth? A review of their microstructure, biomineralization, development and mechanical properties," *Connective Tissue Research* 55 (2014): 41–51.

78. R. G. Bromley, "Comparative analysis of fossil and recent echinoid bioerosion," *Palaeontology* 18 (1975): 725–39; Wilson, "Macroborings and the evolution of bioerosion."

79. O. Mokady et al., "Echinoid bioerosion as a major structuring force of Red Sea coral reefs," *Biological Bulletin* 190 (2006): 367–72.

80. Bromley, "Comparative analysis of fossil and recent echinoid bioerosion"; U. Asgaard and R. G. Bromley, "Echinometrid sea urchins, their trophic styles and corresponding bioerosion," in *Current Developments in Bioerosion*, ed. M. Wisshak and L. Tapanila (Berlin: Springer, 2008), 279–303; A. Santos et al., "Role of environmental change in rock-boring echinoid trace fossils," *Palaeogeography, Palaeoclimatology, Palaeoecology* 432 (2015): 1–14.

81. G. E. Farrow and J. A. Fyfe, "Bioerosion and carbonate mud production on high-latitude shelves," *Sedimentary Geology* 60 (1988): 281–97; M. Wisshak, *High-Latitude Bioerosion: The Kosterfjord Experiment* (Berlin: Springer, 2006).

82. L. A. Buatois et al., "Quantifying ecospace utilization and ecosystem engineering during the early Phanerozoic—The role of bioturbation and bioerosion," *Science Advances* 6 (2020): eabb0618.

83. Brannen, *The Ends of the World*, and references therein.

84. S. Singh et al., *Global Climate Change* (Amsterdam: Elsevier, 2021).

85. M. Wisshak et al., "Ocean acidification accelerates reef bioerosion," *PLOS ONE* 7 (2012): e45124; C. H. L. Schönberg et al., "Bioerosion: the other ocean acidification problem," *ICES Journal of Marine Science* 74 (2017): 895–925.

86. C. H. L. Schönberg and J.-C. Ortiz, "Is sponge bioerosion increasing?," *Proceedings of the 11th International Coral Reef Symposium* (2008): 520–23; Wisshak et al., "Ocean acidification accelerates reef bioerosion"; M. Achtalis et al., "Sponge bioerosion on changing reefs: ocean warming poses physiological constraints to the success of a photosymbiotic excavating sponge," *Scientific Reports* 7 (2017): 10705.

Chapter Four

1. Someone really should do a book-long photographic tribute to parrotfishes, because they richly deserve it. But in the meantime, you can get a visual appreciation for their stunning array of colors and forms via field guidebooks, such as: E. H. Kaplan and S. L. Kaplan, *A Field Guide to Coral Reefs: Caribbean and Florida* (Boston: Houghton Mifflin, 1999).

2. J. T. Streelman et al., "Evolutionary history of the parrotfishes: bio-

geography, ecomorphology, and comparative diversity," *Evolution* 56 (2002): 961–71; J. H. Choat et al., "Patterns and processes in the evolutionary history of parrotfishes (Family Labridae)," *Biological Journal of the Linnean Society* 107 (2012): 529–57.

3. R. M. Bonaldo et al., "The ecosystem roles of parrotfishes on tropical reefs," *Oceanography and Marine Biology: An Annual Review* 52 (2014): 81–132, and references therein; A. C. Siqueira et al., "The evolution of traits and functions in herbivorous coral reef fishes through space and time," *Proceedings of the Royal Society, B* 286 (2019): 20182672.

4. Bonaldo et al., "The ecosystem roles of parrotfishes on tropical reefs."

5. R. R. Warner, "Mating behavior and hermaphroditism in coral reef fishes," *American Scientist* 72 (1984): 128–36; B. M. Taylor, "Drivers of protogynous sex change differ across spatial scales," *Proceedings of the Royal Society, B* 281 (2014): 0132423.

6. A. S. Grutter et al., "Fish mucous cocoons: the 'mosquito nets' of the sea," *Biology Letters* 7 (2010): 292–94.

7. Grutter et al., "Fish mucous cocoons."

8. Grutter et al., "Fish mucous cocoons."

9. Bonaldo et al., "The ecosystem roles of parrotfishes on tropical reefs."

10. D. R. Bellwood and O. Schultz, "A review of the fossil record of the parrotfishes (Labroidei: Scaridae) with a description of a new *Calotomus* species from the Middle Miocene (Badenian) of Austria," *Annals Naturhistorisches Museum Wien* 92 (1991): 55–71; Siqueira et al., "The evolution of traits and functions."

11. R. Syed and S. Sengupta, "First record of parrotfish bite mark on larger foraminifera from the Middle Eocene of Kutch, Gujarat, India," *Current Science* 116 (2019): 363–65.

12. C. Baars et al., "The earliest rugose coral," *Geological Magazine* 150 (2012): 371–80; J. Stolarski et al., "The ancient evolutionary origins of Scleractinia revealed by azooxanthellate corals," *BMC Evolutionary Biology* 11 (2011): 316.

13. R. Osinga et al., "The biology and economics of coral growth," *Marine Biotechnology* 13 (2011): 658–71.

14. Bonaldo et al., "The ecosystem roles of parrotfishes on tropical reefs."

15. Bonaldo et al., "The ecosystem roles of parrotfishes on tropical reefs"; T. C. Adam et al., "Comparative analysis of foraging behavior and bite mechanics reveals complex functional diversity among Caribbean parrotfishes," *Marine Ecology Progress Series* 597 (2018): 207–20.

16. I. D. Lange et al., "Site-level variation in parrotfish grazing and bioerosion as a function of species-specific feeding metrics," *Diversity* 12 (2020): 379.

17. M. A. Marcus et al., "Parrotfish teeth: stiff biominerals whose microstructure makes them tough and abrasion-resistant to bite stony corals," *ACS Nano* 11 (2013): 11858–65.

18. Marcus et al., "Parrotfish teeth."

19. K. W. Gobalet, "Morphology of the parrotfish pharyngeal jaw apparatus," *Integrative and Comparative Biology* 29 (1989): 319–31.

20. Gobalet, "Morphology of the parrotfish."

21. P. Frydl and C. W. Stearn, "Rate of bioerosion by parrotfish in Barbados reef environments," *Journal of Sedimentary Research* 48 (1978): 1149–58.

22. R. T. Yarlett et al., "Constraining species–size class variability in rates of parrotfish bioerosion on Maldivian coral reefs: implications for regional-scale bioerosion estimates," *Marine Ecology Progress Series* 590 (2018): 155–69.

23. Frydl and Stearn, "Rate of bioerosion by parrotfish."

24. Bonaldo et al., "The ecosystem roles of parrotfishes on tropical reefs."

25. M. M. Rice et al., "Macroborer presence on corals increases with nutrient input and promotes parrotfish bioerosion," *Coral Reefs* 39 (2020): 409–18.

26. K. D. Clements et al., "Integrating ecological roles and trophic diversification on coral reefs: multiple lines of evidence identify parrotfishes as microphages," *Biological Journal of the Linnean Society* 120 (2017): 729–51.

27. K. L. Cramer et al., "Prehistorical and historical declines in Caribbean coral reef accretion rates driven by loss of parrotfish," *Nature Communications* 8 (2017): 14160.

28. Bonaldo et al., "The ecosystem roles of parrotfishes on tropical reefs."

29. Bonaldo et al., "The ecosystem roles of parrotfishes on tropical reefs."

30. Bonaldo et al., "The ecosystem roles of parrotfishes on tropical reefs."

31. R. M. Bonaldo and D. R. Bellwood, "Parrotfish predation on massive *Porites* on the Great Barrier Reef," *Coral Reefs* 30 (2011): 259–69; Rice et al., "Macroborer presence on corals."

32. D. R. Bellwood. "Direct estimate of bioerosion by two parrotfish species, *Chlorurus gibbus* and *C. sordidus*, on the Great Barrier Reef, Australia," *Marine Biology* 121 (1995): 419–29.

33. Yarlett et al., "Constraining species–size class variability."

34. C. T. Perry et al., "Linking reef ecology to island building: parrotfish identified as major producers of island-building sediment in the Maldives," *Geology* 43 (2015): 503–6.

35. J. P. Terry and J. Goff, "One hundred and thirty years since Darwin: 'reshaping' the theory of atoll formation," *The Holocene* 23 (2013): 615–19.

36. K. M. Morgan and P. S. Kench, "Parrotfish erosion underpins reef growth, sand talus development and island building in the Maldives," *Sedimentary Geology* 341 (2016): 50–57.

37. Yarlett et al., "Constraining species–size class variability."

38. J. F. Bruno et al., "Climate change, coral loss, and the curious case

of the parrotfish paradigm: why don't marine protected areas improve reef resilience?," *Annual Reviews of Marine Sciences* 11 (2019): 307–34.

39. Cramer et al., "Prehistorical and historical declines." For once, I'm serious: pirates had a negative effect on shallow-marine ecosystems that went far beyond occasional pollution caused by plank-walking or keelhauling.

40. A. E. Berchtold and I. M. Côté, "Effect of early exposure to predation on risk perception and survival of fish exposed to a non-native predator," *Animal Behaviour* 164 (2020): 205–16.

41. Berchtold and I. M. Côté, "Effect of early exposure to predation."

42. I. C. Enochs et al., "Ocean acidification enhances the bioerosion of a common coral reef sponge: implications for the persistence of the Florida Reef Tract," *Bulletin of Marine Sciences* 91 (2015): 271–90; Bruno et al., "Climate change, coral loss."

43. Bonaldo et al., "The ecosystem roles of parrotfishes on tropical reefs"; T. M. Davidson et al., "Bioerosion in a changing world: a conceptual framework," *Ecology Letters* 21 (2018): 422–38.

44. J. M. Sigren et al., "Coastal sand dunes and dune vegetation: restoration, erosion, and storm protection," *Shore & Beach* 82 (2014): 5–12.

45. Martin, *Life Traces of the Georgia Coast*, and references therein.

46. Martin, *Life Traces of the Georgia Coast*.

Chapter Five

1. Clarkson, *Invertebrate Palaeontology and Evolution*.

2. Servais and Harper, "The Great Ordovician Biodiversification Event."

3. R. A. Davis and D. Meyers, *A Sea Without Fish: Life in the Ordovician Sea of the Cincinnati Region* (Bloomington: Indiana University Press, 2009).

4. A. J. Martin, "A Paleoenvironmental Interpretation of the "Arnheim" Micromorph Fossil Assemblage from the Cincinnatian Series (Upper Ordovician), Southeastern Indiana and Southwestern Ohio" (MS thesis, Miami University, 1986); Dattilo et al., "Giants among micromorphs: were Cincinnatian (Ordovician, Katian) small shelly phosphatic faunas dwarfed?," *Palaios* 31 (2016): 55–70.

5. John K. Pope was actually the first paleontologist I ever met, and I will always be grateful to him for that and his mentoring me as a young, naïve graduate student. At first, we didn't get along very well because of our very different backgrounds and personalities, but I'm happy to say that over the course of my three years of working with him, I became a more serious scholar and he warmed to my sense of frivolity. Thank you, John.

6. Hua et al., "Borings in *Cloudina* shells."

7. R. D. C. Bicknell and J. R. Paterson, "Reappraising the early evidence of durophagy and drilling predation in the fossil record: implications for escalation and the Cambrian Explosion," *Biology Reviews* 93 (2018): 754–84.

8. M. G. Mángano et al., "The great Ordovician biodiversification

event," in *The Trace-Fossil Record of Major Evolutionary Events*, ed. M. Mángano and L. Buatois (Dordrecht: Springer, 2016): 127–65; Servais and Harper, "The Great Ordovician Biodiversification Event."

9. L. A. Buatois et al., "Decoupled evolution of soft and hard substrate communities during the Cambrian Explosion and Great Ordovician Biodiversification Event," *Proceedings of the National Academy of Sciences* 113 (2016): 6945–48.

10. Wilson and Palmer, "Patterns and processes"; Mángano, "The great Ordovician biodiversification event"; Buatois et al., "Decoupled evolution of soft and hard substrate communities."

11. C. E. Brett and S. E. Walker, "Predators and predation in Paleozoic marine environments," in *The Fossil Record of Predation*, ed. M. Kowalewski and P. H. Kelley, *Paleontological Society Papers* 8 (2002): 93–118.

12. A. A. Klompmaker et al., "Increase in predator-prey size ratios throughout the Phanerozoic history of marine ecosystems," *Science* 356 (2017): 1178–80.

13. Brett and Walker, "Predators and predation in Paleozoic marine environments"; A. A. Klompmaker and P. H. Kelley, "Shell ornamentation as a likely exaptation: Evidence from predatory drilling on Cenozoic bivalves," *Paleobiology* 41 (2015): 187–201; G. J. Vermeij, "Gastropod skeletal defences: land, freshwater, and sea compared," *Vita Malacologica* 13 (2015): 1–25.

14. *Naticidae: MolluscaBase*, 2021, Naticidae Guilding, 1834, World Register of Marine Species (WoRMS), http://www.marinespecies.org /aphia.php?p=taxdetails&id=145, accessed October 30, 2021; *Muricidae: MolluscaBase*, 2021, MolluscaBase. Muricidae Rafinesque, 1815, World Register of Marine Species (WoRMS) http://www.marinespecies.org/aphia .php?p=taxdetails&id=148, accessed October 30, 2021.

15. S. S. Das et al., "Family Naticidae (Gastropoda) from the Upper Jurassic of Kutch, India and a critical reappraisal of taxonomy and time of origination of the family," *Journal of Paleontology* 93 (2019): 673–84; R. Saha et al., "Gastropod drilling predation in the Upper Jurassic of Kutch, India," *Palaios* 36 (2021): 301–12.

16. Brett and Walker, "Predators and predation in Paleozoic marine environments," and references therein; Paleozoic platyceratids have also been accused of parasitism, particularly on crinoids: A. Nützel, "Gastropods as parasites and carnivorous grazers: a major guild in marine ecosystems," in *The Evolution and Fossil Record of Parasitism*, ed. K. De Baets and J. W. Huntley, *Topics in Geobiology* 49 (Cham: Springer, 2021), 209–29.

17. Witherington and Witherington, *Living Beaches of Georgia and the Carolinas*.

18. Martin, *Life Traces of the Georgia Coast*; Martin, *Tracking the Golden Isles*.

19. Martin, *Life Traces of the Georgia Coast*.

20. M. R. Carriker, "Shell penetration and feeding by naticacean and muricacean predatory gastropods: a synthesis," *Malacologia* 20 (1981): 403–

22; R. L. Kitching and J. Pearson, "Prey localization by sound in a predatory intertidal gastropod," *Marine Biology Letters* 2 (1981): 313–21.

21. E. S. Clelland and N. B. Webster, "Drilling into hard substrate by naticid and muricid gastropods: a chemo-mechanical process involved in feeding," in *Physiology of Molluscs*, ed. S. Saleuddin and S. Mukai (Boca Raton, FL: Apple Academic Press, 2021): 1896–916.

22. Martin, *Life Traces of the Georgia Coast*; Martin, *Tracking the Golden Isles*, and references therein.

23. A.A. Klompmaker et al., "An overview of predation evidence found on fossil decapod crustaceans with new examples of drill holes attributed to gastropods and octopods," *Palaios* 28 (2013): 599–613; J. Villegas-Martín et al., "A small yet occasional meal: predatory drill holes in Paleocene ostracods from Argentina and methods to infer predation intensity," *Palaeontology* 62 (2019): 731–56.

24. R. G. Bromley, "Concepts in ichnotaxonomy illustrated by small round holes in shells," *Acta Geológica Hispánica* 16 (1981): 55–64.

25. *Murex (Murex) pecten* [Lightfoot], 1786, *MolluscaBase*; World Register of Marine Species, http://www.marinespecies.org/aphia .php?p=taxdetails&id=215663, accessed October 30, 2021.

26. *Muricidae [Rafinesque]*, 1815, MolluscaBase, World Register of Marine Species (WoRMS), http://www.marinespecies.org/aphia .php?p=taxdetails&id=148, accessed October 30, 2021.

27. C. L. Garvie, "Two new species of Muricinae from the Cretaceous and Paleocene of the Gulf Coastal Plain, with comments on the genus *Odontopolys* Gabb. 1860," *Tulane Studies in Geology and Paleontology* 24 (1991): 87–92; A. M. Sørensen and F. Surlyk, "Taphonomy and palaeoecology of the gastropod fauna from a Late Cretaceous rocky shore, Sweden," *Cretaceous Research* 32 (2011): 472–79.

28. Bromley, "Concepts in ichnotaxonomy."

29. Carriker, "Shell penetration and feeding"; G. S. Herbert et al., "Behavioural versatility of the giant murex *Muricanthus fulvescens* (Sowerby, 1834) (Gastropoda: Muricidae) in interactions with difficult prey," *Journal of Molluscan Studies* 82 (2016): 357–65.

30. A. A. Klompmaker et al., "The fossil record of drilling predation on barnacles," *Palaeogeography, Palaeoclimatology, Palaeoecology* 426 (2015): 95–111; J. H. Nebelsick and M. Kowalewski. "Drilling predation on Recent clypeasteroid echinoids from the Red Sea," *Palaios* 14 (1999): 127–44; C. A. Meadows et al., "Drill holes in the irregular echinoid, *Fibularia*, from the Oligocene of New Zealand," *Palaios* 30 (2015): 810–17.

31. C. Fouilloux et al., "Cannibalism," *Current Biology* 29 (2019): R1295–97.

32. Fouilloux et al., "Cannibalism."

33. S. Mondal et al., "Naticid confamilial drilling predation through time," *Palaios* 32 (2017): 278–87.

34. D. Chattopadhyay et al., "What controls cannibalism in drilling gas-

tropods? A case study on *Natica tigrine*," *Palaeogeography, Palaeoclimatology, Palaeoecology* 410 (2014): 126–33.

35. A. Pahari et al., "Subaerial naticid gastropod drilling predation by *Natica tigrina* on the intertidal molluscan community of Chandipur, eastern coast of India," *Palaeogeography, Palaeoclimatology, Palaeoecology* 451 (2016): 110–23.

36. M. Nixon and J. Z. Young, *The Brains and Lives of Cephalopods* (Oxford: Oxford University Press, 2003).

37. J. Kluessendorf and P. Doyle, "*Pohlsepia mazonensis*, An early 'octopus' from the Carboniferous of Illinois, USA," *Palaeontology* 43 (2003): 919–26.

38. W. Malik, "Inky's daring escape shows how smart octopuses are," *National Geographic*, April 14, 2016, https://www.nationalgeographic.com /animals/article/160414-inky-octopus-escapes-intelligence.

39. J. K. Fin et al., "Defensive tool use in a coconut-carrying octopus," *Current Biology* 19 (2009): R1069–70.

40. S. L. de Souza Medeiros et al., "Cyclic alternation of quiet and active sleep states in the octopus," *iScience* 24 (2021): 102223.

41. M. E. Q. Pilson and P. B. Taylor, "Hole drilling by *Octopus*," *Science* 134 (1961): 1366–68; J. M. Arnold, "Some aspects of hole-boring predation by *Octopus vulgaris*," *American Zoologist* 9 (1969): 991–96; J. Wodinsky, "Penetration of the shells and feeding on gastropods by *Octopus*," *American Zoologist* 9 (1969): 997–1010.

42. J. Wodinsky, "Penetration of the shells and feeding."

43. W. B. Saunders et al., "*Octopus* predation on *Nautilus*: evidence from Papua New Guinea," *Bulletin of Marine Science* 49 (1991): 280–87; A. A. Klompmaker et al., "First fossil evidence of a drill hole attributed to an octopod in a barnacle," *Lethaia* 47 (2014): 309–12; D. Pech-Puch et al. "Chemical tools of *Octopus maya* during crab predation are also active on conspecifics," *PLOS ONE* 11 (2016): e0148922; A. A. Klompmaker and B. A. Kittle, "Inferring octopodoid and gastropod behavior from their Plio-Pleistocene cowrie prey (Gastropoda: Cypraeidae)," *Palaeogeography, Palaeoclimatology, Palaeoecology* 567 (2021): 110251.

44. Bromley, "Concepts in ichnotaxonomy"; R. G. Bromley, "Predation habits of octopus past and present and a new ichnospecies, *Oichnus ovalis*," *Bulletin of the Geological Society of Denmark* 40 (1993): 167–73.

45. S. K. Donovan, "A plea not to ignore ichnotaxonomy: recognizing and recording *Oichnus* Bromley," *Swiss Journal of Palaeontology* 136 (2017): 369–72.

46. A. A. Klompmaker and N. H. Landman, "Octopodoidea as predators near the end of the Mesozoic Marine Revolution," *Biological Journal of the Linnean Society* 132 (2021): 894–99.

47. Bromley, "Concepts in ichnotaxonomy." This also wasn't the last time that Richard Bromley insulted these cephalopods. While on a field

trip with him and others on Sapelo Island (Georgia) in July 2001, I recall his disparaging octopus intelligence, or at least a specific octopus. After a long, hot day in the field there, and while we were in the air-conditioned comfort of our quarters at the University of Georgia Marine Institute, he picked up a whelk shell from a nearby windowsill and told a marvelous story about that shell. But he began it by pointing to a tiny hole near the whelk's apex and saying, "This was a very stupid octopus."

48. A.-F. Hiemstra, "Recognizing cephalopod boreholes in shells and the northward spread of *Octopus vulgaris* Cuvier, 1797 (Cephalopoda, Octopodoidea)," *Vita Malacologica* 13 (2015): 53–56.

49. P. H. Kelley, "Predation by Miocene gastropods of the Chesapeake Group: stereotyped and predictable," *Palaios* 3 (1988): 436–48; Klompmaker and Kelley, "Shell ornamentation as a likely exaptation."

50. G. P. Dietl, and R. R. Alexander, "Borehole site and prey size stereotypy in naticid predation on *Euspira* (*Lunatia*) *heros* Say and *Neverita* (*Polinices*) *duplicata* Say from the southern New Jersey coast," *Journal of Shellfish Research* 14 (1995): 307–14.

51. E. M. Harper and D. S. Wharton, "Boring predation and Mesozoic articulate brachiopods," *Palaeogeography, Palaeoclimatology, Palaeoecology* 158 (2000): 15–24.

52. S. Mondal et al., "Naticid confamilial drilling predation through time"; S. Mondal et al., "Latitudinal patterns of gastropod drilling predation through time," *Palaios* 34 (2019): 261–70.

53. D. Chattopadhyay and S. Dutta, "Prey selection by drilling predators: A case study from Miocene of Kutch, India," *Palaeogeography, Palaeoclimatology, Palaeoecology* 374 (2013): 187–96; J. A. Hutchings and G. S. Herbert, "No honor among snails: Conspecific competition leads to incomplete drill holes by a naticid gastropod," *Palaeogeography, Palaeoclimatology, Palaeoecology* 379 (2013): 32–38.

54. Klompmaker et al., "Increase in predator-prey size ratios."

55. D. Chattopadhyay et al., "Effectiveness of small size against drilling predation: Insights from lower Miocene faunal assemblage of Quilon Limestone, India," *Palaeogeography, Palaeoclimatology, Palaeoecology* 551 (2020): 109742.

56. P. A. M. Gaemers and B. W. Langeveld, "Attempts to predate on gadid fish otoliths demonstrated by naticid gastropod drill holes from the Neogene of Mill-Langenboom, The Netherlands," *Scripta Geologica* 149 (2015): 159–83.

57. N. C. Chojnacki and L. R. Leighton, "Comparing predatory drillholes to taphonomic damage from simulated wave action on a modern gastropod," *Historical Biology* 26 (2014): 69–79.

58. M. A. Wilson and T. J. Palmer, "Domiciles, not predatory borings: a simpler explanation of the holes in Ordovician shells analyzed by Kaplan and Baumiller, 2000," *Palaios* 16 (2001): 524–25.

59. J. P. Zonnenveld and M. K. Gingras, "*Sedilichnus, Oichnus, Fossichnus,* and *Tremichnus*: 'small round holes in shells' revisited," *Journal of Paleontology* 88 (2015): 895–905.

60. S. K. Donovan and G. Hoare, "Site selection of small round holes in crinoid pluricolumnals, Trearne Quarry SSSI (Mississippian, Lower Carboniferous), north Ayrshire, UK," *Scottish Journal of Geology* 55 (2019): 1–5.

61. D. E. Bar-Yosef Mayer et al., "On holes and strings: earliest displays of human adornment in the Middle Palaeolithic," *PLOS ONE* 15 (2020): e0234924.

62. A. M. Kubicka et al., "A systematic review of animal predation creating pierced shells: implications for the archaeological record of the Old World," *PeerJ* 5 (2017): e2903.

63. Kubicka et al., "A systematic review of animal predation."

64. G. Schweigert et al., "New Early Jurassic hermit crabs from Germany and France," *Journal of Crustacean Biology* 33 (2013): 802–17.

65. A. Mironenko, "A hermit crab preserved inside an ammonite shell from the Upper Jurassic of central Russia: implications to ammonoid palaeoecology," *Palaeogeography, Palaeoclimatology, Palaeoecology* 537 (2020): 109397.

66. J. W. Martin et al., ed., *Decapod Crustacean Phylogenetics* (Boca Raton, FL: CRC Press, 2016).

67. E. C. F. de Souza et al., "Intra-specific competition drives variation in the fundamental and realized niches of the hermit crab, *Pagurus criniticornis*," *Bulletin of Marine Science* 91 (2015): 343–61.

68. J. A. Pechenik and S. Lewis, "Avoidance of drilled gastropod shells by the hermit crab *Pagurus longicarpus* at Nahant, Massachusetts," *Journal of Experimental Marine Biology and Ecology* 253 (2000): 17–32; J. A. Pechenik et al., "Factors selecting for avoidance of drilled shells by the hermit crab *Pagurus longicarpus*," *Journal of Experimental Marine Biology and Ecology* 262 (2001): 75–89.

Chapter Six

1. Brett and Walker, "Predators and predation in Paleozoic marine environments"; S. E. Walker and C. E. Brett, "Post-Paleozoic patterns in marine predation: was there a Mesozoic and Cenozoic marine predatory revolution?," in *The Fossil Record of Predation*, ed. M. Kowalewski and P. H. Kelley, *Paleontological Society Papers* 8 (2002): 119–93.

2. D. M. Rudkin et al., "The oldest horseshoe crab: a new xiphosurid from Late Ordovician konservat-lagerstätten deposits, Manitoba, Canada," *Palaeontology* 51 (2008): 1–9; D. R. Smith et al., *Biology and Conservation of Horseshoe Crabs* (New York: Springer, 2009).

3. Smith, *Biology and Conservation of Horseshoe Crabs*; Martin, *Life Traces of the Georgia Coast*; Martin, *Tracking the Golden Isles*.

4. R. D. C. Bicknell et al., "Computational biomechanical analyses dem-

onstrate similar shell-crushing abilities in modern and ancient arthropods," *Proceedings of the Royal Society, B* 285 (2018): 20181935.

5. Bicknell et al., "Computational biomechanical analyses."

6. Bicknell et al., "Computational biomechanical analyses"; R. S. Bicknell et al., "Biomechanical analyses of Cambrian euarthropod limbs reveal their effectiveness in mastication and durophagy," *Proceedings of the Royal Society of London, B* 288 (2021): 20202075.

7. A. Zacaï et al., "Reconstructing the diet of a 505-million-year-old arthropod: *Sidneyia inexpectans* from the Burgess Shale fauna," *Arthropod Structure and Development* 45 (2016): 200–220.

8. R. Ludvigsen, "Rapid repair of traumatic injury by an Ordovician trilobite," *Lethaia* 10 (1977): 205–7; S. C. Morris and R. J. F. Jenkins, "Healed injuries in Early Cambrian trilobites from South Australia," *Alcheringa* 9 (1985): 167–77; B. Schoenemann et al., "Traces of an ancient immune system—how an injured arthropod survived 465 million years ago," *Scientific Reports* 7 (2017): 40330.

9. C. E. Schweitzer and R. M. Feldmann, "The Decapoda (Crustacea) as predators on Mollusca through geologic time," *Palaios* 25 (2010): 167–82; S. Kuratani, "Evolution of the vertebrate jaw from developmental perspectives," *Evolution and Development* 14 (2012): 76–92.

10. J. Cunningham and N. Herr, *Hands-On Physics Activities and Life Applications: Easy-to-Use Labs and Demonstrations for Grades 8–12* (New York: Wiley, 1994).

11. Cunningham and Herr, *Hands-On Physics Activities*.

12. J. W. Martin et al., *Decapod Crustacean Phylogenetics*; L. M. Tang, "Evolutionary history of true crabs (Crustacea: Decapoda: Brachyura) and the origin of freshwater crabs," *Molecular Biology and Evolution* 31 (2014): 1173–87.

13. C. E. Schweitzer and R. N. Feldmann, "The oldest Brachyura (Decapoda: Homolodromioidea: Glaessneropsoidea) known to date (Jurassic)," *Journal of Crustacean Biology* 30 (2010): 251–56; C. N. de Carvalho et al., "Running crabs, walking crinoids, grazing gastropods: behavioral diversity and evolutionary implications in the Cabeço da Ladeira lagerstätte (Middle Jurassic, Portugal)," *Comunicações Geológicas* 103 (2016): 39–54.

14. J. M. Wolf et al., "How to become a crab: phenotypic constraints on a recurring body plan," *BioEssays* 43 (2021): 2100020.

15. J. S. Weis, *Walking Sideways: The Remarkable World of Crabs* (Ithaca, NY: Cornell University Press, 2012).

16. Weis, *Walking Sideways*.

17. Weis, *Walking Sideways*.

18. E. Zipser and G. J. Vermeij, "Crushing behavior of tropical and temperate crabs," *Journal of Experimental Marine Biology and Ecology* 31 (1978): 155–72; M. D. Bertness and C. Cunningham, "Crab shell-crushing predation and gastropod architectural defense," *Journal of Experimental Marine Biology and Ecology* 50 (1981): 213–30.

19. Bertness and Cunningham, "Crab shell-crushing predation";
M. Ishikawa et al., "Snail versus hermit crabs: a new interpretation of shell-peeling predation on fossil gastropod associations," *Paleontological Research* 8 (2004): 99–108.

20. G. C. Cadée et al., "Gastropod shell repair in the intertidal of Bahía la Choya (N. Gulf of California)," *Palaeogeography, Palaeoclimatology, Palaeoecology* 136 (1997): 67–78.

21. E. S. Stafford et al., "Gastropod shell repair tracks predator abundance," *Marine Ecology* 36 (2015): 1176–84.

22. E. S. Stafford et al., "*Caedichnus*, a new ichnogenus representing predatory attack on the gastropod shell aperture," *Ichnos* 22 (2015): 87–102.

23. M. E. Kosloski, "Recognizing biotic breakage of the hard clam, *Mercenaria mercenaria* caused by the stone crab, *Menippe mercenaria*: An experimental taphonomic approach," *Journal of Experimental Marine Biology and Ecology* 396 (2011): 115–21; M. E. Kosloski and W. D. Allmon, "Macroecology and evolution of a crab 'super predator,' *Menippe mercenaria* (Menippidae), and its gastropod prey," *Biological Journal of the Linnean Society* 116 (2015): 571–81.

24. Martin, *Tracking the Golden Isles*.

25. Martin, *Tracking the Golden Isles*.

26. Bertness and Cunningham, "Crab shell-crushing predation."

27. T. C. Edgell et al., "Simultaneous defense against shell entry and shell crushing in a snail faced with the predatory shorecrab *Carcinus maenas*," *Marine Ecology Progress Series* 371 (2008): 191–98.

28. C. van der Wal et al., "The evolutionary history of Stomatopoda (Crustacea: Malacostraca) inferred from molecular data," *PeerJ* 5 (2017): e3844; C. Huag et al., "New records of Mesozoic mantis shrimp larvae and their implications on modern larval traits in stomatopods," *Palaeodiversity* 8 (2015): 121–33.

29. H. H. Thoen et al., "A different form of color vision in mantis shrimp," *Science* 343 (2015): 411–13.

30. J. S. Harrison et al., "Scaling and development of elastic mechanisms: the tiny strikes of larval mantis shrimp," *Journal of Experimental Biology* 224 (2021): jeb235465.

31. Harrison et al., "Scaling and development of elastic mechanisms."

32. Harrison et al., "Scaling and development of elastic mechanisms."

33. R. L. Crane et al., "Smashing mantis shrimp strategically impact shells," *Journal of Experimental Biology* 221 (2018): jeb176099.

34. Crane et al., "Smashing mantis shrimp."

35. Harrison et al., "Scaling and development of elastic mechanisms."

36. J. Pether, "*Belichnus* new ichnogenus, a ballistic trace on mollusc shells from the Holocene of the Benguela region, South Africa," *Journal of Paleontology* 69 (1995): 171–81.

37. A stingray's "face" is an illusion, as its eyes are on the other side of its body (on the top, or dorsal side), whereas what seem like its "eyes" are

actually its nostrils. Its mouth is in the right place, though, which looks like it is smiling, usually in a *Mona Lisa* sort of way.

38. C. R. L. Amaral et al., "The mitogenomic phylogeny of the Elasmobranchii (Chondrichthyes)," *Mitochondrial DNA Part A* 29 (2018): 867–78.

39. R. M. Kempster et al., "Phylogenetic and ecological factors influencing the number and distribution of electroreceptors in elasmobranchs," *Journal of Fish Biology* 80 (2012): 2055–88.

40. M. R. Gregory et al., "On how some rays (Elasmobranchia) excavate feeding depressions by jetting water," *Journal of Sedimentary Petrology* 49 (1979): 1125–30; O. R. O'Shea et al., "Bioturbation by stingrays at Ningaloo Reef, Western Australia," *Marine and Freshwater Research* 63 (2011): 189–97.

41. A. P. Summers, "Stiffening the stingray skeleton: an investigation of durophagy in myliobatid Stingrays (Chondrichthyes, Batoidea, Myliobatidae)," *Journal of Morphology* 243 (2000): 113–26; K. M. Rutledge et al., "Killing them softly: ontogeny of jaw mechanics and stiffness in mollusk-feeding freshwater stingrays," *Journal of Morphology* 280 (2018): 796–808.

42. Summers, "Stiffening the stingray skeleton."

43. M. A. Kolmann et al., "Intraspecific variation in feeding mechanics and bite force in durophagous stingrays," *Journal of Zoology* 304 (2018): 225–34.

44. Martin, *Life Traces of the Georgia Coast.*

45. M. R. Gregory, "New trace fossils from the Miocene of Northland, New Zealand, *Rosschachichnus amoeba* and *Piscichnus Waitemata*," *Ichnos* 1 (1991): 195–206.

46. J. D. Howard et al., "Biogenic sedimentary structures formed by rays," *Journal of Sedimentary Research* 47 (1977): 339–46; J. Martinell et al., "Cretaceous ray traces? An alternative interpretation for the alleged dinosaur tracks of La Posa, Isona, NE Spain," *Palaios* 16 (2001): 409–16.

47. Brannen, *The Ends of the World.*

48. Z.-Q. Chen et al., "Structural changes of marine communities over the Permian–Triassic transition: ecologically assessing the end-Permian mass extinction and its aftermath," *Global and Planetary Change* 73 (2010): 123–40.

49. D. Chu et al., "Early Triassic wrinkle structures on land: stressed environments and oases for life," *Scientific Reports* 5 (2015): 10109.

50. S. L. A. Cooper and D. M. Martill, "Pycnodont fishes (Actinopterygii, Pycnodontiformes) from the Upper Cretaceous (lower Turonian) Akrabou Formation of Asfla, Morocco," *Cretaceous Research* 116 (2020): 104607.

51. Martin, *Dinosaurs Without Bones.*

52. G. J. Vermeij and E. Zipser, "The diet of *Diodon hystrix* (Teleostei: —Tetraodontiformes): shell-crushing on Guam's reefs," *Bishop Museum Bulletin in Zoology* 9 (2015): 169–75.

53. A. M. Jones et al., "Tool use in the tuskfish *Choerodon schoenleinii*?," *Coral Reef* 30 (2011): 865; G. Bernardi, "The use of tools by wrasses (Labri-

dae)," *Coral Reefs* 31 (2011): 39–39; K. J. Pryor and A. M. Milton, "Tool use by the graphic tuskfish *Choerodon graphicus*," *Journal of Fish Biology* 95 (2019): 663–67.

54. M. J. Benton, *Vertebrate Paleontology* (4th ed.) (Oxford: Wiley-Blackwell, 2014).

55. Benton, *Vertebrate Paleontology*.

56. Benton, *Vertebrate Paleontology*.

57. J. M. Neenan et al., "European origin of placodont marine reptiles and the evolution of crushing dentition in Placodontia," *Nature Communications* 4 (2013): 1621; J. M. Neenan et al., "Unique method of tooth replacement in durophagous placodont marine reptiles, with new data on the dentition of Chinese taxa," *Journal of Anatomy* 224 (2014): 603–13.

58. E. E. Maxwell and M. W. Caldwell, "First record of live birth in Cretaceous ichthyosaurs: closing an 80 million year gap," *Proceedings of the Royal Society, B* 270 (2003): S104–7.

59. J.-D. Huang et al., "Repeated evolution of durophagy during ichthyosaur radiation after mass extinction indicated by hidden dentition," *Scientific Reports* 10 (2020): 7798.

60. M. J. Polcyn et al., "Physical drivers of mosasaur evolution," *Palaeogeography, Palaeoclimatology, Palaeoecology* 400 (2014): 17–27; W. B. Gallagher, "On the last mosasaurs: Late Maastrichtian mosasaurs and the Cretaceous-Paleogene boundary in New Jersey," *Bulletin de la Société Géologique de France* 183 (2012): 145–50.

61. A. S. Gale et al., "Mosasauroid predation on an ammonite—*Pseudaspidoceras*—from the Early Turonian of south-eastern Morocco," *Acta Geologica Polonica* 67 (2017): 31–46; E. G. Kauffman and J. K. Sawdo, "Mosasaur predation on a nautiloid from the Maastrichtian Pierre Shale, Central Colorado, Western Interior Basin, United States," *Lethaia* 46 (2013): 180–87.

62. T. Ikejiri et al., "Two-step extinction of Late Cretaceous marine vertebrates in northern Gulf of Mexico prolonged biodiversity loss prior to the Chicxulub impact," *Scientific Reports* 10 (2020): 4169.

63. D. D. Bermúdez-Rochas et al., "Evidence of predation in Early Cretaceous unionoid bivalves from freshwater sediments in the Cameros Basin, Spain," *Lethaia* 46 (2012): 57–70.

64. C. A. Brochu, "Alligatorine phylogeny and the status of *Allognathosuchus* Mook, 1921," *Journal of Vertebrate Paleontology* 24 (2004): 857–73.

65. In 2015 I was awarded a Fellowship in the Explorers Club, and to celebrate such a career milestone, I went to their annual meeting in New York City. While at the Explorers Club headquarters in Manhattan, I finally met my childhood idol, zoologist Jim Fowler of *Mutual and Omaha's Wild Kingdom* fame. As I reverted to little-kid fandom and thanked him for encouraging my lifelong interest in animals and animal behavior, he graciously listened and was otherwise a very nice fellow. Yes, sometimes you can meet your heroes.

66. G. Sheffield and J. M. Grebmeier, "Pacific walrus (*Odobenus rosmarus*

divergens): differential prey digestion and diet," *Marine Mammal Science* 25 (2009): 761–77.

67. N. Levermann et al., "Feeding behaviour of free-ranging walruses with notes on apparent dextrality of flipper use," *BMC Ecology* 3 (2003): 9.

68. R. A. Kastelein et al., "Oral suction of a Pacific walrus (*Odobenus rosmarus divergens*) in air and under water," *Z. Säugetierkunde* 59 (1994): 105–15.

69. M. K. Gingras et al., "Pleistocene walrus herds in the Olympic peninsula area: trace-fossil evidence of predation by hydraulic jetting," *Palaios* 22 (2007): 539–45.

70. D. W. MacDonald et al. (eds.), *Biology and Conservation of Musteloids* (Oxford: Oxford University Press, 2017).

71. R. G. Kvitek et al., "Sea otter foraging on deep-burrowing bivalves in a California coastal lagoon," *Marine Biology* 98 (1988): 157–67.

72. G. R. VanBlaricom, *Sea Otters* (Stillwater, MN: Voyageur Press, 2001).

73. J.A. Fujii et al., "Ecological drivers of variation in tool-use frequency across sea otter populations," *Behavioral Ecology* 26 (2015): 519–26; M. Haslam et al., "Wild sea otter mussel pounding leaves archaeological traces," *Scientific Reports* 9 (2019): 4417.

74. Haslam et al., "Wild sea otter."

75. Haslam et al., "Wild sea otter."

76. Haslam et al., "Wild sea otter."

77. G. C. Cadée, "Size-selective transport of shells by birds and its palaeoecological implications," *Palaeontology* 32 (1989): 429–37; Martin, *Life Traces of the Georgia Coast*.

78. A. P. Le Rossignol, "Breaking down the mussel (*Mytilus edulis*) shell: which layers affect oystercatchers' (*Haematopus ostralegus*) prey selection?," *Journal of Experimental Marine Biology and Ecology* 405 (2011): 87–92.

79. Like many stories that sound too good to be true, Aeschylus's "death by tortoise" story may have been invented, as Greek-history scholars can find no mention of it. However, the descendants of birds and tortoises accused of killing him are still in the Greek countryside, so perhaps this same fate awaits some follicle-deprived person someday. And if it does, let's just hope a Greek chorus is nearby to provide perspective.

80. C. A. Meyer and B. Thüring, "Dinosaurs of Switzerland," *Comptes Rendus Palevol* 2 (2003): 103–17.

81. C. Püntener et al., "Under the feet of sauropods: a trampled coastal marine turtle from the Late Jurassic of Switzerland?," *Swiss Journal of Geosciences* 112 (2019): 507–15.

82. M. Lockley, *Tracking Dinosaurs: A New Look at an Ancient World* (Cambridge: Cambridge University Press, 1991).

Chapter Seven

1. E. West, *The Last Indian War: The Nez Perce Story* (Oxford: Oxford University Press, 2009); M. Cheater, "Wolf spirit returns to Idaho," *Na-*

tional Wildlife Federation, August 1, 1998, https://www.nwf.org/Magazines /National-Wildlife/1998/Wolf-Spirit-Returns-to-Idaho.

2. L. D. Mech. and R. O. Peterson, "Wolf-prey relations," in *Wolves: Behavior, Ecology, and Conservation*, ed. L. D. Mech and L. Boitani (Chicago: University of Chicago Press, 2010).

3. D. L. Garshelis et al., "Remarkable adaptations of the American black bear help explain why it is the most common bear: a long-term study from the center of its range," in *Bears of the World: Ecology, Conservation and Management*, ed. V. Penteriana and M. Melletti (Cambridge: Cambridge University Press, 2021), 53–62.

4. M. A. Salamon et al., "Putative Late Ordovician land plants," *New Phytologist* 218 (2018): 1305–9.

5. D. Edwards et al., "A vascular conducting strand in the early land plant *Cooksonia*," *Nature* 357 (1992): 683–85; P. Steemans et al. "Origin and radiation of the earliest vascular land plants," *Science* 324 (2009): 353.

6. Hoysted et al., "A mycorrhizal revolution"; A. J. Hetherington and L. Doland, "Stepwise and independent origins of roots among land plants," *Nature* 561 (2018): 235–38.

7. R. Ennos, *The Age of Wood: Our Most Useful Material and the Construction of Civilization* (New York: Scribner, 2020).

8. P. Larson, *The Vascular Cambium: Development and Structure* (Berlin: Springer-Verlag, 2012).

9. S. G. Pallardy, *The Woody Plant Body* (3rd ed.) (Cambridge, MA: Academic Press, 2008).

10. V. Trouet, *Tree Story: The History of the World Written in Rings* (Baltimore: Johns Hopkins University Press, 2020).

11. D. Edwards et al., "Coprolites as evidence for plant-animal interaction in Siluro-Devonian terrestrial ecosystems," *Nature* 377 (1995): 329–31.

12. J. A. Dunlop and R. J. Garwood, "Terrestrial invertebrates in the Rhynie chert ecosystem," *Philosophical Transactions of the Royal Society, B* 373 (2018): 20160493.

13. Edwards et al., "Coprolites as evidence."

14. C. C. Labandeira, "Deep-time patterns of tissue consumption by terrestrial arthropod herbivores," *Naturwissenschaften* 100 (2013): 355–64.

15. C. C. Labandeira, "Middle Devonian liverwort herbivory and anti-herbivory defense," *New Phytologist* 202 (2014): 247–58.

16. Y.-H. Wang et al., "Fossil record of stem groups employed in evaluating the chronogram of insects (Arthropoda: Hexapoda)," *Scientific Reports* 6 (2016): 38939.

17. C. Huag and J. T. Huag, "The presumed oldest flying insect: more likely a myriapod?," *PeerJ* 5 (2017): e3402; Dunlop and Garwood, "Terrestrial invertebrates."

18. B. Misof et al., "Phylogenomics resolves the timing and pattern of insect evolution," *Science* 346 (2014): 6210.

19. Labandeira, "Deep-time patterns"; C. C. Labandeira, "A paleo-

biologic perspective on plant–insect interactions," *Current Opinion in Plant Biology* 16 (2013): 414–21.

20. A. C. Scott, "Trace fossils of plant–arthropod interactions," in *Trace Fossils: Their Paleobiological Aspects*, ed. C. G. Maples and R. R. West, *Paleontological Society Short Course* 5 (1992): 197–223.

21. Labandeira et al., "A paleobiologic perspective."

22. S. K. Turner, "Constraints on the onset duration of the Paleocene-Eocene Thermal Maximum," *Philosophical Transactions of the Royal Society, A* 376 (2018): 20170082.

23. S. A. Marshall, *Beetles: The Natural History and Diversity of Coleoptera* (Richmond Hill, ON: Firefly Books, 2018).

24. F. Vega and R. Hofstetter (eds.), *Bark Beetles: Biology and Ecology of Native and Invasive Species* (Cambridge, MA: Academic Press, 2014).

25. L. R. Kirkendall et al., "Evolution and diversity of bark and ambrosia beetles," in *Bark Beetles: Biology and Ecology of Native and Invasive Species*, ed. F. Vega and R. Hofstetter (Cambridge, MA: Academic Press, 2014), 85–156.

26. Kirkendall et al., "Evolution and diversity."

27. J. P. Audley et al., "Impacts of mountain pine beetle outbreaks on lodgepole pine forests in the Intermountain West, U.S., 2004–2019," *Forest Ecology and Management* 475 (2020): 118403.

28. J. R. Meeker et al., "The Southern pine beetle *Dendroctonus frontalis* Zimmerman (Coleoptera: Scolytidae)," *Florida Department of Agricultural and Consumer Services, Entomology Circular* 369 (1995): 1–4; H. Li and T. Li, "Bark beetle larval dynamics carved in the egg gallery: a study of mathematically reconstructing bark beetle tunnel maps," *Advances in Difference Equations* (2019): 513.

29. K. M. Thompson et al., "Autumn shifts in cold tolerance metabolites in overwintering adult mountain pine beetles," *PLOS ONE* 15 (2020): e0227203.

30. Kirkendall et al., "Evolution and diversity."

31. S.-Q. Zhang et al., "Evolutionary history of Coleoptera revealed by extensive sampling of genes and species," *Nature Communications* 9 (2018): 205; V. Q. P. Turman et al., "A new trace fossil produced by insects in fossil wood of Late Jurassic–Early Cretaceous Missão Velha Formation, Araripe Basin, Brazil," *Journal of South American Earth Sciences* 109 (2021): 103266; D. Peris et al., "Origin and evolution of fungus farming in wood-boring Coleoptera—a palaeontological perspective," *Biological Reviews* 96 (2021): 2476–88.

32. M. V. Walker, "Evidence of Triassic insects in the Petrified Forest National Monument, Arizona," *Proceedings of the United States National Museum* 85 (1938): 137–41.

33. S. T. Hasiotis et al., "Research update on hymenopteran nests and cocoons, Upper Triassic Chinle Formation, Petrified Forest National Park, Arizona," in *National Park Service Paleontological Research, Technical Report*

NPS/NRGRD/GRDTR-98/01, ed. V. L. Santucci and L. McClelland (1998), 116–21.

34. L. Tapanila and E. M. Roberts, "The earliest evidence of holometabolan insect pupation in conifer wood," *PLOS ONE* 7 (2012): e31668.

35. Tapanila and Roberts, "The earliest evidence."

36. Z. Feng et al., "Late Permian wood-borings reveal an intricate network of ecological relationships," *Nature Communications* 8 (2017): 556.

37. F. Legendre et al., "Phylogeny of Dictyoptera: dating the origin of cockroaches, praying mantises and termites with molecular data and controlled fossil evidence," *PLOS ONE* 10 (2015): e0130127; D. A. Evangelista et al., "An integrative phylogenomic approach illuminates the evolutionary history of cockroaches and termites (Blattodea)," *Proceedings of the Royal Society, B* 286 (2019): 20182076.

38. K. Maekawa and C. A. Nalepa, "Biogeography and phylogeny of wood-feeding cockroaches in the genus *Cryptocercus*," *Insects* 2 (2011): 354–68.

39. T. Chouvenc et al., "Termite evolution: mutualistic associations, key innovations, and the rise of Termitidae," *Cellular and Molecular Life Sciences* 78 (2021): 2749–69; C.A. Nalepa, "Origin of termite eusociality: trophallaxis integrates the social, nutritional, and microbial environments," *Ecological Entomology* 40 (2015): 323–35.

40. H. Ritter Jr., "Defense of mate and mating chamber in a wood roach," *Science* 143 (1964): 1459–60; Y. Park et al., "Colony composition, social behavior and some ecological characteristics of the Korean wood-feeding cockroach (*Cryptocercus kyebangensis*)," *Zoological Science* 19 (2002): 1133–39; Y. Park and J. Choe, "Territorial behavior of the Korean wood-feeding cockroach, *Cryptocercus kyebangensis*," *Journal of Ethology* 21 (2003): 79–85.

41. D. E. Bignell et al. (eds.), *Biology of Termites: A Modern Synthesis* (Dordrecht: Springer, 2011).

42. K. J. Howard and B. L. Thorne, "Eusocial evolution in termites and Hymenoptera," in *Biology of Termites: A Modern Synthesis*, ed. D. E. Bignell et al. (Dordrecht: Springer, 2011), 97–132.

43. S. K. Himmi et al., "X-ray tomographic analysis of the initial structure of the royal chamber and the nest-founding behavior of the drywood termite *Incisitermes minor*," *Journal of Wood Science* 60 (2014): 435–60; A. A. Eleuterio et al., "Stem decay in live trees: heartwood hollows and termites in five timber species in Eastern Amazonia," *Forests* 11 (2020): 1087.

44. J. I. Sutherland, "Miocene petrified wood and associated borings and termite faecal pellets from Hukatere Peninsula, Kaipara Harbour, North Auckland, New Zealand," *Journal of the Royal Society of New Zealand* 33 (2003): 395–414.

45. P. Vršanský et al., "Early wood-boring 'mole roach' reveals eusociality 'missing ring,'" *AMBA Projekty* 9 (2019): 1–28; Z. Zhao et al., "Termite colonies from mid-Cretaceous Myanmar demonstrate their early eusocial lifestyle in damp wood," *National Science Review* 7 (2020): 381–90; Z. Zhao

et al., "Termite communities and their early evolution and ecology trapped in Cretaceous amber," *Cretaceous Research* 117 (2021): 104612.

46. Howard and Thorne, "Eusocial evolution."

47. J.-P. Colin et al., "Termite coprolites (Insecta: Isoptera) from the Cretaceous of western France: a palaeoecological insight," *Revue de Micropaléontologie* 54 (2011): 129–39.

48. J. E. Francis and B. M. Harland, "Termite borings in Early Cretaceous fossil wood, Isle of Wight, UK," *Cretaceous Research* 27 (2006): 773–77; D. M. Rohr et al., "Oldest termite nest from the Upper Cretaceous of west Texas," *Geology* 14 (1986): 87–88.

49. E. M. Roberts et al., "Oligocene termite nests with in situ fungus gardens from the Rukwa Rift Basin, Tanzania, support a Paleogene African origin for insect agriculture," *PLOS ONE* 11 (2016): e0156847.

50. T. C. Harrington et al., "Isolations from the redbay ambrosia beetle, *Xyleborus glabratus*, confirm that the laurel wilt pathogen, *Raffaelea lauricola*, originated in Asia," *Mycologia* 103 (2011): 1028–36.

51. B. J. Bentz and A. M. Jönsson, "Modeling bark beetle responses to climate change," in *Bark Beetles: Biology and Ecology of Native and Invasive Species*, ed. F. E. Vega and R. W. Hofstetter (Cambridge, MA: Academic Press, 2014), 533–53.

52. Chouvenc et al., "Termite evolution."

53. P. A. Nauer et al., "Termite mounds mitigate half of termite methane emissions," *Proceedings of the National Academy of Sciences* 115 (2018): 13306–11.

54. K. Chin and B. D. Gill, "Dinosaurs, dung beetles, and conifers: participants in a Cretaceous food web," *Palaios* 11 (1996): 280–85.

55. Chin and Gill, "Dinosaurs, dung beetles, and conifers."

56. K. Chin, "The paleobiological implications of herbivorous dinosaur coprolites from the Upper Cretaceous Two Medicine Formation of Montana: why eat wood?," *Palaios* 22 (2007): 554–66.

57. K. Chin et al., "Consumption of crustaceans by megaherbivorous dinosaurs: dietary flexibility and dinosaur life history strategies," *Scientific Reports* 7 (2017): 11163.

58. Chin et al., "Consumption of crustaceans."

59. For general information about the Piedmont Wildlife Refuge, their website is a good first source: https://www.fws.gov/refuge/Piedmont/.

60. D. A. Sibley, *The Sibley Guide to Birds* (New York: Alfred A. Knopf, 2014).

61. R. N. Conner et al., *The Red-Cockaded Woodpecker: Surviving in a Fire-Maintained Ecosystem* (Austin: University of Texas Press, 2010).

62. D. C. Rudolph and R. N. Conner, "Cavity tree selection by red-cockaded wood-peckers in relation to tree age," *Wilson Bulletin* 103 (1991): 458–67; R. N. Conner et al., "Red-cockaded woodpecker nest-cavity selection: relationships with cavity age and resin production," *The Auk* 115 (1998): 447–54.

63. R. T. Engstrom and F. J. Sanders, "Red-cockaded woodpecker foraging ecology in an old-growth longleaf pine forest," *Wilson Bulletin* 109 (1997): 203–17.

64. B. Finch et al., *Longleaf, as Far as the Eye Can See* (Chapel Hill: University of North Carolina Press, 2012).

65. G. T. Lloyd et al., "Probabilistic divergence time estimation without branch lengths: dating the origins of dinosaurs, avian flight and crown birds," *Biology Letters* 12 (2016): 0160609.

66. C. Foth and O. V. M. Rauhut, "Re-evaluation of the Haarlem *Archaeopteryx* and the radiation of maniraptoran theropod dinosaurs," *BMC Evolutionary Biology* 17 (2017): 236.

67. D. T. Ksepka. "Feathered dinosaurs," *Current Biology* 30 (2020): R1347–53.

68. J. K. O'Connor et al., "A new enantiornithine from the Yixian formation with the first recognized avian enamel specialization," *Journal of Vertebrate Paleontology* 33 (2013): 1–12.

69. L. Xing et al., "A new enantiornithine bird with unusual pedal proportions found in amber," *Current Biology* 29 (2019): 2396–401.

70. D. J. Field et al., "Early evolution of modern birds structured by global forest collapse at the end-Cretaceous mass extinction," *Current Biology* 28 (2018): 1825–31.

71. D. T. Ksepka et al., "Oldest finch-beaked birds reveal parallel ecological radiations in the earliest evolution of passerines," *Current Biology* 29 (2019): 657–63.

72. J. M. McCullough et al., "A Laurasian origin for a pantropical bird radiation is supported by genomic and fossil data (Aves: Coraciiformes)," *Proceedings of the Royal Society, B* 286 (2019): 20190122.

73. V. L. de Pietri et al., "A new species of woodpecker (Aves; Picidae) from the early Miocene of Saulcet (Allier, France)," *Swiss Journal of Paleontology* 130 (2011): 307–14.

74. G. Mayr, "A tiny barbet-like bird from the Lower Oligocene of Germany: the smallest species and earliest substantial fossil record of the Pici (woodpeckers and allies)," *Auk* 122 (2005): 1055–63.

75. A. Manegold and A. Louchart, "Biogeographic and paleoenvironmental implications of a new woodpecker species (Aves, Picidae) from the early Pliocene of South Africa," *Journal of Vertebrate Paleontology* 32 (2012): 926–38.

76. R. C. Laybourne et al., "Feather in amber is earliest new world fossil of Picidae," *Wilson Bulletin* 106 (1994): 18–25.

77. R. Mikuláš and B. Zasadil, "A probable fossil bird nest, ?*Eocavum* isp., from the Miocene wood of the Czech Republic," *4th International Bioerosion Workshop Abstract Book* (Prague, Czech Republic, 2004), 49–51.

78. J.-Y. Jung et al., "A natural stress deflector on the head? Mechanical and functional evaluation of the woodpecker skull bones," *Advanced Theory and Simulation* 2 (2019): 1800152.

79. An annual competition for the world's fastest drummer is measured by a device called a Drumometer™, which counts the number of single drum strokes (but with both hands). The world record as of this writing (November 2021) was by Tom Grosset on July 15, 2013, with 1,208 beats in 60 seconds. This is slightly more than 20 beats/second, which is almost as fast as a woodpecker. See it at *Tom Grosset—Official World's Fastest Drummer—Breaks World Record*, https://youtu.be/Q9FrW-Wr-ds.

80. M. Elbroch and E. Marks, *Bird Tracks and Sign of North America* (Mechanicsburg, PA: Stackpole Books, 2001).

81. L. Wang et al., "Why do woodpeckers resist head impact injury: a biomechanical investigation," *PLOS ONE* 6 (2011): e26490.

82. Wang et al., "Why do woodpeckers."

83. Wang et al., "Why do woodpeckers"; Jung et al., "A natural stress deflector"; S. Van Wassenbergh et al., "Woodpeckers minimize cranial absorption of shocks," *Current Biology* 32 (2022): https://doi.org/10.1016/j.cub.2022.05.052.

84. S. A. Shunk, *Peterson Reference Guide to Woodpeckers of North America* (New York: Houghton Mifflin, 2016).

85. R. D. Magrath et al., "A mutual understanding? Interspecific responses by birds to each other's aerial alarm calls," *Behavioral Ecology* 18 (2007): 944–51; L. I. Hollén and A. N. Radford, "The development of alarm call behaviour in mammals and birds," *Animal Behaviour* 78 (2009): 791–800.

86. R. D. Stark et al., "A quantitative analysis of woodpecker drumming," *The Condor* 100 (1998): 350–56.

87. L. Imbeau and A. Desrochers, "Foraging ecology and use of drumming trees by three-toed woodpeckers," *Journal of Wildlife Management* 66 (2020): 222–31.

88. M. Garcia et al., "Evolution of communication signals and information during species radiation," *Nature Communications* 11 (2020): 4970.

89. Garcia et al., "Evolution of communication signals."

90. C. Reynolds et al., "The role of waterbirds in the dispersal of aquatic alien and invasive species," *Diversity and Distributions* 21 (2015): 744–54.

91. N. R. Johansson et al., "Woodpeckers can act as dispersal vectors for microorganisms," *Ecology and Evolution* 11 (2021): 7154–63.

92. M. A. Jusino et al., "Experimental evidence of a symbiosis between red-cockaded woodpeckers and fungi," *Proceedings of the Royal Society, B* 283 (2016): 20160106.

93. C. K. Vishnudas, "*Crematogaster* ants in shaded coffee plantations: a critical food source for rufous woodpecker *Micropternus brachyurus* and other forest birds," *Indian Birds* 4 (2008): 9–11.

94. Vishnudas, "*Crematogaster* ants."

95. Vishnudas, "*Crematogaster* ants."

96. J. D. Styrsky and M. D. Eubanks, "Ecological consequences of interactions between ants and honeydew-producing insects," *Proceedings of the Royal Society, B* 274 (2007): 151–64.

97. L. A. Blanc and J. R. Walters, "Cavity-nest webs in a longleaf pine ecosystem," *The Condor* 110 (2008): 80–92.

98. A. B. Edworthy et al., "Tree cavity occupancy by nesting vertebrates across cavity age," *Journal of Wildlife Management* 82 (2018): 639–48.

99. K. L. Cockle and K. Martin, "Temporal dynamics of a commensal network of cavity-nesting vertebrates: increased diversity during an insect outbreak," *Ecology* 96 (2015): 1093–104.

100. Sibley, *Sibley Guide to Birds*.

101. S. Barve et al., "Lifetime reproductive benefits of cooperative polygamy vary for males and females in the acorn woodpecker (*Melanerpes formicivorus*)," *Proceedings of the Royal Society, B* 288 (2021): 20210579.

102. R. E. Harness and E. L. Walters, "Woodpeckers and utility pole damage," *IEEE Industry Applications Magazine* 11 (2005): 68–73. Also, I don't know if acorn woodpeckers have ever drilled into wooden Sasquatch sculptures, but I hope they have, because woodpecker traces actually represent a real animal.

Chapter Eight

1. R. Hutchinson, *The Spanish Armada* (New York: Thomas Dunne Books, 2013).

2. P. Palma and L. N. Samthakumaran, *Shipwrecks and Global "Worming"* (Oxford: Archaeopress, 2014).

3. J. A. Sokolow, *The Great Encounter: Native Peoples and European Settlers in the Americas, 1492–1800* (Abingdon: Taylor & Francis, 2016).

4. Although the name "Bahamas" lends itself well to tales that it is derived from Spanish (where *baha* = "shallow" and *mar* = "sea"), it may actually be from the original Lucayan name for Grand Bahama Island, which was simply *Bahama*: W. P. Aren, "Naming the Bahamas islands: history and folk etymology," in *Names and Their Environment*, ed. C. Hough and D. Izdebska (Glasgow: *Proceedings of the 25th International Congress of Onomastic Sciences*, 2014), 42–49.

5. D. Méndez, "Shipwrecked by worms, saved by canoe: the last voyage of Columbus," in *The Ocean Reader: History, Culture, Politics*, ed. E. P. Roordia (Durham, NC: Duke University Press, 2020), 297–304.

6. C. C. Mann, *1493: Uncovering the New World Columbus Created* (New York: Vintage Books, 2012).

7. A. Sundberg, "Molluscan explosion: the Dutch shipworm epidemic of the 1730s," *Environment & Society Portal, Arcadia* 14 (2015): 1–6; D. L. Nelson, "The ravages of *Teredo*: the rise and fall of shipworm in US history, 1860–1940," *Environmental History* 21 (2016): 100–124.

8. Palma and Samthakumaran, *Shipwrecks and Global "Worming"*; S. Gilman, "The clam that sank a thousand ships," *Hakai Magazine*, December 5, 2016, https://www.hakaimagazine.com/features/clam-sank-thousand-ships/.

9. Herman Melville's classic novel *Moby-Dick; or, The Whale* was originally published in the UK on October 18, 1851, by Richard Bentley, then on November 14, 1851, in the US by Harper & Brothers. Project Gutenberg has a free ebook edition of this classic novel at https://www.gutenberg.org /files/2701/2701-h/2701-h.htm.

10. D. L. Distel, "The biology of marine wood boring bivalves and their bacterial endosymbionts," *Wood Deterioration and Preservation* 845 (2003): 253–71; J. R. Voight, "Xylotrophic bivalves: aspects of their biology and the impacts of humans," *Journal of Molluscan Studies* 81 (2015): 175–86.

11. Distel, "The biology of marine wood boring bivalves."

12. A. G. Steinmayer and J. M. Turfa, "Effects of shipworm on the performance of ancient Mediterranean warships," *International Journal of Nautical Archaeology* 25 (1996): 104–21; C. A. Rayes et al., "Boring through history: an environmental history of the extent, impact and management of marine woodborers in a global and local context, 500 BCE to 1930s CE," *Environment and History* 21 (2015): 477–512.

13. L. M. S. Borges et al., "Diversity, environmental requirements, and biogeography of bivalve wood-borers (Teredinidae) in European coastal waters," *Frontiers in Zoology* 11 (2014): 13.

14. Nelson, "The ravages of *Teredo*."

15. S. P. Ville and J. Kearney, *Transport and the Development of the European Economy, 1750–1918* (London: Palgrave Macmillan, 1990).

16. Steemans et al., "Origin and radiation."

17. N. S. Davies and M. R. Gibling, "Paleozoic vegetation and the Siluro-Devonian rise of fluvial lateral accretion sets," *Geology* 38 (2010): 51–54.

18. S. I. Kaiser et al., "The global Hangenberg Crisis (Devonian–Carboniferous transition): review of a first-order mass extinction," *Geological Society, London, Special Publications* 423 (2015): 387–437.

19. Kaiser et al., "The global Hangenberg Crisis."

20. Kaiser et al., "The global Hangenberg Crisis."

21. Brannen, *The Ends of the World*.

22. T. Martin et al., "Triassic-Jurassic biodiversity, ecosystems, and climate in the Junggar Basin, Xinjiang, Northwest China," *Palaeobiodiversity and Palaeoenvironments* 90 (2010): 171–73.

23. W. I. Ausich et al., "Early phylogeny of crinoids within the pelmatozoan clade," *Palaeontology* 58 (2015): 937–52.

24. Clarkson, *Invertebrate Palaeontology and Evolution*.

25. Clarkson, *Invertebrate Palaeontology and Evolution*.

26. One of the better swimming-crinoid videos with contextual information is by National Geographic, posted February 17, 2017: *Watch: Entrancing Sea Creature Glides through Water*, https://youtu.be/_u6lJ7EEzak.

27. K. R. Brom et al., "Experimental neoichnology of crawling stalked crinoids," *Swiss Journal of Palaeontology* 137 (2018): 197–203; Carvalho et al., "Running crabs."

28. T. K. Baumiller et al., "Post-Paleozoic crinoid radiation in response

to benthic predation preceded the Mesozoic marine revolution," *Proceedings of the National Academy of Sciences* 107 (2010): 5893–96.

29. H. Hess, "Lower Jurassic Posidonia Shale of southern Germany," in *Fossil Crinoids*, ed. H. Hess et al. (Cambridge: Cambridge University Press, 1999), 183–96; R. B. Hauff and U. Joger, "Holzmaden: prehistoric museum Hauff: a fossil museum since 4 generations (Urweltmuseum Hauff)," in *Paleontological Collections of Germany, Austria and Switzerland*, ed. L. Beck and U. Joger (Cham: Springer, 2018), 325–29.

30. Hauff and Joger, "Holzmaden."

31. A. J. van Loo, "Ichthyosaur embryos outside the mother body: not due to carcass explosion but to carcass implosion," *Palaeobiology and Palaeoenvironments* 93 (2013): 103–9.

32. H. W. Rasmussen, "Function and attachment of the stem in Isocrinidae and Pentacrinitidae: review and interpretation," *Lethaia* 10 (1977): 51–57.

33. H.-J. Röhl et al., "The Posidonia Shale (Lower Toarcian) of SW-Germany: an oxygen-depleted ecosystem controlled by sea level and palaeoclimate," *Palaeogeography, Palaeoclimatology, Palaeoecology* 165 (2001): 27–52.

34. A. W. Hunter et al., "Reconstructing the ecology of a Jurassic pseudoplanktonic raft colony," *Royal Society Open Science* 7 (2020): 200142.

35. Hunter et al., "Reconstructing the ecology."

36. Hunter et al., "Reconstructing the ecology." But also note that master ichnologist and paleo-detective Dolf Seilacher worked out a similar solution using finely tuned observations, not math. For that, please read A. Seilacher, "Developmental transformations in Jurassic driftwood crinoids," *Swiss Journal of Palaeontology* 130 (2011): 129–41.

37. *Teredo* Linnaeus, 1758, MolluscaBase, World Register of Marine Species (WoRMS), http://www.marinespecies.org/aphia.php?p=taxdetails&id=138539, accessed November 7, 2021.

38. D. L. Distel et al., "Molecular phylogeny of Pholadoidea Lamarck, 1809 supports a single origin for xylotrophy (wood feeding) and xylotrophic bacterial endosymbiosis in Bivalvia," *Molecular Phylogenetics and Evolution* 61 (2011): 245–54.

39. Distel et al., "Molecular phylogeny."

40. D. L. Distel et al., "Discovery of chemoautotrophic symbiosis in the giant shipworm *Kuphus polythalamia* (Bivalvia: Teredinidae) extends wooden-steps theory," *Proceedings of the National Academy of Science* 114 (2017): E3652–58.

41. M. Velásquez and R. Shipway, "A new genus and species of deep-sea wood-boring shipworm (Bivalvia: Teredinidae) *Nivanteredo coronata* n. sp. from the Southwest Pacific," *Marine Biology Research* 14 (2018): 808–15.

42. D. L. Distel and S. J. Roberts, "Bacterial endosymbionts in the gills of the deep-sea wood-boring bivalves *Xylophaga atlantica* and *Xylophaga washingtona*," *Biological Bulletin* 192 (1997): 253–61.

43. A. D. Ansell and N. B. Nair, "Shell movements of a wood-boring

bivalve," *Nature* 216 (1967): 595; A. D. Ansell and N. B. Nair, "The mechanisms of boring in *Martesia striata* Linne (Bivalvia: Pholadidae) and *Xylophaga dorsalis* Turton (Bivalvia: Xylophaginidae)," *Proceedings of the Royal Society of London, B* 174 (1969): 123–33.

44. Ansell and Nair, "The mechanisms of boring."

45. R. G. Bromley et al., "A Cretaceous woodground: the *Teredolites* ichnofacies," *Journal of Paleontology* 58 (1984): 488–98.

46. L. Brodsley et al., "Prince Rupert's drops," *Notes and Records of the Royal Society of London* 41 (1986): 1–26.

47. D. J. Amon et al., "Burrow forms, growth rates and feeding rates of wood-boring Xylophagaidae bivalves revealed by micro-computed tomography," *Frontiers in Marine Science* 2 (2015): 1–10.

48. Amon et al., "Burrow forms."

49. I. Guarneri et al., "A simple method to calculate the volume of shipworm tunnels from radiographs," *International Biodeterioration & Biodegradation* 156 (2021): 105109.

50. Rayes et al., "Boring through history."

51. Distel et al., "Molecular phylogeny."

52. Distel et al., "Molecular phylogeny."

53. Distel et al., "Molecular phylogeny."

54. A. Kaim, "Non-actualistic wood-fall associations from Middle Jurassic of Poland," *Lethaia* 44 (2011): 109–24.

55. J. Villegas-Martín et al., "Jurassic *Teredolites* from Cuba: new trace fossil evidence of early wood-boring behavior in bivalves," *Journal of South American Earth Sciences* 38 (2012): 123–28.

56. Bromley et al., "A Cretaceous woodground."

57. Z. Belaústegui et al., "Ichnogeny and bivalve bioerosion: examples from shell and wood substrates," *Ichnos* 27 (2020): 277–83.

58. Belaústegui et al., "Ichnogeny and bivalve bioerosion."

59. R. H. B. Fraaije et al., "The oldest record of galatheoid anomurans (Decapoda, Crustacea) from Normandy, northwest France," *Neues Jahrbuch für Geologie und Paläontologie—Abhandlungen Band* 292 (2019): 291–97.

60. S. Melnyk et al., "A new marine woodground ichnotaxon from the Lower Cretaceous Mannville Group, Saskatchewan, Canada," *Journal of Paleontology* 95 (2020): 162–69.

61. Martin, *Life Traces of the Georgia Coast.*

62. Bromley et al., "A Cretaceous woodground."

63. MacEachern et al., "The ichnofacies paradigm."

64. C. E. Savrda et al., "Log-grounds and *Teredolites* in transgressive deposits, Eocene Tallahatta Formation (southern Alabama, USA)," *Ichnos* 12 (2005): 47–57; P. Monaco et al., "First documentation of wood borings (*Teredolites* and insect larvae) in Early Pleistocene lower shoreface storm deposits (Orvieto area, central Italy)," *Bollettino della Società Paleontologica Italiana* 50 (2011): 55–63; C. I. Serrano-Brañas et al., "*Teredolites* trace fossils in log-grounds from the Cerro del Pueblo Formation (Upper Cretaceous) of

the state of Coahuila, Mexico," *Journal of South American Earth Sciences* 95 (2019): 102316.

65. M. K. Gingras et al., "Modern perspectives on the *Teredolites* ichnofacies: observations from Willapa Bay, Washington," *Palaios* 19 (2004): 79–88; MacEachern et al., "The ichnofacies paradigm."

66. J. R. Shipway et al., "A rock-boring and rock-ingesting freshwater bivalve (shipworm) from the Philippines," *Proceedings of the Royal Society, B* 286 (2019): 20190434.

67. I. N. Bolotov et al., "Discovery of a silicate rock-boring organism and macrobioerosion in fresh water," *Nature Communications* 9 (2018): 2882.

68. Shipway et al., "A rock-boring and rock-ingesting freshwater bivalve."

Chapter Nine

1. M. S. Savoca et al., "Baleen whale prey consumption based on high-resolution foraging measurements," *Nature* 599 (2021): 85–90.

2. The first known observation of a whale fall in the deep sea was on February 19, 1977, by a US Navy submersible crew: T. Vetter, *30,000 Leagues Undersea: True Tales of a Submariner and Deep Submergence Pilot* (self-published, Tom Vetter Books). However, the first scientific discovery and assessment of a whale fall was in 1987, with results reported in C. R. Smith et al., "Vent fauna on whale remains," *Nature* 341 (1989): 27–28.

3. C. R. Smith et al., "Whale-fall ecosystems: recent insights into ecology, paleoecology, and evolution," *Annual Review Marine Sciences* 7 (2015): 571–96.

4. Smith et al., "Whale-fall ecosystems."

5. S. Danise and S. Dominici, "A record of fossil shallow-water whale falls from Italy," *Lethaia* 47 (2014): 229–43.

6. M. J. Moore et al., "Dead cetacean? Beach, bloat, float, sink," *Frontiers in Marine Science* 7 (2020): 333.

7. J. C. Mallon et al., "A 'bloat-and-float' taphonomic model best explains the upside-down preservation of ankylosaurs," *Palaeogeography, Palaeoclimatology, Palaeoecology* 497 (2018): 117–27.

8. Moore et al., "Dead cetacean?"

9. N. D. Higgs et al., "The morphological diversity of *Osedax* worm borings (Annelida: Siboglinidae)," *Journal of the Marine Biological Association of the United Kingdom* 94 (2014): 1429–39.

10. Distel et al., "Molecular phylogeny of Pholadoidea Lamarck."

11. Y. Li et al., "Phylogenomics of tubeworms (Siboglinidae, Annelida) and comparative performance of different reconstruction methods," *Zoologica Scripta* 46 (2016): 200–213.

12. Y. Li et al., "Phylogenomics of tubeworms."

13. G. W. Rouse et al., "*Osedax*: bone-eating marine worms with dwarf males," *Science* 305 (2004): 668–71.

14. Rouse et al., "Osedax: bone-eating marine worms."

15. M. Tresguerres et al., "How to get into bones: proton pump and carbonic anhydrase in *Osedax* boneworms," *Proceedings of the Royal Society, B* 280 (2013): 20130625.

16. Tresguerres et al., "How to get into bones."

17. N. D. Higgs et al., "Bone-boring worms: characterizing the morphology, rate, and method of bioerosion by *Osedax mucofloris* (Annelida, Siboglinidae)," *Biological Bulletin* 221 (2011): 307–16.

18. Higgs et al., "Bone-boring worms."

19. G. W. Rouse et al., "A dwarf male reversal in bone-eating worms," *Current Biology* 25 (2015): 236–41.

20. Rouse et al., "A dwarf male reversal."

21. Rouse et al., "A dwarf male reversal."

22. Higgs et al., "The morphological diversity of *Osedax* worm borings."

23. Higgs et al., "The morphological diversity of *Osedax* worm borings."

24. Higgs et al., "The morphological diversity of *Osedax* worm borings."

25. S. Kiel et al., "Fossil traces of the bone-eating worm *Osedax* in early Oligocene whale bones," *Proceedings of the National Academy of Sciences* 107 (2010): 8656–59.

26. Benton, *Vertebrate Paleontology.*

27. D. J. E. Murdock, "The 'biomineralization toolkit' and the origin of animal skeletons," *Biological Reviews* 95 (2020): 1372–92.

28. S. G. Platt et al., "Diet of the American crocodile (*Crocodylus acutus*) in marine environments of coastal Belize," *Journal of Herpetology* 47 (2013): 1–10.

29. Benton, *Vertebrate Paleontology.*

30. Benton, *Vertebrate Paleontology.*

31. Benton, *Vertebrate Paleontology.*

32. P. M. Sander et al., "Short-snouted toothless ichthyosaur from China suggests Late Triassic diversification of suction feeding ichthyosaurs," *PLOS ONE* 6 (2011): e19480.

33. R. B. J. Benson et al., "A giant pliosaurid skull from the Late Jurassic of England," *PLoS ONE* 8 (2013): e65989; V. Fischer et al. "Peculiar microphagous adaptations in a new Cretaceous pliosaurid," *Royal Society of Open Science* 2 (2015): 150552.

34. A. S. Schulp et al., "On diving and diet: resource partitioning in   . type-Maastrichtian mosasaurs," *Netherlands Journal of Geosciences— Geologie En Mijnbouw* 92 (2014): 165–70.

35. Everhart, *Oceans of Kansas.*

36. A. P. Cossette and C. A. Brochu, "A systematic review of the giant alligatoroid *Deinosuchus* from the Campanian of North America and its implications for the relationships at the root of Crocodylia," *Journal of Vertebrate Paleontology* 40 (2020): e1767638.

37. A. Berta, *Whales, Dolphins, and Porpoises: A Natural History and Species Guide* (Chicago: University of Chicago Press, 2015).

38. H. G. Ferrón et al., "Assessing metabolic constraints on the maximum body size of actinopterygians: locomotion energetics of *Leedsichthys problematicus* (Actinopterygii, Pachycormiformes)," *Palaeontology* 61 (2018): 775–83.

39. S. Danise et al., "Ecological succession of a Jurassic shallow-water ichthyosaur fall," *Nature Communications* 5 (2014): 4789.

40. Danise et al., "Ecological succession."

41. Danise et al., "Ecological succession."

42. Danise et al., "Ecological succession."

43. A. Kaim, "Chemosynthesis-based associations on Cretaceous plesiosaurid carcasses," *Acta Palaeontologica Polonica* 53 (2008): 97–104.

44. S. B. Johnson et al., "*Rubyspira*, new genus and two new species of bone-eating deep-sea snails with ancient habits," *Biological Bulletin* 219 (2010): 166–77.

45. Johnson et al., "*Rubyspira.*"

46. S. Danise and N. D. Higgs, "Bone-eating *Osedax* worms lived on Mesozoic marine reptile deadfalls," *Biology Letters* 11 (2015): 20150072.

47. M. Talevi and S. Brezina, "Bioerosion structures in a Late Cretaceous mosasaur from Antarctica," *Facies* 65 (2019): 1–5.

48. J. W. M. Jagt et al., "Episkeletozoans and bioerosional ichnotaxa on isolated bones of Late Cretaceous mosasaurs and cheloniid turtles from the Maastricht area, the Netherlands," *Geologos* 26 (2020): 39–49.

49. W. J. Jones et al., "Marine worms (genus *Osedax*) colonize cow bones," *Proceedings of the Royal Society, B* 275 (2008): 387–91.

50. G. W. Rouse et al., "Not whale-fall specialists, *Osedax* worms also consume fishbones," *Biology Letters* 7 (2011): 736–39.

51. S. Kiel et al., "*Osedax* borings in fossil marine bird bones," *Naturwissenschaften* 98 (2011): 51–55.

52. S. Kiel et al., "Traces of the bone-eating annelid *Osedax* in Oligocene whale teeth and fish bones," *Paläontologische Zeitschrift* 87 (2013): 161–67.

53. N. D. Higgs et al., "Evidence of *Osedax* worm borings in Pliocene (~3 Ma) whale bone from the Mediterranean," *Historical Biology* 24 (2012): 269–77.

54. C. R. McClain et al., "Alligators in the abyss: The first experimental reptilian food fall in the deep ocean," *PLOS ONE* 14 (2019): e0225345.

55. McClain et al., "Alligators in the abyss."

56. Smith et al., "Whale-fall ecosystems."

Chapter Ten

1. R. W. Thorington et al., *Squirrels: The Animal Answer Guide* (Baltimore: Johns Hopkins University Press, 2006).

2. A. Luebke et al., "Optimized biological tools: ultrastructure of rodent and bat teeth compared to human teeth," *Bioinspired, Biomimetic and Nanobiomaterials* 8 (2019): 247–53.

3. M. Elbroch, *Mammal Tracks and Sign: A Guide to North American Species* (Mechanicsburg, PA: Stackpole Books, 2003).

4. C. Rindali and T. M. Cole III, "Environmental seasonality and incremental growth rates of beaver (*Castor canadensis*) incisors: implications for palaeobiology," *Palaeogeography, Palaeoclimatology, Palaeoecology* 206 (2004): 289–301.

5. W. E. Klippel and J. A. Synstelien, "Rodents as taphonomic agents: bone gnawing by brown rats and gray squirrels," *Journal of Forensic Science* 52 (2007): 765–73, and references therein.

6. Klippel and Synstelien, "Rodents as taphonomic agents."

7. J. H. Frank et al., *American Beetles, Volume II: Polyphaga: Scarabaeoidea through Curculionoidea* (Boca Raton, FL: CRC Press, 2002).

8. Frank et al., *American Beetles.*

9. "Dermestarium," by Stephen H. Hinshaw of the University of Michigan Museum of Zoology, explains how dermestid beetles are used to prepare skeletons for the museum: https://webapps.lsa.umich.edu/ummz/mammals/dermestarium/default.asp.

10. B. B. Britt et al., "A suite of dermestid beetle traces on dinosaur bone from the Upper Jurassic Morrison Formation, Wyoming, USA," *Ichnos* 15 (2008): 59–71.

11. L. D. Martin and D. L. West, "The recognition and use of dermestid (Insecta, Coleoptera) pupation chambers in paleoecology," *Palaeogeography, Palaeoclimatology, Palaeoecology* 113 (1995): 303–10.

12. Britt et al., "A suite of dermestid beetle traces"; K. S. Bader et al., "Application of forensic science techniques to trace fossils on dinosaur bones from a quarry in the Upper Jurassic Morrison Formation, northeastern Wyoming," *Palaios* 24 (2009): 140–58.

13. L. R. Backwell et al., "Criteria for identifying bone modification by termites in the fossil record," *Palaeogeography, Palaeoclimatology, Palaeoecology* 337 (2012): 72–87.

14. M. Tappen, "Bone weathering in the tropical rain forest," *Journal of Archaeological Science* 21 (1994): 667–73.

15. Blackwell et al., "Criteria."

16. B. P. Freymann et al., "Termites of the genus *Odontotermes* are optionally keratophagous," *Ecotropica* 13 (2007): 143–47.

17. M. C. Go, "A case of human bone modification by ants (Hymenoptera: Formicidae) in the Philippines," *Forensic Anthropology* 1 (2018): 116–23.

18. Go, "A case of human bone modification."

19. V. D. P. Neto et al., "Oldest evidence of osteophagic behavior by insects from the Triassic of Brazil," *Palaeogeography, Palaeoclimatology, Palaeoecology* 453 (2016): 30–41.

20. O. Fejar and T. M. Kaiser, "Insect bone-modification and paleoecology of Oligocene mammal-bearing sites in the Doupov Mountains, northwestern Bohemia," *Palaeontologia Electronica* 8 (2005): Article 8.1.8A.

21. E. J. Odes et al., "Osteopathology and insect traces in the Australo-

pithecus africanus skeleton StW 431," *South African Journal of Science* 113 (2017): 1 7.

22. Dinosaur National Monument (US National Park Service), https://www.nps.gov/dino/index.htm.

23. The Late Jurassic dinosaur *Stegosaurus* is well known for its distinctive bony plates arranged along its back, but also for four formidable spikes on its tail, which paleontologists assume were used to defend against predatory dinosaurs. In 1982, cartoonist Gary Larson applied the term "thagomizer" to these tail spikes in a comic, in which an (anachronistic) caveman is teaching other cavemen by pointing to an illustration of a *Stegosaurus* tail and says, "Now this end is called the thagomizer . . . after the late Thag Simmons." The term has persisted with mirthful paleontologists since.

24. K. Carpenter, "Rocky start of Dinosaur National Monument (USA), the world's first dinosaur geoconservation site," *Geoconservation Research* 1 (2018): 1–20.

25. Britt et al., "A suite of dermestid beetle traces."

26. Britt et al., "A suite of dermestid beetle traces."

27. Britt et al., "A suite of dermestid beetle traces."

28. J. Foster et al., "Paleontology, taphonomy, and sedimentology of the Mygatt-Moore Quarry, a large dinosaur bonebed in the Morrison Formation, western Colorado: implications for Upper Jurassic dinosaur preservation modes," *Geology of the Intermountain West* 5 (2018): 23–93.

29. S. K. Drumheller et al., "High frequencies of theropod bite marks provide evidence for feeding, scavenging, and possible cannibalism in a stressed Late Jurassic ecosystem," *PLOS ONE* 15 (2020): e0233115.

30. J. B. McHugh et al., "Decomposition of dinosaurian remains inferred by invertebrate traces on vertebrate bone reveal new insights into Late Jurassic ecology, decay, and climate in western Colorado," *PeerJ* 8 (2020): e9510.

31. McHugh et al., "Decomposition of dinosaurian remains."

32. C. McNassor, *Images of America: Los Angeles's La Brea Tar Pits and Hancock Park* (Charleston, SC: Arcadia Publishing, 2011).

33. C. Stock (7th ed., revised by J. M. Harris), *Rancho La Brea: A Record of Pleistocene Life in California Science Series* 37 (Los Angeles: Natural History Museum of Los Angeles County, 1992).

34. The La Brea Tar Pits and Museum website provides a list of mammals discovered from the site: https://tarpits.org/research-collections/tar-pits-collections/mammal-collections.

35. B. K. McHorse et al., "The carnivoran fauna of Rancho La Brea: Average or aberrant?," *Palaeogeography, Palaeoclimatology, Palaeoecology* 329–30 (2012): 118–23.

36. McHorse et al., "The carnivoran fauna of Rancho La Brea."

37. A. R. Holden et al., "Paleoecological and taphonomic implications of insect-damaged Pleistocene vertebrate remains from Rancho La Brea, Southern California," *PLOS ONE* 8 (2013): e67119.

38. Frank et al., *American Beetles*.

39. Holden et al., "Paleoecological and taphonomic implications."

40. F. J. Augustin et al., "The smallest eating the largest: the oldest mammalian feeding traces on dinosaur bone from the Late Jurassic of the Junggar Basin (northwestern China)," *The Science of Nature* 107 (2020): 32.

41. C. S. Ozeki et al., "Biological modification of bones in the Cretaceous of North Africa," *Cretaceous Research* 114 (2020): 104529; F. J. Augustin et al., "Dinosaur taphonomy of the Jurassic Shishugou Formation (Northern Junggar Basin, NW China): insights from bioerosional trace fossils on bone," *Ichnos* 28 (2021): 87–96.

42. Klippel and Synstelian, "Rodents as taphonomic agents"; J. T. Pokines et al., "The taphonomic effects of eastern gray squirrels (*Sciurus carolinensis*) gnawing on bone," *Journal of Forensic Identification* 66 (2016): 349–75.

43. J. M. Hutson et al., "Osteophagia and bone modifications by giraffes and other large ungulates," *Journal of Archaeological Science* 40 (2013): 4139–49.

44. L. A. Meckel et al., "White-tailed deer as a taphonomic agent: photographic evidence of white-tailed deer gnawing on human bone," *Journal of Forensic Sciences* 63 (2018): 292–94.

45. M. Agaba et al., "Giraffe genome sequence reveals clues to its unique morphology and physiology," *Nature Communications* 7 (2016): 11519; B. Shorrocks, *The Giraffe: Biology, Ecology, Evolution and Behaviour* (West Sussex: John Wiley & Sons, 2016).

46. Hutson et al., "Osteophagia and bone modifications."

47. Hutson et al., "Osteophagia and bone modifications."

48. Hutson et al., "Osteophagia and bone modifications."

49. M. M. Selvaggio, "Carnivore tooth marks and stone tool butchery marks on scavenged bones: archaeological implications," *Journal of Human Evolution* 27 (1994): 215–28.

50. The Petrified Forest National Park (US National Park Service) website is at https://www.nps.gov/pefo/index.htm.

51. W. G. Parker et al. (eds.), *A Century of Research at Petrified Forest National Park* (Flagstaff: Museum of Northern Arizona Bulletin No. 62, 2006).

52. A. J. Martin and S. T. Hasiotis, "Vertebrate tracks and their significance in the Chinle Formation (Late Triassic), Petrified Forest National Park, Arizona," in *National Park Service Paleontological Research* 3 (1998): 38–143; A. Wahl et al., "Vertebrate coprolites and coprophagy traces, Chinle Formation (Late Triassic), Petrified Forest National Park, Arizona," *National Park Service Paleontological Research* 3 (1998): 144–48.

53. Wahl et al., "Vertebrate coprolites and coprophagy traces."

54. Parker et al., *A Century of Research*.

55. S. K. Drumheller et al., "Direct evidence of trophic interactions among apex predators in the Late Triassic of western North America," *Naturwissenschaften* 101: (2014) 975–87.

56. Drumheller et al., "Direct evidence of trophic interactions."

57. Drumheller et al., "Direct evidence of trophic interactions."

58. Brett and Walker, "Predators and predation."

59. M. W. Hardisty, *Lampreys: Life Without Jaws* (London: Forrest Text, 2006).

60. Benton, *Vertebrate Paleontology*.

61. Kuratani, "Evolution of the vertebrate jaw."

62. I. J. Sansom et al., "Chondrichthyan-like scales from the Middle Ordovician of Australia," *Palaeontology* 55 (2012): 243–47.

63. P. Gorzelak et al., "Inferred placoderm bite marks on Devonian crinoids from Poland," *Neues Jahrbuch für Geologie und Paläontologie* 259 (2011): 105–12.

64. Gorzelak et al., "Inferred placoderm bite marks."

65. O. A. Lebedev et al., "Bite marks as evidence of predation in early vertebrates," *Acta Zoologica* 90 (2009): 344–56.

66. Lebedev et al., "Bite marks as evidence."

67. J. Pier, "The Devonian monster of the deep," *Palaeontologia Electronica* blog post, related to Z. Johanson et al., "Fusion in the vertebral column of the pachyosteomorph arthrodire *Dunkleosteus terrelli* ('Placodermi')," *Palaeontologia Electronica* 22.2.20 (2019), https://palaeo-electronica.org /content/2011-11-30-22-01-23/2528-the-devonian-monster-of-the-deep.

68. P. S. L. Anderson and M. W. Westneat, "Feeding mechanics and bite force modelling of the skull of *Dunkleosteus terrelli*, an ancient apex predator," *Biology Letters* 3 (2007): 76–79; P. S. L. Anderson and M. W. Westneat, "A biomechanical model of feeding kinematics for *Dunkleosteus terrelli* (Arthrodira, Placodermi)," *Paleobiology* 35 (2009): 251–69.

69. Anderson and Westneat, "Feeding mechanics and bite force modelling."

70. Anderson and Westneat, "A biomechanical model."

71. Anderson and Westneat, "A biomechanical model."

72. L. Hall et al., "Possible evidence for cannibalism in the giant arthrodire *Dunkleosteus*, the apex predator of the Cleveland Shale Member (Fammenian) of the Ohio Shale," *Journal of Vertebrate Paleontology, Programs and Abstracts Book* (2016): 148.

73. Hall et al., "Possible evidence for cannibalism."

74. D. A. Ebert et al., *Sharks of the World: A Complete Guide* (Princeton, NJ: Princeton University Press, 2013).

75. J. Nielsen et al., "Eye lens radiocarbon reveals centuries of longevity in the Greenland shark (*Somniosus microcephalus*)," *Science* 353 (2016): 702–4.

76. B. Pobiner, "Paleoecological information in predator tooth marks," *Journal of Taphonomy* 6 (2008): 373–97.

77. S. J. Godfrey and J. B. Smith, "Shark-bitten vertebrate coprolites from the Miocene of Maryland," *Naturwissenschaften* 97 (2010): 461–67.

78. J. I. Castro, *The Sharks of North America* (Oxford: Oxford University Press, 2011).

79. Castro, *The Sharks of North America*.

80. S. Wroe et al., "Three-dimensional computer analysis of white shark jaw mechanics: how hard can a great white bite?," *Journal of Zoology* 276 (2008): 336–42.

81. Wroe et al., "Three-dimensional computer analysis."

82. R. W. Boessenecker et al., "The Early Pliocene extinction of the mega-toothed shark *Otodus megalodon*: a view from the eastern North Pacific," *PeerJ* 7 (2019): e6088.

83. J. A. Cooper et al., "Body dimensions of the extinct giant shark *Otodus megalodon*: a 2D reconstruction," *Scientific Reports* 10 (2020): 14596.

84. Cooper et al., "Body dimensions."

85. Wroe et al., "Three-dimensional computer analysis."

86. A. Ballell and H. G. Ferrón, "Biomechanical insights into the dentition of megatooth sharks (Lamniformes: Otodontidae)," *Scientific Reports* 11 (2021): 1232.

87. O. A. Aguilera et al., "Giant-toothed white sharks and cetacean trophic interaction from the Pliocene Caribbean Paraguaná Formation," *Paläontologische Zeitschrift* 82 (2008): 204–8.

88. S. J. Godfrey et al., "*Carcharocles*-bitten odontocete caudal vertebrae from the Coastal Eastern United States," *Acta Palaeontologica Polonica* 63 (2018): 463–68.

89. S. J. Godfrey et al., "*Otodus*-bitten sperm whale tooth from the Neogene of the Coastal Eastern United States," *Acta Paleontologica Polonica* 66 (2021): 1–5.

90. G. Grigg, *Biology and Evolution of Crocodylians* (Ithaca, NY: Cornell University Press, 2015).

91. D. R. Schwimmer, *King of the Crocodylians: The Paleobiology of* Deinosuchus (Bloomington: Indiana University Press, 2002).

92. C. A. Boyd et al., "Crocodyliform feeding traces on juvenile ornithischian dinosaurs from the Upper Cretaceous (Campanian) Kaiparowits Formation, Utah," *PLOS ONE* 8 (2013): e57605.

93. C. D. Brownstein, "Trace fossils on dinosaur bones reveal ecosystem dynamics along the coast of eastern North America during the latest Cretaceous," *PeerJ* 6 (2018): e4973.

94. G. M. Erickson et al., "Insights into the ecology and evolutionary success of crocodilians revealed through bite-force and tooth-pressure experimentation," *PLOS ONE* 7 (2012): e31781.

95. Erickson et al., "Insights into the ecology."

96. Erickson et al., "Insights into the ecology."

97. G. M. Erickson and K. H. Olson, "Bite marks attributable to *Tyrannosaurus rex*: preliminary description and implications," *Journal of Vertebrate Paleontology* 16 (1996): 175–78.

98. G. M. Erickson et al., "Bite-force estimation for *Tyrannosaurus rex* from tooth-marked bones," *Nature* 382 (1996): 706–8.

99. Erickson et al., "Bite-force estimation."

100. K. Chin et al. "A king-sized theropod coprolite," *Nature* 393 (1998): 680–82.

101. P. M. Gignac and G. M. Erickson, "The biomechanics behind extreme osteophagy in *Tyrannosaurus rex*," *Scientific Reports* 7 (2017): 2012.

102. K. Carpenter, "Evidence of predatory behavior by carnivorous dinosaurs," *Gaia* 15 (1998): 135–44.

103. J. R. Horner and D. Lessem, *The Complete T. Rex* (New York: Simon & Schuster, 1993).

104. R. A. DePalma et al., "Physical evidence of predatory behavior in *Tyrannosaurus rex*," *Proceedings of the National Academy of Science* 110 (2013): 12560–64.

105. C. M. Brown et al., "Intraspecific facial bite marks in tyrannosaurids provide insight into sexual maturity and evolution of bird-like intersexual display," *Paleobiology* 48 (2021): doi: https://doi.org/10.1017/pab.2021.29.

106. D. H. Tanke and P. J. Currie, "Head-biting behavior in theropod dinosaurs: paleopathological evidence," *Gaia* 15 (1998): 167–84.

107. Y. Hu et al., "Large Mesozoic mammals fed on young dinosaurs," *Nature* 433 (2005): 149–52.

108. T. R. Lyson et al., "Exceptional continental record of biotic recovery after the Cretaceous–Paleogene mass extinction," *Science* 366 (2019): 977–83.

109. A. K. Behrensmeyer, "Terrestrial vertebrate accumulations," in *Taphonomy: Releasing the Data Locked in the Fossil Record*, ed. P. A. Allison and D. E. G. Briggs (New York: Plenum Press, 1991), 291–327.

110. N. D. Pyenson, "The ecological rise of whales chronicled by the fossil record," *Current Biology* 27 (2017): R558–64.

111. J. G. M. Thewissen, *The Emergence of Whales: Evolutionary Patterns in the Origin of Cetacea* (New York: Springer, 2013).

112. E. Snively et al., "Bone-breaking bite force of *Basilosaurus isis* (Mammalia, Cetacea) from the Late Eocene of Egypt estimated by finite element analysis," *PLOS ONE* 10 (2015): e0118380.

113. Snively et al., "Bone-breaking bite force."

Chapter Eleven

1. R. N. Scoon, "Mount Elgon National Park(s)," in *Geology of National Parks of Central/Southern Kenya and Northern Tanzania*, ed. R. N. Scoon (Cham: Springer, 2018), 81–90.

2. R. J. Bowell et al., "Formation of cave salts and utilization by elephants in the Mount Elgon region, Kenya," in *Environmental Geochemistry and Health*, ed. J. D. Appleton et al. (London: Geological Society Special

Publication No. 113, 1996), 63–79; J. Lundberg and D. A. McFarlane, "Speleogenesis of the Mount Elgon elephant caves," in *Perspectives on Karst Geomorphology, Hydrology, and Geochemistry*, ed. R. S. Harmon and C. Wicks (Boulder, CO: Geological Society of America, 2006), 51–63.

3. C. J. Moss et al. (eds.), *The Amboseli Elephants: A Long-Term Perspective on a Long-Lived Mammal* (Chicago: University of Chicago Press, 2011).

4. C. A. Lundquist and W. W. Varnedoe Jr., "Salt ingestion caves," *International Journal of Speleology* 35 (2005): 13–18.

5. Lundquist and Varnedoe, "Salt ingestion caves."

6. L. A. Borrero and F. M. Martin, "Taphonomic observations on ground sloth bone and dung from Cueva del Milodón, Ultima Esperanza, Chile: 100 years of research history," *Quaternary International* 278 (2012): 3–11; B. van Geel et al., "Diet and environment of *Mylodon darwinii* based on pollen of a Late-Glacial coprolite from the Mylodon Cave in southern Chile," *Review of Palaeobotany and Palynology* 296 (2021): 104549.

7. E. Hadjisterkotis and D. S. Reese, "Considerations on the potential use of cliffs and caves by the extinct endemic late Pleistocene hippopotami and elephants of Cyprus," *European Journal of Wildlife Research* 54 (2008): 122–33.

8. P. Davies and A. M. Lister, "*Palaeoloxodon cypriotes*, the dwarf elephant of Cyprus: size and scaling comparisons with *P. falconeri* (Sicily-Malta) and mainland *P. antiquus*," in *The World of Elephants* (Rome: International Congress Proceedings, 2001), 479–80.

9. Hadjisterkotis and Reese, "Considerations on the potential use of cliffs and caves."

10. Martin, *The Evolution Underground*, and references therein.

11. Martin, *The Evolution Underground*.

12. The Lakota Sioux name Tȟuŋkášila Šákpe ("Six Grandfathers") for the granite outcrop in the Black Hills—renamed "Mount Rushmore"—comes from a vision related by Lakota medicine man Nicolas Black Elk to John G. Neihardt. The transcript was published as a book in 1931, and has been reprinted many times since: B. Elk and J. G. Neihardt, *Black Elk Speaks: Being the Life Story of a Holy Man of the Ogala Sioux* (Albany: State University of New York Press, 2008).

13. E. P. Kiver and D. V. Harris, *Geology of U.S. Parklands* (5th ed.) (New York: John Wiley & Sons, 1999).

14. Of the four US presidents represented on Mount Rushmore, the first born was George Washington (1732) and the last dead was Theodore Roosevelt (1919). Atmospheric carbon dioxide levels in 1732–1735 were 277 parts per million (ppm) and 302 ppm in 1919. This translates to an 8 percent increase during the presidents' cumulative lifespans. When the carving was completed in 1941, carbon dioxide was 310 ppm. Since 1941 it has increased 25 percent (413 ppm), and 37 percent since the death of Theodore Roosevelt. Data from https://www.co2levels.org/.

15. G. Haynes, *Mammoths, Mastodons, and Elephants* (Cambridge: Cambridge University Press, 1993).

16. F. Bibi et al., "Early evidence for complex social structure in Proboscidea from a late Miocene trackway site in the United Arab Emirates," *Biology Letters* 8 (2012): 670–73.

17. Haynes, *Mammoths, Mastodons, and Elephants*.

18. *Mammoth Rocks*, California Department of Parks and Recreation, https://www.parks.ca.gov/?page_id=23566.

19. E. B. Parkman, "Rancholabrean rubbing rocks on California's north coast," *California State Parks, Science Notes Number* 72 (2007): 1–32; E. B. Parkman et al., "Extremely high polish on the rocks of uplifted sea stacks along the north coast of Sonoma County, California, USA," *Mammoth Rocks and the Geology of the Sonoma Coast, Northern California Geological Society Guidebook* (2010).

20. D. D. Alt and D. W. Hyndman, *Roadside Geology of Northern and Central California* (Missoula, MT: Mountain Press Publishing, 2000).

21. Alt and Hyndman, *Roadside Geology*.

22. Alt and Hyndman, *Roadside Geology*.

23. Parkman et al., "Extremely high polish."

24. Parkman et al., "Extremely high polish."

25. G. T. Jefferson, "A catalogue of late Quaternary vertebrates from California: part two, mammals," *Technical Reports Number* 7 (Los Angeles: Natural History Museum of Los Angeles County, 1991); Parkman et al., "Extremely high polish."

26. B. R. Coppedge et al., "Grassland soil depressions: Relict bison wallows or inherent landscape heterogeneity?," *American Midland Naturalist* 142 (1999): 382–92.

27. For a video depicting this behavior, and in juvenile elephants no less, please watch "Baby elephants enjoying a mud bath" (2021): https://vimeo.com/547960736.

28. Parkman, "Rancholabrean rubbing rocks"; Parkman et al., "Extremely high polish."

29. J. E. Lewis and S. Harmand, "An earlier origin for stone tool making: implications for cognitive evolution and the transition to *Homo*," *Philosophical Transactions of the Royal Society, B* 371 (2016): 20150233.

30. M. Domínguez-Rodrigo et al., "Artificial intelligence provides greater accuracy in the classification of modern and ancient bone surface modifications," *Scientific Reports* 10 (2020): 18862.

31. T. A. Surovell et al., "Test of Martin's overkill hypothesis using radiocarbon dates on extinct megafauna," *Proceedings of the National Academy of Sciences* 113 (2016): 886–91.

32. P. Villa et al., "Elephant bones for the Middle Pleistocene toolmaker," *PLOS ONE* 16 (2021): e0256090.

33. K. Sano et al., "A 1.4-million-year-old bone handaxe from Konso,

Ethiopia, shows advanced tool technology in the early Acheulean," *Proceedings of the National Academy of Science* 117 (2020): 18393–400.

34. M. R. Bennett et al., "Evidence of humans in North America during the Last Glacial Maximum," *Science* 373 (2021): 1528–31; R. A. Fariña, "Bone surface modifications, reasonable certainty, and human antiquity in the Americas: the case of the Arroyo Del Vizcaíno site," *American Antiquity* 80 (2017): 193–200.

35. Carvalho et al., "First tracks of newborn straight-tusked elephants (*Palaeoloxodon antiquus*)," *Scientific Reports* 11 (2021): 17311.

36. D. M. Fragaszy et al., "Wild bearded capuchin monkeys (*Sapajus libidinosus*) strategically place nuts in a stable position during nut-cracking," *PLOS ONE* 8 (2013a): e56182; D. M. Fragaszy et al., "The fourth dimension of tool use: temporally enduring artefacts aid primates learning to use tools," *Philosophical Transactions of the Royal Society, B* 368 (2013b): 20120410.

37. Fragaszy et al., "Wild bearded capuchin monkeys."

38. Fragaszy et al., "Wild bearded capuchin monkeys."

39. Fragaszy et al. "Wild bearded capuchin monkeys."

40. T. Falótico et al., "Three thousand years of wild capuchin stone tool use," *Nature Ecology & Evolution* 3: 1034–38.

41. Fragaszy et al., "The fourth dimension of tool use."

42. Martin, *Dinosaurs Without Bones*.

43. Lewis and Harmand, "An earlier origin for stone tool making"; S. P. McPherron et al., "Evidence for stone-tool-assisted consumption of animal tissues before 3.39 million years ago at Dikika, Ethiopia," *Nature* 466 (2010): 857–60. However, some researchers have pointed out that presumed "tool marks" may actually have been from non-hominin causes, such as crocodile bites, trampling, and abrasion. For alternative interpretations, read M. Domínguez-Rodrigo and L. Alcalá, "3.3-million-year-old stone tools and butchery traces? More evidence needed," *PaleoAnthropology* 2016 (2016): 46–53; S. D. Domínguez-Solera et al., "Equids can also make stone artefacts," *Journal of Archaeological Science: Reports* 40 (2021): 103260.

44. G. D. Bader et al., "The forgotten kingdom: new investigations in the prehistory of Eswatini," *Journal of Global Archaeology* 2021 (2021): 1–8.

45. T. Kerig, "Prehistoric mining," *Antiquity* 94 (2020): 802–5.

46. B. C. Black, *Crude Reality: Petroleum in World History* (2nd ed.) (Lanham, MD: Rowman and Littlefield, 2021).

47. P. M. Varley and C. D. Warren, "History of the geological investigations for the Channel Tunnel," in *Engineering Geology of the Channel Tunnel*, ed. C. S. Harris et al. (London: Thomas Telford, 1996), 5–18.

48. Varley and Warren, "History of the geological investigations."

49. C. D. Warren et al., "UK tunnels: geotechnical monitoring and encountered conditions," in *Engineering Geology of the Channel Tunnel*, ed. C. S. Harris et al. (London: Thomas Telford, 1996), 219–43.

50. P. Magron, "General geology and geotechnical considerations," in

Engineering Geology of the Channel Tunnel, ed. C. S. Harris et al. (London: Thomas Telford, 1996), 57–63.

51. "Channel tunnel," Geological Society, https://www.geolsoc.org.uk /GeositesChannelTunnel, accessed November 10, 2021.

52. H. P. Horton et al., "Expert assessment of sea-level rise by AD 2100 and AD 2300," *Quaternary Science Reviews* 84 (2014): 1–6.

Bibliography

Print Sources

Achtalis, M., R. M. van der Zande, C. H. L. Schönberg, J. K. H. Fang, O. Hoegh-Guldberg, et al. "Sponge bioerosion on changing reefs: ocean warming poses physiological constraints to the success of a photosymbiotic excavating sponge," *Scientific Reports* 7 (2017): 10705.

Adam, T. C., A. Duran, C. E. Fuchs, M. V. Roycroft, M. C. Rojas, B. I. Ruttenberg, et al. "Comparative analysis of foraging behavior and bite mechanics reveals complex functional diversity among Caribbean parrotfishes," *Marine Ecology Progress Series* 597 (2018): 207–20.

Agaba, M., E. Ishengoma, W. C. Miller, B. C. McGrath, C. N. Hudson, O. C. B. Reina, et al. "Giraffe genome sequence reveals clues to its unique morphology and physiology," *Nature Communications* 7 (2016): 11519.

Aguilera, O. A., L. Garcia, and M. A. Cozzuol. "Giant-toothed white sharks and cetacean trophic interaction from the Pliocene Caribbean Paraguaná Formation," *Paläontologische Zeitschrift* 82 (2008): 204–8.

Ahrens, W. P. "Naming the Bahamas islands: history and folk etymology," in *Names and Their Environment*, ed. C. Hough and D. Izdebska (Glasgow, *Proceedings of the 25th International Congress of Onomastic Sciences*, 2014): 42–49.

Allwood, A. C., M. R. Walter, B. S. Kamber, C. P. Marshall, and I. W. Burch. "Stromatolite reef from the Early Archaean era of Australia," *Nature* 441 (2006): 714–18.

Alt, D. D., and D. W. Hyndman. *Roadside Geology of Northern and Central California* (Missoula, MT: Mountain Press Publishing, 2000).

Amaral, C. R. L., F. Pereira, D. A. Silva, A. Amorim, and E. F. de Carvalho. "The mitogenomic phylogeny of the Elasmobranchii (Chondrichthyes)," *Mitochondrial DNA Part A* 29 (2018): 867–78.

Amon, D. J., D. Sykes, F. Ahmed, J. T. Copley, K. M. Kemp, P. A. Tyler, et al. "Burrow forms, growth rates and feeding rates of wood-boring Xylophagaidae bivalves revealed by micro-computed tomography," *Frontiers in Marine Science* 2 (2015): 1–10.

Anderson, J. B., J. N. Bruhn, D. Kasimer, H. Wang, N. Rodrigue, and M. L. Smith. "Clonal evolution and genome stability in a 2500-year-old fungal individual," *Proceedings of the Royal Society of London, B* 285 (2018): 20182233.

Anderson, P. S. L., and M. W. Westneat. "Feeding mechanics and bite force modelling of the skull of *Dunkleosteus terrelli*, an ancient apex predator," *Biology Letters* 3 (2007): 76–79.

Anderson, P. S. L., and M. W. Westneat. "A biomechanical model of feeding kinematics for *Dunkleosteus terrelli* (Arthrodira, Placodermi)," *Paleobiology* 35 (2009): 251–69.

Ansell, A. D., and N. B. Nair. "Shell movements of a wood-boring bivalve," *Nature* 216 (1967): 595.

Ansell, A. D., and N. B. Nair. "The mechanisms of boring in *Martesia striata* Linne (Bivalvia: Pholadidae) and *Xylophaga dorsalis* Turton (Bivalvia: Xylophaginidae)," *Proceedings of the Royal Society of London, B* 174 (1969): 123–33.

Armstrong, R. A. "The use of the lichen genus *Rhizocarponin* in lichneometric dating with special reference to Holocene glacial events," in *Advances in Environmental Research*, ed. J. A. Daniels (Hauppauge, NY: Nova Science Publishers, 2015), 23–50.

Arnold, J. M. "Some aspects of hole-boring predation by *Octopus vulgaris*," *American Zoologist* 9 (1969): 991–96.

Asadzadeh, S. S., T. Kiørboe, P. S. Larsen, S. P. Leys, G. Yahel, and J. H. Walther. "Hydrodynamics of sponge pumps and evolution of the sponge body plan," *eLife* 9 (2020): e61012.

Asgaard, U., and R. G. Bromley. 2008. "Echinometrid sea urchins, their trophic styles and corresponding bioerosion," in *Current Developments in Bioerosion*, ed. M. Wisshak and L. Tapanila (Berlin: Springer, 2008), 279–303.

Audley, J. P., C. J. Fettig, A. S. Munson, J. B. Runyon, L. A. Mortenson, B. E. Steed, et al. "Impacts of mountain pine beetle outbreaks on lodgepole pine forests in the Intermountain West, U.S., 2004–2019," *Forest Ecology and Management* 475 (2020): 118403.

Augustin, F. J., A. T. Matzke, M. W. Maisch, J. K. Hinz, and H.-U. Pfretzschner. "The smallest eating the largest: the oldest mammalian feeding traces on dinosaur bone from the Late Jurassic of the Junggar Basin (northwestern China)," *The Science of Nature* 107 (2020): 32.

Augustin, F. J., A. T. Matzke, M. W. Maisch, J. K. Hinz, and H.-U. Pfretzschner. "Dinosaur taphonomy of the Jurassic Shishugou Formation (Northern Junggar Basin, NW China): insights from bioerosional trace fossils on bone," *Ichnos* 28 (2021): 87–96.

Ausich, W. I., T. W. Kammer, E. C. Rhenberg, and D. F. Wright. "Early phylogeny of crinoids within the pelmatozoan clade," *Palaeontology* 58 (2015): 937–52.

Ausich, W. I., and G. D. Webster (eds.). *Echinoderm Paleobiology* (Bloomington: Indiana University Press, 2008).

Baars, C., M. G. Pour, and R. C. Atwood. "The earliest rugose coral," *Geological Magazine* 150 (2012): 371–80.

Backwell, L. R., A. H. Parkinson, E. M. Roberts, F. d'Erricoc, and J.-B. Huchete. "Criteria for identifying bone modification by termites in the fossil record," *Palaeogeography, Palaeoclimatology, Palaeoecology* 337 (2012): 72–87.

Bader, G. D., B. Forrester, L. Ehlers, E. Velliky, B. L. MacDonald, and J. Linstädter. "The forgotten kingdom: new investigations in the prehistory of Eswatini," *Journal of Global Archaeology* 2021 (2021): 1–8.

Bader, K. S., S. T. Hasiotis, and L. D. Martin. "Application of forensic science techniques to trace fossils on dinosaur bones from a quarry in the Upper Jurassic Morrison Formation, northeastern Wyoming," *Palaios* 24 (2009): 140–58.

de Bakker, D. M., A. E. Webb, L. A. van den Bogaart, S. M. A. C. van Heuven, E. H. Meesters, and F. C. van Duyl. "Quantification of chemical and mechanical bioerosion rates of six Caribbean excavating sponge species found on the coral reefs of Curaçao," *PLOS ONE* 13 (2018): e0197824.

Ballell, A., and H. G. Ferrón. "Biomechanical insights into the dentition of megatooth sharks (Lamniformes: Otodontidae)," *Scientific Reports* 11 (2021): 1232.

Balog, S.-J. "Boring thallophytes in some Permian and Triassic reefs: bathymetry and bioerosion," in *Global and Regional Controls on Biogenic Sedimentation. I. Reef Evolution*, ed. J. Reitner et al. (*Göttinger Arbeiten Geologic Paläontologie*, 2, 1996): c305–9.

Bar-Yosef Mayer, D. E., I. Groman-Yaroslavski, O. Bar-Yosef, I. Hershkovitz, A. Kampen-Hasday, B. Vandermeersch, et al. "On holes and strings: earliest displays of human adornment in the Middle Palaeolithic," *PLOS ONE* 15 (2020): e0234924.

Barve, S., C. Riehl, E. L. Walters, J. Haydock, H. L. Dugdale, and W. D. Koenig. "Lifetime reproductive benefits of cooperative polygamy vary for males and females in the acorn woodpecker (*Melanerpes formicivorus*)," *Proceedings of the Royal Society, B* 288 (2021): 20210579.

Baumiller, T. K., M. A. Salamon, P. Gorzelak, R. Mooi, C. G. Messing, and F. J. Gahn. "Post-Paleozoic crinoid radiation in response to benthic predation preceded the Mesozoic marine revolution," *Proceedings of the National Academy of Sciences* 107 (2010): 5893–96.

Behrensmeyer, A. K. "Terrestrial vertebrate accumulations," in *Taphonomy: Releasing the Data Locked in the Fossil Record*, ed. P. A. Allison and D. E. G. Briggs (New York: Plenum Press, 1991), 291–327.

Belaústegui, Z., F. Muñiz, R. Domènech, and J. Martinell. "Ichnogeny and bivalve bioerosion: examples from shell and wood substrates," *Ichnos* 27 (2020): 277–83.

Bellwood, D. R. "Direct estimate of bioerosion by two parrotfish species, *Chlorurus gibbus* and *C. sordidus*, on the Great Barrier Reef, Australia," *Marine Biology* 121 (1995): 419–29.

Bellwood, D. R., and O. Schultz. "A review of the fossil record of the parrotfishes (Labroidei: Scaridae) with a description of a new *Calotomus* species from the Middle Miocene (Badenian) of Austria," *Annals Naturhistorisches Museum Wien* 92 (1991): 55–71.

Bengston, S. "Fungus-like mycelial fossils in 2.4-billion-year-old vesicular basalt," *Nature Ecology & Evolution* 1 (2017): 0141.

Benner, J. B., A. A. Ekdale, and J. M. de Gibert. "Macroborings (*Gastrochaenolites*) in Lower Ordovician hardgrounds of Utah: sedimentologic, paleoecologic, and evolutionary implications," *Palaios* 19 (2004): 543–50.

Bennett, M. R., D. Bustos, J. S. Pigati, K. B. Springer, T. M. Urban, V. T. Holliday, et al. "Evidence of humans in North America during the Last Glacial Maximum," *Science* 373 (2021): 1528–31.

Benson, R. B. J., M. Evans, A. S. Smith, J. Sassoon, S. Moore-Faye, H. F. Ketchum, et al. "A giant pliosaurid skull from the Late Jurassic of England," *PLOS ONE* 8 (2013): e65989.

Benton, M. J. *Vertebrate Paleontology* (4th ed.) (Oxford: Wiley-Blackwell, 2014).

Bentz, B. J., and A. M. Jönsson. 2014. "Modeling bark beetle responses to climate change," in *Bark Beetles: Biology and Ecology of Native and Invasive Species*, ed. F. E. Vega and R. W. Hofstetter (Cambridge, MA: Academic Press, 2014), 533–53.

Berchtold, A. E., and I. M. Côté. "Effect of early exposure to predation on risk perception and survival of fish exposed to a non-native predator," *Animal Behaviour* 164 (2020): 205–16.

Bermúdez-Rochas, D. D., G. Delvene, and J. I. Ruiz-Omeñaca. "Evidence of predation in Early Cretaceous unionoid bivalves from freshwater sediments in the Cameros Basin, Spain," *Lethaia* 46 (2012): 57–70.

Bernardi, G. "The use of tools by wrasses (Labridae)," *Coral Reefs* 31 (2011): 39–39.

Berta, A. *Whales, Dolphins, and Porpoises: A Natural History and Species Guide* (Chicago: University of Chicago Press, 2015).

Bertness, M. D., and C. Cunningham. "Crab shell-crushing predation and gastropod architectural defense," *Journal of Experimental Marine Biology and Ecology* 50 (1981): 213–30.

Bibi, F., B. Kraatz, N. Craig, M. Beech, M. Schuster, and A. Hill. "Early evidence for complex social structure in Proboscidea from a late Miocene trackway site in the United Arab Emirates," *Biology Letters* 8 (2012): 670–73.

Bicknell, R. D. C., J. D. Holmes, G. D. Edgecombe, S. R. Losso, J. Ortega-Hernández, S. Wroe, et al. "Biomechanical analyses of Cambrian euarthropod limbs reveal their effectiveness in mastication and durophagy," *Proceedings of the Royal Society of London, B* 288 (2021): 20202075.

Bicknell, R. D. C., J. A. Ledogar, S. Wroe, B. C. Gutzler, W. H. Watson, and J. R. Paterson. "Computational biomechanical analyses demonstrate similar shell-crushing abilities in modern and ancient arthropods," *Proceedings of the Royal Society, B* 285 (2018): 20181935.

Bicknell, R. D. C., and J. R. Paterson. "Reappraising the early evidence of durophagy and drilling predation in the fossil record: implications for escalation and the Cambrian Explosion," *Biology Reviews* 93 (2018): 754–84.

Bignell, D. E., Y. Roisin, and N. Lo (eds.). *Biology of Termites: A Modern Synthesis* (Dordrecht: Springer, 2011).

Bisno, A. L., G. S. Peter, and E. L. Kaplan. "Diagnosis of strep throat in adults: are clinical criteria really good enough?," *Clinical Infectious Diseases* 35 (2002): 126–29.

Black, B. C. *Crude Reality: Petroleum in World History* (2nd ed.) (Lanham, MD: Rowman and Littlefield, 2021).

Black, R. *The Last Days of the Dinosaurs: An Asteroid, Extinction, and the Beginning of Our World* (New York: St. Martin's Press, 2022).

Blanc, L. A., and J. R. Walters. "Cavity-nest webs in a longleaf pine ecosystem," *The Condor* 110 (2008): 80–92.

Bobrovskiy, I., J. M. Hope, B. J. Nettersheim, J. K. Volkman, C. Hallmann, and J. J. Brocks. "Algal origin of sponge sterane biomarkers negates the oldest evidence for animals in the rock record," *Nature Ecology & Evolution* 5 (2021): 165–68.

Bodył, A., P. Mackiewicz, and P. Gagat. "Organelle evolution: *Paulinella* breaks a paradigm," *Current Biology* 22 (2012): R304R306.

Boessenecker, R. W., D. J. Ehret, D. J. Long, M. Churchill, E. Martin, and S. J. Boessenecker. "The Early Pliocene extinction of the mega-toothed shark *Otodus megalodon*: a view from the eastern North Pacific," *PeerJ* 7 (2019): e6088.

Bolotov, I. N., O. V. Aksenova, T. Bakken, C. J. Glasby, M. Y. Gofarov, A. V. Kondakov, et al. "Discovery of a silicate rock-boring organism and macrobioerosion in fresh water," *Nature Communications* 9 (2018): 2882.

Bonaldo, R. M., and D. R. Bellwood. "Parrotfish predation on massive *Porites* on the Great Barrier Reef," *Coral Reefs* 30 (2011): 259–69.

Bonaldo, R. M., A. S. Hoey, and D. R. Bellwood. "The ecosystem roles of parrotfishes on tropical reefs," *Oceanography and Marine Biology: An Annual Review* 52 (2014): 81–132.

Borges, L. M. S., L. M. Merckelbach, Í. Sampaio, and S. M. Cragg. "Diversity, environmental requirements, and biogeography of bivalve wood-borers (Teredinidae) in European coastal waters," *Frontiers in Zoology* 11 (2014): 13.

Borrero, L. A., and F. M. Martin. "Taphonomic observations on ground sloth bone and dung from Cueva del Milodón, Ultima Esperanza, Chile: 100 years of research history," *Quaternary International* 278 (2012): 3–11.

Bowell, R. J., A. Warren, and I. Redmond. "Formation of cave salts and utilization by elephants in the Mount Elgon region, Kenya," in *Environmental Geochemistry and Health*, ed. J. D. Appleton, R. Fuge, and G. J. H. McCall (London: Geological Society Special Publication No. 113, 1996), 63–79.

Boyd, C. A., S. K. Drumheller, and T. A. Gates. "Crocodyliform feeding traces on juvenile ornithischian dinosaurs from the Upper Cretaceous (Campanian) Kaiparowits Formation, Utah," *PLOS ONE* 8 (2013): e57605.

Brannen, P. *The Ends of the World: Volcanic Apocalypses, Lethal Oceans, and Our Quest to Understand Earth's Past Mass Extinctions* (New York: Ecco, 2017).

Brett, C. E., and S. E. Walker. "Predators and predation in Paleozoic marine environments," in *The Fossil Record of Predation*, ed. M. Kowalewski and P. H. Kelley, *Paleontological Society Papers* 8 (2002): 93118.

Briggs, D. E. G. "The Cambrian explosion," *Current Biology* 25 (2015): R864–68.

Britt, B. B., R. D. Scheetz, and A. Dangerfield. "A suite of dermestid beetle traces on dinosaur bone from the Upper Jurassic Morrison Formation, Wyoming, USA," *Ichnos* 15 (2008): 59–71.

Brochu, C. A. "Alligatorine phylogeny and the status of *Allognathosuchus* Mook, 1921," *Journal of Vertebrate Paleontology* 24 (2004): 857–73.

Brodsley, L., C. Frank, and J. W. Steeds. "Prince Rupert's drops," *Notes and Records of the Royal Society of London* 41 (1986): 1–26.

Brom, K. R., K. Oguri, T. Oji, M. A. Salamon, and P. Gorzelak. "Experimental neoichnology of crawling stalked crinoids," *Swiss Journal of Palaeontology* 137 (2018): 197–203.

Bromley, R. G. "Comparative analysis of fossil and recent echinoid bioerosion," *Palaeontology* 18 (1975): 725–39.

Bromley, R. G. "Concepts in ichnotaxonomy illustrated by small round holes in shells," *Acta Geológica Hispánica* 16 (1981): 55–64.

Bromley, R. G. "Predation habits of octopus past and present and a new ichnospecies, *Oichnus ovalis*," *Bulletin of the Geological Society of Denmark* 40 (1993): 167–73.

Bromley, R. G., and A. D'Alessandro. "Bioerosion in the Pleistocene of southern Italy: ichnogenera *Caulostrepsis* and *Maeandropolydora*," *Rivista Italiana di Paleontologia e Stratigrafia* 89 (1983): 283–309.

Bromley, R. G., and A. D'Alessandro. "Ichnological study of shallow marine endolithic sponges from the Italian coast," *Rivista Italiana di Paleontologia e Stratigrafia* 95 (1989): 279–340.

Bromley, R. G., S. G. Pemberton, and R. A. Rahmani. "A Cretaceous woodground: the *Teredolites* ichnofacies," *Journal of Paleontology* 58 (1984): 488–98.

Bronn, H. G. "*Lethaea geognostica, 2: Das Kreide und Molassen–Gebirge*" (1837–1838), 545–1350.

Brown, C. M., P. J. Currie, and F. Therrien. "Intraspecific facial bite marks in tyrannosaurids provide insight into sexual maturity and evolution of bird-like intersexual display," *Paleobiology* 48 (2021): doi: https://doi .org/10.1017/pab.2021.29.

Brown, M., T. Johnson, and N. J. Gardiner. "Plate tectonics and the Archean Earth," *Annual Review of Earth and Planetary Sciences* 48 (2020): 291–320.

Brownstein, C. D. "Trace fossils on dinosaur bones reveal ecosystem dynamics along the coast of eastern North America during the latest Cretaceous," *PeerJ* 6 (2018): e4973.

Bruno, J. F., I. M. Côté, and L. T. Toth. "Climate change, coral loss, and the curious case of the parrotfish paradigm: why don't marine protected areas improve reef resilience?," *Annual Reviews of Marine Sciences* 11 (2019): 307–34.

Buatois, L. A., M. G. Mángano, N. J. Minter, K. Zhou, M. Wisshak, M. A. Wilson, et al. "Quantifying ecospace utilization and ecosystem engineering during the early Phanerozoic—the role of bioturbation and bioerosion," *Science Advances* 6 (2020): eabb0618.

Buatois, L. A., M. G. Mángano, R. A. Olea, and M. A. Wilson. "Decoupled evolution of soft and hard substrate communities during the Cambrian Explosion and Great Ordovician Biodiversification Event," *Proceedings of the National Academy of Sciences* 113 (2016): 6945–48.

Cadée, G. C. "Size-selective transport of shells by birds and its palaeoecological implications," *Palaeontology* 32 (1989): 429–37.

Cadée, G. C., S. E. Walker, and K. W. Flessa. "Gastropod shell repair in the intertidal of Bahía la Choya (N. Gulf of California)," *Palaeogeography, Palaeoclimatology, Palaeoecology* 136 (1997): 67–78.

Carpenter, K. "Evidence of predatory behavior by carnivorous dinosaurs," *Gaia* 15 (1998): 135–44.

Carpenter, K. "Rocky start of Dinosaur National Monument (USA), the world's first dinosaur geoconservation site," *Geoconservation Research* 1 (2018): 1–20.

Carriker, M. R. "Shell penetration and feeding by naticacean and muricacean predatory gastropods: a synthesis," *Malacologia* 20 (1981): 403–22.

de Carvalho, C. N. (ed.). *Icnologia de Portugal e Transfronteriça* [*Ichnology of Portugal and Cross-Border*], *Comunicações Geológicas* 103 (2016).

de Carvalho, C. N., Z. Belaústegui, A. Toscano, F. Muñiz, J. Belo, J. María Galán, et al. "First tracks of newborn straight-tusked elephants (*Palaeoloxodon antiquus*)," *Scientific Reports* 11 (2021): 17311.

de Carvalho, C. N., B. Pereira, A. Klompmaker, A. Baucon, J. A. Moita, P. Pereira, et al. "Running crabs, walking crinoids, grazing gastropods: behavioral diversity and evolutionary implications in the Cabeço da

Ladeira lagerstätte (Middle Jurassic, Portugal)," *Comunicações Geológicas* 103 (2016): 39–54.

Castro, J. I. *The Sharks of North America* (Oxford: Oxford University Press, 2011).

Catling, D. C., and K. J. Zahnle. "The Archean atmosphere," *Science Advances* 6 (2020): eaax1420.

Chan, B. K. K., N. Dreyer, A. S. Gale, H. Glenner, C. Ewers-Saucedo, M. Pérez-Losada, et al. "The evolutionary diversity of barnacles, with an updated classification of fossil and living forms," *Zoological Journal of the Linnean Society* 193 (2021): 789–846.

Chattopadhyay, D., and S. Dutta. "Prey selection by drilling predators: a case study from Miocene of Kutch, India," *Palaeogeography, Palaeoclimatology, Palaeoecology* 374 (2013): 187–96.

Chattopadhyay, D., V. Gopal, S. Kella, and D. Chattopadhyay. "Effectiveness of small size against drilling predation: insights from lower Miocene faunal assemblage of Quilon Limestone, India," *Palaeogeography, Palaeoclimatology, Palaeoecology* 551 (2020): 109742.

Chattopadhyay, D., D. Sarkar, S. Dutta, and S. R. Prasanjit. "What controls cannibalism in drilling gastropods? A case study on *Natica tigrine*," *Palaeogeography, Palaeoclimatology, Palaeoecology* 410 (2014): 126–33.

Chen, J., H.-P. Blume, and L. Beyer. "Weathering of rocks induced by lichen colonization: a review," *Catena* 39 (2000): 121–46.

Chen, Z.-Q., J. Tong, Z.-T. Liao, and J. Chen. "Structural changes of marine communities over the Permian–Triassic transition: ecologically assessing the end-Permian mass extinction and its aftermath," *Global and Planetary Change* 73 (2010): 123–40.

Cherchi, A., C. Buosi, P. Zuddas, and G. De Giudici. "Bioerosion by microbial euendoliths in benthic foraminifera from heavy metal-polluted coastal environments of Portovesme (south-western Sardinia, Italy)," *Biogeosciences* 9 (2012): 4607–20.

Chiappe, L. M., and M. Qingjin. *Birds of Stone: Chinese Avian Fossils* (Baltimore: Johns Hopkins University Press, 2016).

Chin, K. "The paleobiological implications of herbivorous dinosaur coprolites from the Upper Cretaceous Two Medicine Formation of Montana: why eat wood?," *Palaios* 22 (2007): 554–66.

Chin, K., and B. D. Gill. "Dinosaurs, dung beetles, and conifers: participants in a Cretaceous food web," *Palaios* 11 (1996): 280–85.

Chin, K., R. M. Feldmann, and J. N. Tashman. "Consumption of crustaceans by megaherbivorous dinosaurs: dietary flexibility and dinosaur life history strategies," *Scientific Reports* 7 (2017): 11163.

Chin, K., T. T. Tokaryk, G. M. Erickson, and L. C. Calk. "A king-sized theropod coprolite," *Nature* 393 (1998): 680–82.

Choat, J. H., O. S. Klanten, L. van Herwerden, D. R. Robertson, and K. D. Clements. "Patterns and processes in the evolutionary history of parrotfishes (Family Labridae)," *Biological Journal of the Linnean Society* 107 (2012): 529–57.

Chojnacki, N. C., and L. R. Leighton. "Comparing predatory drillholes to taphonomic damage from simulated wave action on a modern gastropod," *Historical Biology* 26 (2014): 69–79.

Chouvenc, T., J. Šobotník, M. S. Engel, and T. Bourguignon. "Termite evolution: mutualistic associations, key innovations, and the rise of Termitidae," *Cellular and Molecular Life Sciences* 78 (2021): 274969.

Chu, D., J. Tong, H. Song, M. J. Benton, D. J. Bottjer, H. Song, et al. "Early Triassic wrinkle structures on land: stressed environments and oases for life," *Scientific Reports* 5 (2015): 10109.

Çinar, M., and E. Dagli. "Bioeroding (boring) polychaete species (Annelida: Polychaeta) from the Aegean Sea (eastern Mediterranean)," *Journal of the Marine Biological Association of the United Kingdom* 101 (2021): 309–18.

Clarkson, E. N. K. *Invertebrate Palaeontology and Evolution* (4th ed.) (Oxford: Wiley-Blackwell, 2013).

Clelland, E. S., and N. B. Webster. "Drilling into hard substrate by naticid and muricid gastropods: a chemo-mechanical process involved in feeding," in *Physiology of Molluscs*, ed. S. Saleuddin and S. Mukai (Boca Raton, FL: Apple Academic Press, 2021), 1896–916.

Clements, K. D., D. P. German, J. Piché, A. Tribollet, and J. H. Choat 2017. "Integrating ecological roles and trophic diversification on coral reefs: multiple lines of evidence identify parrotfishes as microphages," *Biological Journal of the Linnean Society* 120 (2017): 729–51.

Cockle, K. L., and K. Martin. "Temporal dynamics of a commensal network of cavity-nesting vertebrates: increased diversity during an insect outbreak," *Ecology* 96 (2015): 1093–104.

Colin, J.-P., D. Néraudeau, A. Nel, and V. Perrichot. "Termite coprolites (Insecta: Isoptera) from the Cretaceous of western France: a palaeoecological insight," *Revue de Micropaléontologie* 54 (2011): 129–39.

Conner, R. N., D. C. Rudolph, and J. R. Walters. *The Red-Cockaded Woodpecker: Surviving in a Fire-Maintained Ecosystem* (Austin: University of Texas Press, 2010).

Conner, R. N., D. Saenz, D. C. Rudolph, W. G. Ross, and D. L. Kulhavy. "Red-cockaded woodpecker nest-cavity selection: relationships with cavity age and resin production," *The Auk* 115 (1998): 447–54.

Cooper, J. A., C. Pimiento, H. G. Ferrón, and M. J. Benton. "Body dimensions of the extinct giant shark *Otodus megalodon*: a 2D reconstruction," *Scientific Reports* 10 (2020): 14596.

Cooper, S. L. A., and D. M. Martill. "Pycnodont fishes (Actinopterygii, Pycnodontiformes) from the Upper Cretaceous (lower Turonian) Akrabou Formation of Asfla, Morocco," *Cretaceous Research* 116 (2020): 104607.

Coppedge, B. R., S. D. Fuhlendorf, D. M. Engle, B. J. Carter, and J. H. Shaw. "Grassland soil depressions: relict bison wallows or inherent landscape heterogeneity?," *American Midland Naturalist* 142 (1999): 382–92.

Cossette, A. P., and C. A. Brochu. "A systematic review of the giant alligatoroid *Deinosuchus* from the Campanian of North America and its implications for the relationships at the root of Crocodylia," *Journal of Vertebrate Paleontology* 40 (2020): e1767638.

Cramer, K. L., A. O'Dea, T. R. Clark, J.-X. Zhao, and R. D. Norris. "Prehistorical and historical declines in Caribbean coral reef accretion rates driven by loss of parrotfish," *Nature Communications* 8 (2017): 14160.

Crane, R. L., S. M. Cox, S. A. Kisare, and S. N. Patek. "Smashing mantis shrimp strategically impact shells," *Journal of Experimental Biology* 221 (2018): jeb176099.

Cunningham, J., and N. Herr. *Hands-On Physics Activities and Life Applications: Easy-to-Use Labs and Demonstrations for Grades 8–12* (New York: Wiley, 1994).

Curran, H. A., and B. White. "Ichnology of Holocene carbonate eolianites on San Salvador Island, Bahamas: diversity and significance," in *Proceedings of the 9th Symposium on the Geology of the Bahamas and other Carbonate Regions*, ed. H. A. Curran and J. E. Mylroie (San Salvador, Bahamas: Gerace Research Centre, 1999), 22–35.

Danise, S., and S. Dominici. "A record of fossil shallow-water whale falls from Italy," *Lethaia* 47 (2014): 229–43.

Danise, S., and N. D. Higgs. "Bone-eating *Osedax* worms lived on Mesozoic marine reptile deadfalls," *Biology Letters* 11 (2015): 20150072.

Danise, S., R. J. Twitchett, and K. Matts. "Ecological succession of a Jurassic shallow-water ichthyosaur fall," *Nature Communications* 5 (2014): 4789.

Darwin, C. R. *Living Cirripedia, A Monograph on the Sub-class Cirripedia, with Figures of All the Species. The Lepadidæ; or, Pedunculated Cirripedes* (London: The Ray Society, vol. 1, 1851–1852).

Darwin, C. R. *Living Cirripedia, The Balanidæ, (or Sessile Sirripedes); the Verrucidæ* (London: The Ray Society, vol. 2, 1854).

Das, S. S., S. Mondal, S. Saha, S. Bardhan, and Ranita Saha. "Family Naticidae (Gastropoda) from the Upper Jurassic of Kutch, India and a critical reappraisal of taxonomy and time of origination of the family," *Journal of Paleontology* 93 (2019): 673–84.

Dashper, S. G., and E. C. Reynolds. "Lactic acid excretion by *Streptococcus mutans*," *Microbiology* 142 (1996): 33–39.

Dattilo, B. F., R. L. Freeman, W. Peters, B. Heimbrock, B. Deline, A. J. Martin, et al. "Giants among micromorphs: were Cincinnatian (Ordovician, Katian) small shelly phosphatic faunas dwarfed?," *Palaios* 31 (2016): 55–70.

Davidson, T. M., A. H. Altieri, G. M. Ruiz, and M. E. Torchin. "Bioerosion in a changing world: a conceptual framework," *Ecology Letters* 21 (2018): 422–38.

Davies, N. S., and M. R. Gibling. "Paleozoic vegetation and the Siluro-Devonian rise of fluvial lateral accretion sets," *Geology* 38 (2010): 51–54.

Davies, P., and A. M. Lister. "*Palaeoloxodon cypriotes*, the dwarf elephant of Cyprus: size and scaling comparisons with *P. falconeri* (Sicily-Malta) and mainland *P. antiquus*," in *The World of Elephants* (Rome: International Congress Proceedings, 2001), 479–80.

Davis, R. A., and D. Meyers. *A Sea Without Fish: Life in the Ordovician Sea of the Cincinnati Region* (Bloomington: Indiana University Press, 2009).

Deline, B., J. R. Thompson, N. S. Smith, S. Zamora, I. A. Rahman, S. L. Sheffield, et al. "Evolution and development at the origin of a phylum," *Current Biology* 30 (2020): 1672–79.

Demoulin, C. F., Y. J. Lara, L. Cornet, C. François, D. Baurain, A. Wilmotte, et al. "Cyanobacteria evolution: insight from the fossil record," *Free Radical Biology & Medicine* 140 (2019): 206–23.

DePalma, R. A., D. A. Burnham, L. D. Martin, B. M. Rothschild, and P. L. Larson. "Physical evidence of predatory behavior in *Tyrannosaurus rex*," *Proceedings of the National Academy of Science* 110 (2013): 12560–64.

de Pietri, V. L., A. Manegold, L. Costeur, and G. Mayr. "A new species of woodpecker (Aves; Picidae) from the early Miocene of Saulcet (Allier, France)," *Swiss Journal of Paleontology* 130 (2011): 307–14.

de Souza, E. C. F., C. F. Estevão, A. Turra, F. P. P. Leite, and D. Gorman. "Intraspecific competition drives variation in the fundamental and realized niches of the hermit crab, *Pagurus criniticornis*," *Bulletin of Marine Science* 91 (2015): 343–61.

de Souza Medeiros, S. L., M. M. M. de Paiva, P. H. Lopes, W. Blanco, F. D. de Lima, J. B. C. de Oliveira, et al. "Cyclic alternation of quiet and active sleep states in the octopus," *iScience* 24 (2021): 102223.

Dietl, G. P., and R. R. Alexander. "Borehole site and prey size stereotypy in naticid predation on *Euspira* (*Lunatia*) *heros* Say and *Neverita* (*Polinices*) *duplicata* Say from the southern New Jersey coast," *Journal of Shellfish Research* 14 (1995): 307–14.

Distel, D. L. "The biology of marine wood boring bivalves and their bacterial endosymbionts," *Wood Deterioration and Preservation* 845 (2003): 253–71.

Distel, D. L., M. A. Altamia, Z. Lin, J. R. Shipway, A. Han, I. Forteza, et al. "Discovery of chemoautotrophic symbiosis in the giant shipworm *Kuphus polythalamia* (Bivalvia: Teredinidae) extends wooden-steps theory," *Proceedings of the National Academy of Science* 114 (2017): E3652–58.

Distel, D. L., M. Amin, A. Burgoyne, E. Linton, G. Mamangkey, W. Morrill, et al. "Molecular phylogeny of Pholadoidea Lamarck, 1809 supports a single origin for xylotrophy (wood feeding) and xylotrophic bacterial endosymbiosis in Bivalvia," *Molecular Phylogenetics and Evolution* 61 (2011): 245–54.

Distel, D. L., and S. J. Roberts. "Bacterial endosymbionts in the gills of the deep-sea wood-boring bivalves *Xylophaga atlantica* and *Xylophaga washingtona*," *Biological Bulletin* 192 (1997): 253–61.

Dodd, M. S., D. Papineau, T. Grenne, J. F. Slack, M. Rittner, F. Pirajno, et al. "Evidence for early life in Earth's oldest hydrothermal vent precipitates," *Nature* 543 (2017): 60–64.

Dohrmann, M., and G. Wörheide. "Dating early animal evolution using phylogenomic data," *Scientific Reports* 7 (2017): 3599.

Domínguez-Rodrigo, M., and L. Alcalá. "3.3-million-year-old stone tools and butchery traces? More evidence needed," *PaleoAnthropology* (2016): 46–53.

Domínguez-Rodrigo, M., G. Cifuentes-Alcobendas, B. Jiménez-García, N. Abellán, M. Pizarro-Monzo, E. Organista, et al. "Artificial intelligence provides greater accuracy in the classification of modern and ancient bone surface modifications," *Scientific Reports* 10 (2020): 18862.

Domínguez-Solera, S. D., J.-M. Maíllo-Fernández, E. Baquedano, and M. Domínguez-Rodrigo. "Equids can also make stone artefacts," *Journal of Archaeological Science: Reports* 40 (2021): 103260.

Donovan, S. K. "A plea not to ignore ichnotaxonomy: recognizing and recording *Oichnus* Bromley," *Swiss Journal of Palaeontology* 136 (2017): 369–72.

Donovan, S. K., and G. Hoare. "Site selection of small round holes in crinoid pluricolumnals, Trearne Quarry SSSI (Mississippian, Lower Carboniferous), north Ayrshire, UK," *Scottish Journal of Geology* 55 (2019): 1–5.

Donn, T. F., and M. R. Boardman. "Bioerosion of rocky carbonate coastlines on Andros Island, Bahamas," *Journal of Coastal Research* 4 (1988): 381–94.

Drumheller, S. K., M. R. Stocker, and S. J. Nesbitt. "Direct evidence of trophic interactions among apex predators in the Late Triassic of western North America," *Naturwissenschaften* 101: (2014) 975–87.

Drumheller, S. K., M. R. Stocker, and S. J. Nesbitt. "High frequencies of theropod bite marks provide evidence for feeding, scavenging, and possible cannibalism in a stressed Late Jurassic ecosystem," *PLOS ONE* 15 (2020): e0233115.

Dunlop, J. A., and R. J. Garwood. "Terrestrial invertebrates in the Rhynie chert ecosystem," *Philosophical Transactions of the Royal Society, B* 373 (2018): 20160493.

Ebert, D. A., M. Dando, and S. Fowler. *Sharks of the World: A Complete Guide* (Princeton, NJ: Princeton University Press, 2013).

Edgell, T. C., C. Brazeau, J. W. Grahame, and R. Rochette. "Simultaneous defense against shell entry and shell crushing in a snail faced with the predatory shorecrab *Carcinus maenas*," *Marine Ecology Progress Series* 371 (2008): 191–98.

Edwards, D., K. L. Davies, and L. Axe. "A vascular conducting strand in the early land plant *Cooksonia*," *Nature* 357 (1992): 683–85.

Edwards, D., P. A. Selden, J. B. Richardson, and L. Axe. "Coprolites as evidence for plant-animal interaction in Siluro-Devonian terrestrial ecosystems," *Nature* 377 (1995): 329–31.

Edworthy, A. B., M. K. Trzcinski, K. L. Cockle, K. L. Wiebe, and K. Martin. "Tree cavity occupancy by nesting vertebrates across cavity age," *Journal of Wildlife Management* 82 (2018): 639–48.

El-Hedeny, M., A. El-Sabbagh, and S. A. Farraj. "Bioerosion and encrustation: evidences from the Middle–Upper Jurassic of central Saudi Arabia," *Journal of African Earth Sciences* 134 (2017): 466–75.

Elbroch, M. *Mammal Tracks and Sign: A Guide to North American Species* (Mechanicsburg, PA: Stackpole Books, 2003).

Elbroch, M., and E. Marks. *Bird Tracks and Sign of North America* (Mechanicsburg, PA: Stackpole Books, 2001).

Eleuterio, A. A., M. A. de Jesus, and F. E. Putz. "Stem decay in live trees: heartwood hollows and termites in five timber species in Eastern Amazonia," *Forests* 11 (2020): 1087.

Elk, B., and J. G. Neihardt. *Black Elk Speaks: Being the Life Story of a Holy Man of the Ogala Sioux* (Albany: State University of New York Press, 2008).

Engstrom, R. T., and F. J. Sanders. "Red-cockaded woodpecker foraging ecology in an old-growth longleaf pine forest," *Wilson Bulletin* 109 (1997): 203–17.

Ennos, R. *The Age of Wood: Our Most Useful Material and the Construction of Civilization* (New York: Scribner, 2020).

Enochs, I. C., D. P. Manzello, R. D. Carlton, D. M. Graham, R. Ruzicka, and M. A. Colella. "Ocean acidification enhances the bioerosion of a common coral reef sponge: implications for the persistence of the Florida Reef Tract," *Bulletin of Marine Sciences* 91 (2015): 271–90.

Erickson, G. M., P. M. Gignac, S. J. Steppan, A. K. Lappin, K. A. Vliet, J. D. Brueggen, et al. "Insights into the ecology and evolutionary success of crocodilians revealed through bite-force and tooth-pressure experimentation," *PLOS ONE* 7 (2012): e31781.

Erickson, G. M., and K. H. Olson. "Bite marks attributable to *Tyrannosaurus rex*: preliminary description and implications," *Journal of Vertebrate Paleontology* 16 (1996): 175–78.

Erickson, G. M., S. D. van Kirk, J. Su, M. E. Levenston, W. E. Caler, and D. R. Carter. "Bite-force estimation for *Tyrannosaurus rex* from tooth-marked bones," *Nature* 382 (1996): 706–8.

Erwin, D., and J. Valentine. *The Cambrian Explosion: The Construction of Animal Biodiversity* (Greenwood Village, CO: Roberts and Company, 2013).

Evangelista, D. A., B. Wipfler, O. Béthoux, A. Donath, M. Fujita, M. K. Kohli, et al. "An integrative phylogenomic approach illuminates the evolutionary history of cockroaches and termites (Blattodea)," *Proceedings of the Royal Society, B* 286 (2019): 20182076.

Everhart, M. J. *Oceans of Kansas: A Natural History of the Western Interior Sea* (Bloomington: Indiana University Press, 2017).

Falini, G., S. Albeck, S. Weiner, and L. Aaddadi. "Control of aragonite or calcite polymorphism by mollusk shell macromolecules," *Science* 271 (1996): 67–69.

Falótico, T., T. Proffitt, E. B. Ottoni, R. A. Staff, and M. Haslam. "Three thousand years of wild capuchin stone tool use," *Nature Ecology & Evolution* 3 (2019): 1034–38.

Fang, L. S., and P. Shen. "A living mechanical file: the burrowing mechanism of the coral-boring bivalve *Lithophaga nigra*," *Marine Biology* 97 (1988): 349–54.

Fariña, R. A. "Bone surface modifications, reasonable certainty, and human antiquity in the Americas: the case of the Arroyo Del Vizcaíno site," *American Antiquity* 80 (2017): 193–200.

Farrow, G. E., and J. A. Fyfe. "Bioerosion and carbonate mud production on high-latitude shelves," *Sedimentary Geology* 60 (1988): 281–97.

Fedonkin, M. A., J. G. Gehling, K. Grey, G. M. Narbonne, and P. Vickers-Rich. *The Rise of Animals: Evolution and Diversification of the Kingdom Animalia* (Baltimore: Johns Hopkins University Press, 2007).

Fedonkin, M. A., A. Simonetta, and A. Y. Ivantsov. "New data on *Kimberella*, the Vendian mollusc-like organism (White Sea region, Russia): palaeoecological and evolutionary implications," in *The Rise and Fall of the Ediacaran Biota*, ed. P. Vickers-Rich and P. Komarower (London: *Geological Society of London, Special Publications* 286, 2007), 157–79.

Fedonkin, M. A., and B. M. Waggoner. "The Late Precambrian fossil *Kimberella* is a mollusc-like bilaterian organism," *Nature* 388 (1997): 868–71.

Fejar, O., and T. M. Kaiser. "Insect bone-modification and paleoecology of Oligocene mammal-bearing sites in the Doupov Mountains, northwestern Bohemia," *Palaeontologia Electronica* 8 (2005): Article 8.1.8A.

Feng, Z., J. Wang, R. Rößler, A. Ślipiński, and C. Labandeira. "Late Permian wood-borings reveal an intricate network of ecological relationships," *Nature Communications* 8 (2017): 556.

Ferrón, H. G., B. Holgado, J. J. Liston, C. Martínez-Pérez, and H. Botella. "Assessing metabolic constraints on the maximum body size of actinopterygians: locomotion energetics of *Leedsichthys problematicus* (Actinopterygii, Pachycormiformes)," *Palaeontology* 61 (2018): 775–83.

Field, D. J., A. Bercovici, J. S. Berv, R. Dunn, D. E. Fastovsky, T. R. Lyson, et al. "Early evolution of modern birds structured by global forest collapse at the end-Cretaceous mass extinction," *Current Biology* 28 (2018): 1825–31.

Fin, J. K., T. Tregenza, and M. D. Norman. "Defensive tool use in a coconut-carrying octopus," *Current Biology* 19 (2009): R1069–70.

Finch, B., B. M. Young, R. Johnson, and J. C. Hall. *Longleaf, as Far as the Eye Can See* (Chapel Hill: University of North Carolina Press, 2012).

Fischer, V., M. S. Arkhangelsky, I. M. Stenshin, G. N. Uspensky, N. G. Zverkov, and R. B. J. Benson. "Peculiar macrophagous adaptations in a new Cretaceous pliosaurid," *Royal Society of Open Science* 2 (2015): 150552.

Foster, J., R. K. Hunt-Foster, M. A. Gorman II, K. C. Trujillo, C. Suarez, J. B. McHugh, et al. "Paleontology, taphonomy, and sedimentology of the Mygatt-Moore Quarry, a large dinosaur bonebed in the Morrison

Formation, western Colorado: implications for Upper Jurassic dinosaur preservation modes," *Geology of the Intermountain West* 5 (2018): 23–93.

Foth, C., and O. V. M. Rauhut. "Re-evaluation of the Haarlem *Archaeopteryx* and the radiation of maniraptoran theropod dinosaurs," *BMC Evolutionary Biology* 17 (2017): 236.

Fouilloux, C., E. Ringler, and B. Rojas. "Cannibalism," *Current Biology* 29 (2019): R1295–97.

Fraaije, R. H. B., B. W.M. van Bakel, J. W. M. Jagt, S. Charbonnier, and J.-P. Pezy. "The oldest record of galatheoid anomurans (Decapoda, Crustacea) from Normandy, northwest France," *Neues Jahrbuch für Geologie und Paläontologie—Abhandlungen Band* 292 (2019): 291–97.

Fragaszy, D. M., D. Biro, Y. Eshchar, T. Humle, P. Izar, B. Resende, et al. "The fourth dimension of tool use: temporally enduring artefacts aid primates learning to use tools," *Philosophical Transactions of the Royal Society, B* 368 (2013): 20120410.

Fragaszy, D. M., Q. Liu, B. W. Wright, A. Allen, C. W. Brown, and E. Visalberghi. "Wild bearded capuchin monkeys (*Sapajus libidinosus*) strategically place nuts in a stable position during nut-cracking," *PLOS ONE* 8 (2013): e56182.

Francis, J. E., and B. M. Harland. "Termite borings in Early Cretaceous fossil wood, Isle of Wight, UK," *Cretaceous Research* 27 (2006): 773–77.

Frank, J. H., P. E. Skelley, J. H. Frank, J. R. Ross, and H. Arnett (eds.). *American Beetles, Volume II: Polyphaga: Scarabaeoidea through Curculionoidea* (Boca Raton, FL: CRC Press, 2002).

Freymann, B. P., S. N. de Visser, E. P. Mayemba, and H. Olff. "Termites of the genus *Odontotermes* are optionally keratophagous," *Ecotropica* 13 (2007): 143–47.

Frydl, P., and C. W. Stearn. "Rate of bioerosion by parrotfish in Barbados reef environments," *Journal of Sedimentary Research* 48 (1978): 114958.

Fujii, J. A., K. Ralls, and M. T. Tinker. "Ecological drivers of variation in tool-use frequency across sea otter populations," *Behavioral Ecology* 26 (2015): 519–26.

Gacesa, R., W. C. Dunlap, D. J. Barlow, R. A. Laskowski, and P. F. Long. "Rising levels of atmospheric oxygen and evolution of Nrf2," *Scientific Reports* 6 (2016): 27740.

Gaemers, P. A. M., and B. W. Langeveld. "Attempts to predate on gadid fish otoliths demonstrated by naticid gastropod drill holes from the Neogene of Mill-Langenboom, The Netherlands," *Scripta Geologica* 149 (2015): 159–83.

Gale, A. S., W. J. Kennedy, and D. Martill, "Mosasauroid predation on an ammonite—*Pseudaspidoceras*—from the Early Turonian of southeastern Morocco," *Acta Geologica Polonica* 67 (2017): 31–46.

Gallagher, W. B. "On the last mosasaurs: Late Maastrichtian mosasaurs and the Cretaceous-Paleogene boundary in New Jersey," *Bulletin de la Société Géologique de France* 183 (2012): 145–50.

Gan, T., K. Pang, C. Zhou, G. Zhou, B. Wan, G. Li, et al. "Cryptic terrestrial fungus-like fossils of the early Ediacaran Period," *Nature Communications* 12 (2021): 641.

Garcia, M., F. Theunissen, F. Sèbe, J. Clavel, A. Ravignani, T. Marin-Cudraz, et al. "Evolution of communication signals and information during species radiation," *Nature Communications* 11 (2020): 4970.

Garshelis, D. L., K. V. Noyce, M. A. Ditmer, P. L. Coy, A. N. Tri, T. G. Laske, et al. "Remarkable adaptations of the American black bear help explain why it is the most common bear: a long-term study from the center of its range," in *Bears of the World: Ecology, Conservation and Management*, ed. V. Penteriana and M. Melletti (Cambridge: Cambridge University Press, 2021), 53–62.

Garvie, C. L. "Two new species of Muricinae from the Cretaceous and Paleocene of the Gulf Coastal Plain, with comments on the genus *Odontopolys* Gabb. 1860," *Tulane Studies in Geology and Paleontology* 24 (1991): 87–92.

Gehling, J. G., B. N. Runnegar, and M. L. Droser. "Scratch traces of large Ediacara bilaterian animals," *Journal of Paleontology* 88 (2015): 284–98.

Gerace, D. T. *Life Quest: Building the Gerace Research Centre, San Salvador, Bahamas* (Port Charlotte, FL: Book-Broker Publishers, 2011).

Gerberich, W. W., R. Ballarini, E. D. Hintsala, M. Mishra, J.-F. Molinari, and I. Szlufarska. "Toward demystifying the Mohs hardness scale," *Journal of the American Chemical Society* 98 (2015): 2681–88.

Gignac, P. M., and G. M. Erickson. "The biomechanics behind extreme osteophagy in *Tyrannosaurus rex*," *Scientific Reports* 7 (2017): 2012.

Gilman, S. "The clam that sank a thousand ships," *Hakai Magazine*, December 5, 2016, https://www.hakaimagazine.com/features/clam-sank -thousand-ships/.

Gingras, M. K., I. A. Armitage, S. G. Pemberton, and H. E. Clifton. "Pleistocene walrus herds in the Olympic Peninsula area: trace-fossil evidence of predation by hydraulic jetting," *Palaios* 22 (2007): 539–45.

Gingras, M. K., J. W. Hagadorn, A. Seilacher, S. V. Lalonde, E. Pecoits, D. Petrash, et al. "Possible evolution of mobile animals in association with microbial mats," *Nature Geoscience* 4 (2011): 372–75.

Gingras, M. K., J. A. Maceachern, and R. K. Pickerill. "Modern perspectives on the *Teredolites* ichnofacies: observations from Willapa Bay, Washington," *Palaios* 19 (2004): 79–88.

Glaub, I., S. Golubic, M. Gektidis, G. Radtke, G. Radtke, and K. Vogel. "Microborings and microbial endoliths: geological implications," in *Trace Fossils: Concepts, Problems, Prospects*, ed. W. Miller III (Amsterdam: Elsevier, 2007), 368–81.

Glaub, I., and K. Vogel. "The stratigraphic record of microborings," *Fossils & Strata* 51 (2004): 126–35.

Glaub, I., K. Vogel, and M. Gektidis. "The role of modern and fossil cyanobacterial borings in bioerosion and bathymetry," *Ichnos* 8 (2001): 185–95.

Gleason, F. H., G. M. Gadd, J. I. Pitt, and A. W. D Larkum. "The roles of endolithic fungi in bioerosion and disease in marine ecosystems. I. General concepts," *Mycology* 8 (2017): 205–15.

Gleason, F. H., G. M. Gadd, J. I. Pitt, and A. W. D Larkum. "The roles of endolithic fungi in bioerosion and disease in marine ecosystems. II. Potential facultatively parasitic anamorphic ascomycetes can cause disease in corals and molluscs," *Mycology* 8 (2017): 216–27.

Go, M. C. "A case of human bone modification by ants (Hymenoptera: Formicidae) in the Philippines," *Forensic Anthropology* 1 (2018): 116–23.

Gobalet, K. W. "Morphology of the parrotfish pharyngeal jaw apparatus," *Integrative and Comparative Biology* 29 (1989): 319–31.

Godfrey, S. J., M. Ellwood, S. Groff, and M. S. Verdin. "*Carcharocles*-bitten odontocete caudal vertebrae from the Coastal Eastern United States," *Acta Palaeontologica Polonica* 63 (2018): 463–68.

Godfrey, S. J., J. R. Nance, and N. L. Riker. "*Otodus*-bitten sperm whale tooth from the Neogene of the Coastal Eastern United States," *Acta Paleontologica Polonica* 66 (2021): 1–5.

Godfrey, S. J., and J. B. Smith. "Shark-bitten vertebrate coprolites from the Miocene of Maryland," *Naturwissenschaften* 97 (2010): 461–67.

Gold, D. A., J. Grabenstatter, A. de Mendoza, A. Riesgo, I. Ruiz-Trillo, and R. E. Summons. "Sterol and genomic analyses validate the sponge biomarker hypothesis," *Proceedings of the National Academy of Sciences* 113 (2016): 2684–89.

Golubic, S., T. Le Campion-Alsumard, and S. E. Campbell. "Diversity of marine cyanobacteria," in *Marine Cyanobacteria*, ed. L. Charpy and A. W. D. Larkum (Monaco: *Bulletin de l'Institut Oceanographique* Special Issue 19, 1999): 53–76.

Golubic, S., G. Radtke, and T. Le Campion-Alsumard. "Endolithic fungi in marine ecosystems," *Trends in Microbiology* 13 (2005): 229–35.

Gorzelak, P., L. Rakowicz, M. A. Salamon, and P. Szrek. "Inferred placoderm bite marks on Devonian crinoids from Poland," *Neues Jahrbuch für Geologie und Paläontologie* 259 (2011): 105–12.

Gould, S. J. *Wonderful Life: The Burgess Shale and the Nature of History* (New York: W. W. Norton, 1990).

Gray, M. W. "Lynn Margulis and the endosymbiont hypothesis: 50 years later," *Molecular Biology of the Cell* 28 (2017): 1285–87.

Gregory, M. R. "New trace fossils from the Miocene of Northland, New Zealand, *Rosschachichnus amoeba* and *Piscichnus waitemata*," *Ichnos* 1 (1991): 195–206.

Gregory, M. R., P. F. Balance, G. W. Gibson, and A. M. Ayling. "On how some rays (Elasmobranchia) excavate feeding depressions by jetting water," *Journal of Sedimentary Petrology* 49 (1979): 1125–30.

Grigg, G. *Biology and Evolution of Crocodylians* (Ithaca, NY: Cornell University Press, 2015).

Grutter, A. S., J. G. Rumney, T. Sinclair-Taylor, P. Waldie, and C. E. Franklin.

"Fish mucous cocoons: the 'mosquito nets' of the sea," *Biology Letters* 7 (2010): 292–94.

Guarneri, I., M. Sigovini, and D. Tagliapietra. "A simple method to calculate the volume of shipworm tunnels from radiographs," *International Biodeterioration & Biodegradation* 156 (2021): 105109.

Guida, B. S., and F. Garcia-Pichel. "Extreme cellular adaptations and cell differentiation required by a cyanobacterium for carbonate excavation," *Proceedings of the National Academy of Sciences* 113 (2016): 5712–17.

Gumsley, A. P. "Timing and tempo of the Great Oxidation Event," *Proceedings of the National Academy of Sciences* 114 (2017): 1811–16.

Hadjisterkotis, E., and D. S. Reese. "Considerations on the potential use of cliffs and caves by the extinct endemic late Pleistocene hippopotami and elephants of Cyprus," *European Journal of Wildlife Research* 54 (2008): 122–33.

Haigler, S. M. "Boring mechanism of *Polydora websteri* inhabiting *Crassostrea virginica*," *American Zoologist* 9 (1969): 821.

Hall, L., M. Ryan, and E. Scott. "Possible evidence for cannibalism in the giant arthrodire *Dunkleosteus*, the apex predator of the Cleveland Shale Member (Famennian) of the Ohio Shale," *Journal of Vertebrate Paleontology, Programs and Abstracts Book* (2016): 148.

Hamada, S., and H. D. Slade. "Biology, immunology, and cariogenicity of *Streptococcus mutans*," *Microbiological Review* 44 (1980): 331–84.

Hardisty, M. W. *Lampreys: Life Without Jaws* (London: Forrest Text, 2006).

Harness, R. E., and E. L. Walters. "Woodpeckers and utility pole damage," *IEEE Industry Applications Magazine* 11 (2005): 68–73.

Harper, E. M., and D. S. Wharton. "Boring predation and Mesozoic articulate brachiopods," *Palaeogeography, Palaeoclimatology, Palaeoecology* 158 (2000): 15–24.

Harrington, T. C., H. Y. Yun, S.-S. Lu, H. Goto, and D. N. Aghayeva. "Isolations from the redbay ambrosia beetle, *Xyleborus glabratus*, confirm that the laurel wilt pathogen, *Raffaelea lauricola*, originated in Asia," *Mycologia* 103 (2011): 1028–36.

Harrison, J. S., M. L. Porter, M. J. McHenry, H. E. Robinson, and S. N. Patek. "Scaling and development of elastic mechanisms: the tiny strikes of larval mantis shrimp," *Journal of Experimental Biology* 224 (2021): jeb235465.

Hasiotis, S. T., R. F. Dubiel, P. T. Kay, T. M. Demko, K. Kowalska, and D. McDaniel. "Research update on hymenopteran nests and cocoons, Upper Triassic Chinle Formation, Petrified Forest National Park, Arizona," in *National Park Service Paleontological Research, Technical Report* NPS/—NRGRD/GRDTR-98/01, ed. V. L. Santucci and L. McClelland (1998), 116–21.

Haslam, M., J. Fujii, S. Espinosa, K. Mayer, K. Ralls, M. T. Tinker, et al. "Wild sea otter mussel pounding leaves archaeological traces," *Scientific Reports* 9 (2019): 4417.

Hauff, R. B., and U. Joger. "Holzmaden: prehistoric museum Hauff: a fossil museum since 4 generations (Urweltmuseum Hauff)," in *Paleontological Collections of Germany, Austria and Switzerland*, ed. L. Beck and U. Joger (Cham: Springer, 2018), 325–29.

Hay, W. W. "Toward understanding Cretaceous climate—an updated review," *Science China Earth Sciences* 60 (2017): 5–19.

Haynes, G. *Mammoths, Mastodons, and Elephants* (Cambridge: Cambridge University Press, 1993).

Herbert, G. S., L. B. Whitenack, and J. Y. McKnight. "Behavioural versatility of the giant murex *Muricanthus fulvescens* (Sowerby, 1834) (Gastropoda: Muricidae) in interactions with difficult prey," *Journal of Molluscan Studies* 82 (2016): 357–65.

Hess, H. 1999. "Lower Jurassic Posidonia Shale of southern Germany," in *Fossil Crinoids*, ed. H. Hess, W. I. Ausich, C. E. Brett, and Michaels J. Simms (Cambridge: Cambridge University Press, 1999), 183–96.

Hetherington, A. J., and L. Doland. "Stepwise and independent origins of roots among land plants," *Nature* 561 (2018): 235–38.

Hiemstra, A.-F. "Recognizing cephalopod boreholes in shells and the northward spread of *Octopus vulgaris* Cuvier, 1797 (Cephalopoda, Octopodoidea)," *Vita Malacologica* 13 (2015): 53–56.

Higgs, N. D., A. G. Glover, T. G. Dahlgren, and C. T. S. Little. "Bone-boring worms: characterizing the morphology, rate, and method of bioerosion by *Osedax mucofloris* (Annelida, Siboglinidae)," *Biological Bulletin* 221 (2011): 307–16.

Higgs, N. D., A. G. Glover, T. G. Dahlgren, C. R. Smith, Y. Fujiwara, F. Pradillon, et al. "The morphological diversity of *Osedax* worm borings (Annelida: Siboglinidae)," *Journal of the Marine Biological Association of the United Kingdom* 94 (2014): 1429–39.

Higgs, N. D., C. T. S. Little, A. G. Glover, T. G. Dahlgren, C. R. Smith, and S. Dominici. "Evidence of *Osedax* worm borings in Pliocene (~3 Ma) whale bone from the Mediterranean," *Historical Biology* 24 (2012): 269–77.

Himmi, S. K., T. Yoshimura, Y. Yanase, M. Oya, T. Torigoe, and S. Imazu. "X-ray tomographic analysis of the initial structure of the royal chamber and the nest-founding behavior of the drywood termite *Incisitermes minor*," *Journal of Wood Science* 60 (2014): 435–60.

Holden, A. R., J. M. Harris, and R. M. Timm. "Paleoecological and taphonomic implications of insect-damaged Pleistocene vertebrate remains from Rancho La Brea, Southern California," *PLOS ONE* 8 (2013): e67119.

Hollén, L. I., and A. N. Radford. "The development of alarm call behaviour in mammals and birds," *Animal Behaviour* 78 (2009): 791–800.

Honegger, R., D. Edwards, L. Axe, and C. Strullu-Derrien. "Fertile *Prototaxites taiti*: a basal ascomycete with inoperculate, polysporous asci lacking croziers," *Philosophical Transactions of the Royal Society, B* 373 (2018): 20170146.

Horner, J. R., and D. Lessem. *The Complete T. Rex* (New York: Simon & Schuster, 1993).

Horton, H. P., S. Rahmstorf, S. E. Engelhart, and A. C. Kemp. "Expert assessment of sea-level rise by AD 2100 and AD 2300," *Quaternary Science Reviews* 84 (2014): 1–6.

Howard, J. D., T. V. Mayou, and R. W. Heard. "Biogenic sedimentary structures formed by rays," *Journal of Sedimentary Research* 47 (1977): 339–46.

Howard, K. J., and B. L. Thorne. "Eusocial evolution in termites and Hymenoptera," in *Biology of Termites: A Modern Synthesis*, ed. D. E. Bignell et al. (Dordrecht: Springer, 2011), 97–132.

Hoysted, G. A., J. Kowal, A. Jacob, W. R. Rimington, J. G. Duckett, S. Pressel, et al. "A mycorrhizal revolution," *Current Opinion in Plant Biology* 44 (2018): 1–6.

Hu, Y., J. Meng, Y. Wang, and C. Li. "Large Mesozoic mammals fed on young dinosaurs," *Nature* 433 (2005): 149–52.

Hua, H., B. R. Pratt, and L.-Y. Zhang. "Borings in *Cloudina* shells: complex predator-prey dynamics in the terminal Neoproterozoic," *Palaios* 18 (2003): 454–59.

Huag, C., and J. T. Huag. "The presumed oldest flying insect: more likely a myriapod?," *PeerJ* 5 (2017): e3402.

Huag, C., J. Wiethase, and J. Haug. "New records of Mesozoic mantis shrimp larvae and their implications on modern larval traits in stomatopods," *Palaeodiversity* 8 (2015): 121–33.

Huang, J.-D., R. Motani, D.-Y. Jiang, X.-X. Ren, A. Tintori, O. Rieppel, et al. "Repeated evolution of durophagy during ichthyosaur radiation after mass extinction indicated by hidden dentition," *Scientific Reports* 10 (2020): 7798.

Huber, B. T., K. G. MacLeod, D. K. Watkins, and M. F. Coffin. 2018. "The rise and fall of the Cretaceous hot greenhouse climate," *Global and Planetary Change* 167 (2018): 1–23.

Hunter, A. W., D. Casenove, E. G. Mitchell, and C. Mayers. "Reconstructing the ecology of a Jurassic pseudoplanktonic raft colony," *Royal Society Open Science* 7 (2020): 200142.

Hutchings, J. A., and G. S. Herbert. "No honor among snails: Conspecific competition leads to incomplete drill holes by a naticid gastropod," *Palaeogeography, Palaeoclimatology, Palaeoecology* 379 (2013): 32–38.

Hutchinson, R. *The Spanish Armada* (New York: Thomas Dunne Books, 2013).

Hutson, J. M., C. C. Burke, and G. Haynes. "Osteophagia and bone modifications by giraffes and other large ungulates," *Journal of Archaeological Science* 40 (2013): 4139–49.

Ikejiri, T., Y. Lu, and B. Zhang. "Two-step extinction of Late Cretaceous marine vertebrates in northern Gulf of Mexico prolonged biodiversity loss prior to the Chicxulub impact," *Scientific Reports* 10 (2020): 4169.

Imbeau, L., and A. Desrochers. "Foraging ecology and use of drumming trees by three-toed woodpeckers," *Journal of Wildlife Management* 66 (2020): 222–31.

Irisarri, I., J. E. Uribe, D. J. Eernisse, and R. Zardoya. "A mitogenomic phylogeny of chitons (Mollusca: Polyplacophora)," *BMC Ecology and Evolution* 20 (2020): 22.

Ishikawa, M., T. Kase, H. Tsutsui, and B. Tojo. "Snail versus hermit crabs: a new interpretation of shell-peeling predation on fossil gastropod associations," *Paleontological Research* 8 (2004): 99–108.

Jagt, J. W. M., M. J. M. Deckers, M. De Leebeeck, S. K. Donovan, and E. Nieuwenhuis. "Episkeletozoans and bioerosional ichnotaxa on isolated bones of Late Cretaceous mosasaurs and cheloniid turtles from the Maastricht area, the Netherlands," *Geologos* 26 (2020): 39–49.

Jefferson, G. T. "A catalogue of late Quaternary vertebrates from California: part two, mammals," *Technical Reports Number 7* (Los Angeles: Natural History Museum of Los Angeles County, 1991).

Joester, D., and L. R. Brooker. 2016. "The chiton radula: a model system for versatile use of iron oxides," in *Iron Oxides: From Nature to Applications*, ed. D. Faivre (Hoboken, NJ: Wiley, 2016), 177–206.

Johansson, N. R., U. Kaasalainen, and J. Rikkinen. "Woodpeckers can act as dispersal vectors for microorganisms," *Ecology and Evolution* 11 (2021): 7154–63.

Johnson, S. B., A. Warén, R. W. Lee, Y. Kano, A. Kaim, A. Davis, et al. "*Rubyspira*, new genus and two new species of bone-eating deep-sea snails with ancient habits," *Biological Bulletin* 219 (2010): 166–77.

Jones, A. M., C. Brown, and S. Gardner. "Tool use in the tuskfish *Choerodon schoenleinii*?," *Coral Reef* 30 (2011): 865.

Jones, W. J., S. B. Johnson, G. W. Rouse, and R. C. Vrijenhoek. "Marine worms (genus *Osedax*) colonize cow bones," *Proceedings of the Royal Society, B* 275 (2008): 387–91.

Jung, J.-Y., A. Pissarenko, A. A. Trikanad, D. Restrepo, F. Y. Su, A. Marquez, et al. "A natural stress deflector on the head? Mechanical and functional evaluation of the woodpecker skull bones," *Advanced Theory and Simulation* 2 (2019): 1800152.

Jusino, M. A., D. L. Lindner, M. T. Banik, K. R. Rose, and J. R. Walters. "Experimental evidence of a symbiosis between red-cockaded woodpeckers and fungi," *Proceedings of the Royal Society, B* 283 (2016): 20160106.

Kaim, A. "Chemosynthesis-based associations on Cretaceous plesiosaurid carcasses," *Acta Palaeontologica Polonica* 53 (2008): 97–104.

Kaim, A. "Non-actualistic wood-fall associations from Middle Jurassic of Poland," *Lethaia* 44 (2011): 109–24.

Kaiser, S. I., M. Aretz, and R. T. Becker. "The global Hangenberg Crisis (Devonian–Carboniferous transition): review of a first-order mass extinction," *Geological Society, London, Special Publications* 423 (2015): 387–437.

Kaplan, E. H., and S. L. Kaplan. *A Field Guide to Coral Reefs: Caribbean and Florida* (Boston: Houghton Mifflin, 1999).

Kastelein, R. A., M. Muller, and A. Terlouw. "Oral suction of a Pacific walrus (*Odobenus rosmarus divergens*) in air and under water," *Z. Säugetierkunde* 59 (1994): 105–15.

Kauffman, E. G., and J. K. Sawdo. "Mosasaur predation on a nautiloid from the Maastrichtian Pierre Shale, Central Colorado, Western Interior Basin, United States," *Lethaia* 46 (2013): 180–87.

Kelley, P. H. "Predation by Miocene gastropods of the Chesapeake Group: stereotyped and predictable," *Palaios* 3 (1988): 436–48.

Kempster, R. M., I. D. McCarthy, and S. P. Collin. "Phylogenetic and ecological factors influencing the number and distribution of electroreceptors in elasmobranchs," *Journal of Fish Biology* 80 (2012): 2055–88.

Kerig, T. "Prehistoric mining," *Antiquity* 94 (2020): 802–5.

Kiel, S., J. L. Goedert, W.-A. Kahl, and G. W. Rouse. "Fossil traces of the bone-eating worm *Osedax* in early Oligocene whale bones," *Proceedings of the National Academy of Sciences* 107 (2010): 8656–59.

Kiel, S., W.-A. Kahl, and J. L. Goedert. "*Osedax* borings in fossil marine bird bones," *Naturwissenschaften* 98 (2011): 51–55.

Kiel, S., W.-A. Kahl, and J. L. Goedert. "Traces of the bone-eating annelid *Osedax* in Oligocene whale teeth and fish bones," *Paläontologische Zeitschrift* 87 (2013): 161–67.

Kirkendall, L. R., P. H. W. Biedermann, and B. H. Jordal. "Evolution and diversity of bark and ambrosia beetles," in *Bark Beetles: Biology and Ecology of Native and Invasive Species*, ed. F. Vega and R. Hofstetter (Cambridge, MA: Academic Press, 2014), 85–156.

Kitching, R. L., and J. Pearson. "Prey localization by sound in a predatory intertidal gastropod," *Marine Biology Letters* 2 (1981): 313–21.

Kiver, E. P., and D. V. Harris. *Geology of U.S. Parklands* (5th ed.) (New York: John Wiley & Sons, 1999).

Kleeman, K. "Biocorrosion by bivalves," *Marine Ecology* 17 (1996): 145–58.

Klinger, T. S., and J. M. Lawrence. "The hardness of the teeth of five species of echinoids (Echinodermata)," *Journal of Natural History* 19 (1984): 917–20.

Klippel, W. E., and J. A. Synstelien. "Rodents as taphonomic agents: bone gnawing by brown rats and gray squirrels," *Journal of Forensic Science* 52 (2007): 765–73.

Klompmaker, A. A., H. Karasawa, R. W. Portell, R, H. B. Fraaije, and Y. Ando. "An overview of predation evidence found on fossil decapod crustaceans with new examples of drill holes attributed to gastropods and octopods," *Palaios* 28 (2013): 599–613.

Klompmaker, A. A., and P. H. Kelley. "Shell ornamentation as a likely exaptation: evidence from predatory drilling on Cenozoic bivalves," *Paleobiology* 41 (2015): 187–201.

Klompmaker, A. A., and B. A. Kittle. "Inferring octopodoid and gastropod

behavior from their Plio-Pleistocene cowrie prey (Gastropoda: Cypraeidae)," *Palaeogeography, Palaeoclimatology, Palaeoecology* 567 (2021):
110251.

Klompmaker, A. A., M. Kowalewski, J. W. Huntley, and S. Finnegan. "Increase in predator-prey size ratios throughout the Phanerozoic history
of marine ecosystems," *Science* 356 (2017): 1178–80.

Klompmaker, A. A., and N. H. Landman. "Octopodoidea as predators near
the end of the Mesozoic Marine Revolution," *Biological Journal of the
Linnean Society* 132 (2021): 894–99.

Klompmaker, A. A., R. W. Portell, and H. Karasawa. "First fossil evidence of
a drill hole attributed to an octopod in a barnacle," *Lethaia* 47 (2014):
309–12.

Klompmaker, A. A., R. W. Portell, S. E. Lad, and M. Kowalewski. "The fossil
record of drilling predation on barnacles," *Palaeogeography, Palaeoclimatology, Palaeoecology* 426 (2015): 95–111.

Kluessendorf, J., and P. Doyle. "*Pohlsepia mazonensis*, An early 'octopus' from
the Carboniferous of Illinois, USA," *Palaeontology* 43 (2003): 919–26.

Kolmann, M. A., R. D. Grubbs, D. R. Huber, R. Fisher, N. R. Lovejoy, and G. M.
Erickson. "Intraspecific variation in feeding mechanics and bite force in
durophagous stingrays," *Journal of Zoology* 304 (2018): 225–34.

Kosloski, M. E. "Recognizing biotic breakage of the hard clam, *Mercenaria
mercenaria* caused by the stone crab, *Menippe mercenaria*: An experimental taphonomic approach," *Journal of Experimental Marine Biology
and Ecology* 396 (2011): 115–21.

Kosloski, M. E., and W. D. Allmon. "Macroecology and evolution of a crab
'super predator,' *Menippe mercenaria* (Menippidae), and its gastropod
prey," *Biological Journal of the Linnean Society* 116 (2015): 571–81.

Kroh, A. "Phylogeny and classification of echinoids," *Developments in Aquaculture and Fisheries Science* 43 (2020): 1–17.

Ksepka, D. T. "Feathered dinosaurs," *Current Biology* 30 (2020): R1347–53.

Ksepka, D. T., L. Grande, and G. Mayr. "Oldest finch-beaked birds reveal
parallel ecological radiations in the earliest evolution of passerines,"
Current Biology 29 (2019): 657–63.

Kubicka, A. M., Z. M. Rosin, P. Tryjanowski, and E. Nelson. "A systematic
review of animal predation creating pierced shells: implications for the
archaeological record of the Old World," *PeerJ* 5 (2017): e2903.

Kuratani, S. "Evolution of the vertebrate jaw from developmental perspectives," *Evolution and Development* 14 (2012): 76–92.

Kvitek, R. G., A. K. Fukayama, B. S. Anderson, and B. K. Grimm. "Sea otter
foraging on deep-burrowing bivalves in a California coastal lagoon,"
Marine Biology 98 (1988): 157–67.

Labandeira, C. C. "Deep-time patterns of tissue consumption by terrestrial
arthropod herbivores," *Naturwissenschaften* 100 (2013): 355–64.

Labandeira, C. C. "A paleobiologic perspective on plant–insect interactions,"
Current Opinion in Plant Biology 16 (2013): 414–21.

Labandeira, C. C., S. L. Tremblay, K. E. Bartowski, and L. V. Hernick. "Middle Devonian liverwort herbivory and antiherbivory defense," *New Phytologist* 202 (2014): 247–58.

Lachat, J., and D. Haag-Wackernagel. "Novel mobbing strategies of a fish population against a sessile annelid predator," *Scientific Reports* 6 (2016): 33187.

Lange, I. D., C. T. Perry, K. M. Morgan, R. Roche, C. E. Benkwitt, and N. A. J. Graham. "Site-level variation in parrotfish grazing and bioerosion as a function of species-specific feeding metrics," *Diversity* 12 (2020): 379.

Larson, P. *The Vascular Cambium: Development and Structure* (Berlin: Springer-Verlag, 2012).

Lawson, B. M. *Shelling San Sal: An Illustrated Guide to Common Shells of San Salvador Island, Bahamas*. (San Salvador Island, Bahamas: Gerace Research Centre, 1993).

Laybourne, R. C., D. W. Deedrick, and F. M. Hueber. "Feather in amber is earliest new world fossil of Picidae," *Wilson Bulletin* 106 (1994): 18–25.

Lebedev, O. A., E. Mark-Kurik, V. N. Karatajūtė-Talimaa, E. Lukševičs, and A. Ivanov. "Bite marks as evidence of predation in early vertebrates," *Acta Zoologica* 90 (2009): 344–56.

Legendre, F., A. Nel, G. J. Svenson, T. Robillard, R. Pellens, and P. Grandcolas. "Phylogeny of Dictyoptera: dating the origin of cockroaches, praying mantises and termites with molecular data and controlled fossil evidence," *PLOS ONE* 10 (2015): e0130127.

Lenton, T. M., and S. J. Daines. "Matworld—the biogeochemical effects of early life on land," *New Phytologist* 215 (2016): 531–37.

Le Rossignol, A. P. "Breaking down the mussel (*Mytilus edulis*) shell: which layers affect oystercatchers' (*Haematopus ostralegus*) prey selection?," *Journal of Experimental Marine Biology and Ecology* 405 (2011): 87–92.

Levermann, N., A. Galatius, G. Ehlme, S. Rysgaard, and E. W. Born. "Feeding behaviour of free-ranging walruses with notes on apparent dextrality of flipper use," *BMC Ecology* 3 (2003): 9.

Levi-Setti, R. *The Trilobite Book: A Visual Journey* (Chicago: University of Chicago Press, 2014).

Lewis, J. E., and S. Harmand. "An earlier origin for stone tool making: implications for cognitive evolution and the transition to *Homo*," *Philosophical Transactions of the Royal Society, B* 371 (2016): 20150233.

Li, H., and T. Li. "Bark beetle larval dynamics carved in the egg gallery: a study of mathematically reconstructing bark beetle tunnel maps," *Advances in Difference Equations* 2019 (2019): 513.

Li, Y., K. M. Kocot, N. V. Whelan, S. R. Santos, D. S. Waits, D. J. Thornhill, et al. "Phylogenomics of tubeworms (Siboglinidae, Annelida) and comparative performance of different reconstruction methods," *Zoologica Scripta* 46 (2016): 200–213.

Lloyd, G. T., D. W. Bapst, M. Friedman, and K. E. Davis. "Probabilistic divergence time estimation without branch lengths: dating the origins of dinosaurs, avian flight and crown birds," *Biology Letters* 12 (2016): 0160609.

Lockley, M. *Tracking Dinosaurs: A New Look at an Ancient World* (Cambridge: Cambridge University Press, 1991).

Lopes, R. P., and J. C. Pereira. "Molluskan grazing traces (ichnogenus *Radulichnus* Voigt, 1977) on a Pleistocene bivalve from southern Brazil, with the proposal of a new ichnospecies with *Radulichnus tranversus* connected to chitons (vs. *R. inopinatus*)," *Ichnos* 26 (2019): 141–57.

Loron, C. C., C. François, R. H. Rainbird, E. C. Turner, S. Borensztajn, and E. J. Javaux. "Early fungi from the Proterozoic Era in Arctic Canada," *Nature* 570 (2019): 232–35.

Love, G. D., E. Grosjean, C. Stalvies, D. A. Fike, J. P. Grotzinger, A. S. Bradley, et al. "Fossil steroids record the appearance of Demospongiae during the Cryogenian Period," *Nature* 457 (2008): 718–21.

Ludvigsen, R. "Rapid repair of traumatic injury by an Ordovician trilobite," *Lethaia* 10 (1977): 205–7.

Luebke, A., K. Loza, O. Prymak, P. Dammann, H. O. Fabritius, and M. Epple. "Optimized biological tools: ultrastructure of rodent and bat teeth compared to human teeth," *Bioinspired, Biomimetic and Nanobiomaterials* 8 (2019): 247–53.

Lundberg, J., and D. A. McFarlane. 2006. "Speleogenesis of the Mount Elgon elephant caves," in *Perspectives on Karst Geomorphology, Hydrology, and Geochemistry*, ed. R. S. Harmon and C. Wicks (Boulder, CO: Geological Society of America, 2006), 51–63.

Lundquist, C. A., and W. W. Varnedoe Jr. "Salt ingestion caves," *International Journal of Speleology* 35 (2005): 13–18.

Lyson, T. R., I. M. Miller, A. D. Bercovich, K. Weissenburger, J. Fuentes, C. Clyde, et al. "Exceptional continental record of biotic recovery after the Cretaceous–Paleogene mass extinction," *Science* 366 (2019): 977–83.

MacDonald, D. W., C. Newman, and L. A. Harrington (eds.). *Biology and Conservation of Musteloids* (Oxford: Oxford University Press, 2017).

MacEachern, J. A., S. G. Pemberton, M. K. Gingras, and K. L. Bann. "The ichnofacies paradigm: a fifty-year perspective," in *Trace Fossils: Concepts, Problems, Prospects*, ed. W. M. Miller III (Amsterdam: Elsevier, 2007), 52–77.

MacEachern, J. A., S. G. Pemberton, M. K. Gingras, K. L. Bann, and L. T. Dafoe. "Uses of trace fossils in genetic stratigraphy," in *Trace Fossils: Concepts, Problems, Prospects*, ed. W. M. Miller III (Amsterdam: Elsevier, 2007), 110–34.

Maekawa, K., and C. A. Nalepa. "Biogeography and phylogeny of wood-feeding cockroaches in the genus *Cryptocercus*," *Insects* 2 (2011): 354–68.

Magrath, R. D., B. J. Pitcher, and J. L. Gardner. "A mutual understanding? Interspecific responses by birds to each other's aerial alarm calls," *Behavioral Ecology* 18 (2007): 944–51.

Magron, P. "General geology and geotechnical considerations," in *Engineer-*

ing Geology of the Channel Tunnel, ed. C. S. Harris et al. (London: Thomas Telford, 1996), 57–63.

Malik, W. "Inky's daring escape shows how smart octopuses are," *National Geographic*, April 14, 2016, https://www.nationalgeographic.com /animals/article/160414-inky-octopus-escapes-intelligence.

Mallon, J. C., D. M. Henderson, C. M. McDonough, and W. J. Loughry. "A 'bloat-and-float' taphonomic model best explains the upside-down preservation of ankylosaurs," *Palaeogeography, Palaeoclimatology, Palaeoecology* 497 (2018): 117–27.

Manegold, A., and A. Louchart. "Biogeographic and paleoenvironmental implications of a new woodpecker species (Aves, Picidae) from the early Pliocene of South Africa," *Journal of Vertebrate Paleontology* 32 (2012): 926–38.

Mángano, M. G., L. A. Buatois, M. A. Wilson, and M. L. Droser. "The great Ordovician biodiversification event," in *The Trace-Fossil Record of Major Evolutionary Events*, ed. M. Mángano and L. Buatois (Dordrecht: Springer, 2016), 127–65.

Mann, C. C. *1493: Uncovering the New World Columbus Created* (New York: Vintage Books, 2012).

Marcus, M. A., S. Amini, C. A. Stifler, C.-Y. Sun, N. Tamura, H. A. Bechtel, et al. "Parrotfish teeth: stiff biominerals whose microstructure makes them tough and abrasion-resistant to bite stony corals," *ACS Nano* 11 (2013): 11858–65.

Marsh, P. D. "Dental plaque as a biofilm: the significance of pH in health and caries," *Compendium of Continuing Education in Dentistry* 30 (2009): 76–78.

Marshall, S. A. *Beetles: The Natural History and Diversity of Coleoptera* (Richmond Hill, ON: Firefly Books, 2018).

Martin, A. J. "A Paleoenvironmental Interpretation of the 'Arnheim' Micromorph Fossil Assemblage from the Cincinnatian Series (Upper Ordovician), Southeastern Indiana and Southwestern Ohio" (MS thesis, Miami University, 1986).

Martin, A. J. *Trace Fossils of San Salvador* (San Salvador Island, Bahamas: Gerace Research Centre, 2006).

Martin, A. J. *Life Traces of the Georgia Coast: Revealing the Unseen Lives of Plants and Animals* (Bloomington: Indiana University Press, 2013).

Martin, A. J. *Dinosaurs Without Bones: Dinosaur Lives Revealed by Their Trace Fossils* (New York: Pegasus Books, 2014).

Martin, A. J. *The Evolution Underground: Burrows, Bunkers, and the Marvelous Subterranean World Beneath Our Feet* (New York: Pegasus Books, 2017).

Martin, A. J. *Tracking the Golden Isles: The Natural and Human Histories of the Georgia Coast* (Athens: University of Georgia Press, 2020).

Martin, A. J., and S. T. Hasiotis. "Vertebrate tracks and their significance in the Chinle Formation (Late Triassic), Petrified Forest National Park, Arizona," *National Park Service Paleontological Research* 3 (1998): 38–143.

Martin, J. W., Crandell, K. A., and Felder, D. L. (eds.). *Decapod Crustacean Phylogenetics* (Boca Raton, FL: CRC Press, 2016).

Martin, L. D., and D. L. West. "The recognition and use of dermestid (Insecta, Coleoptera) pupation chambers in paleoecology," *Palaeogeography, Palaeoclimatology, Palaeoecology* 113 (1995): 303–10.

Martin, T., G. Sun, and V. Mosbrugger. "Triassic-Jurassic biodiversity, ecosystems, and climate in the Junggar Basin, Xinjiang, Northwest China," *Palaeobiodiversity and Palaeoenvironments* 90 (2010): 171–73.

Martinell, J., J. M. de Gibert, R. Domènech, A. A. Ekdale, and P. Steen. "Cretaceous ray traces? an alternative interpretation for the alleged dinosaur tracks of La Posa, Isona, NE Spain," *Palaios* 16 (2001): 409–16.

Maxwell, E. E., and M. W. Caldwell. "First record of live birth in Cretaceous ichthyosaurs: closing an 80 million year gap," *Proceedings of the Royal Society, B* 270 (2003): S104–7.

Mayr, G. "A tiny barbet-like bird from the Lower Oligocene of Germany: the smallest species and earliest substantial fossil record of the Pici (woodpeckers and allies)," *Auk* 122 (2005): 1055–63.

McClain, C. R., C. Nunnally, R. Dixon, G. W. Rouse, and M. Benfield. "Alligators in the abyss: the first experimental reptilian food fall in the deep ocean," *PLOS ONE* 14 (2019): e0225345.

McCollum, T. M. "Miller-Urey and beyond: what have we learned about prebiotic organic synthesis reactions in the past 60 years?," *Annual Review of Earth and Planetary Sciences* 41 (2013): 207–29.

McCullough, J. M., R. G. Moyle, B. T. Smith, and M. J. Andersen. "A Laurasian origin for a pantropical bird radiation is supported by genomic and fossil data (Aves: Coraciiformes)," *Proceedings of the Royal Society, B* 286 (2019): 20190122.

McHorse, B. K., J. D. Orcutt, and E. B. Davis. "The carnivoran fauna of Rancho La Brea: average or aberrant?," *Palaeogeography, Palaeoclimatology, Palaeoecology* 329–30 (2012): 118–23.

McHugh, J. B., S. K. Drumheller, A. Riedel, and M. Kane. "Decomposition of dinosaurian remains inferred by invertebrate traces on vertebrate bone reveal new insights into Late Jurassic ecology, decay, and climate in western Colorado," *PeerJ* 8 (2020): e9510.

McKinney, F. K., and J. B. C. Jackson. *Bryozoan Evolution* (Chicago: University of Chicago Press, 1989).

McNassor, C. *Images of America: Los Angeles's La Brea Tar Pits and Hancock Park* (Charleston, SC: Arcadia Publishing, 2011).

McPherron, S. P., Z. Alemseged, C. W. Marean, J. G. Wynn, D. Reed, D. Geraads, et al. "Evidence for stone-tool-assisted consumption of animal tissues before 3.39 million years ago at Dikika, Ethiopia," *Nature* 466 (2010): 857–60.

Meadows, C. A., R. E. Fordyce, and T. Baumiller. "Drill holes in the irregular echinoid, *Fibularia*, from the Oligocene of New Zealand," *Palaios* 30 (2015): 810–17.

Mech, L. D., and R. O. Peterson. "Wolf-prey relations," in *Wolves: Behavior, Ecology, and Conservation*, ed. L. D. Mech and L. Boitani (Chicago: University of Chicago Press, 2010).

Meckel, L. A., C. P. McDaneld, and D. J. Wescott. "White-tailed deer as a taphonomic agent: photographic evidence of white-tailed deer gnawing on human bone," *Journal of Forensic Sciences* 63 (2018): 292–94.

Meeker, J. R., W. N. Dixon, J. L. Foltz, and T. R. Fasulo. "The Southern pine beetle *Dendroctonus frontalis* Zimmerman (Coleoptera: Scolytidae)," *Florida Department of Agricultural and Consumer Services, Entomology Circular* 369 (1995): 1–4.

Melnyk, S., S. Packer, J.-P. Zonneveld, and M. K. Gingras. "A new marine woodground ichnotaxon from the Lower Cretaceous Mannville Group, Saskatchewan, Canada," *Journal of Paleontology* 95 (2020): 162–69.

Méndez, D. "Shipwrecked by worms, saved by canoe: the last voyage of Columbus," in *The Ocean Reader: History, Culture, Politics*, ed. E. P. Roordia (Durham, NC: Duke University Press, 2020), 297–304.

Meyer, C. A., and B. Thüring. "Dinosaurs of Switzerland," *Comptes Rendus Palevol* 2 (2003): 103–17.

Meyer, N., M. Wisshak, and A. Freiwald. "Ichnodiversity and bathymetric range of microbioerosion traces in polar barnacles of Svalbard," *Polar Research* 39 (2020): 3766.

Mikuláš, R., and B. Zasadil. "A probable fossil bird nest, ?*Eocavum* isp., from the Miocene wood of the Czech Republic," *4th International Bioerosion Workshop Abstract Book* (Prague, Czech Republic, 2004), 49–51.

Mironenko, A. "A hermit crab preserved inside an ammonite shell from the Upper Jurassic of central Russia: implications to ammonoid palaeoecology," *Palaeogeography, Palaeoclimatology, Palaeoecology* 537 (2020): 109397.

Misof, B., S. Liu, K. Meusmann, R. S. Peters, A. Donath, C. Mayer, et al. "Phylogenomics resolves the timing and pattern of insect evolution," *Science* 346 (2014): 6210.

Mokady, O., B. Lazar, and Y. Loya. "Echinoid bioerosion as a major structuring force of Red Sea coral reefs," *Biological Bulletin* 190 (2006): 367–72.

Monaco, P., F. FaMiani, R. Bizzarri, and A. Baldanza. "First documentation of wood borings (*Teredolites* and insect larvae) in Early Pleistocene lower shoreface storm deposits (Orvieto area, central Italy)," *Bollettino della Società Paleontologica Italiana* 50 (2011): 55–63.

Mondal, S., H. Chakraborty, and S. Paul. "Latitudinal patterns of gastropod drilling predation through time," *Palaios* 34 (2019): 261–70.

Mondal, S., P. Goswami, and S. Bardhan. "Naticid confamilial drilling predation through time," *Palaios* 32 (2017): 278–87.

Moore, M. J., G. H. Mitchell, T. K. Rowles, and G. Early. "Dead cetacean? Beach, bloat, float, sink," *Frontiers in Marine Science* 7 (2020): 333.

Morgan, K. M., and P. S. Kench. "Parrotfish erosion underpins reef growth, sand talus development and island building in the Maldives," *Sedimentary Geology* 341 (2016): 50–57.

Morris, S. C., and R. J. F. Jenkins. "Healed injuries in Early Cambrian trilobites from South Australia," *Alcheringa* 9 (1985): 167–77.

Morrison, S. M., S. E. Runyon, and R. M. Hazen. "The paleomineralogy of the Hadean Eon revisited," *Life* 8 (2018): 64.

Moss, C. J., H. Croze, and P. C. Lee (eds.). *The Amboseli Elephants: A Long-Term Perspective on a Long-Lived Mammal* (Chicago: University of Chicago Press, 2011).

Mounika, S., and J. M. Nithya. "Association of *Streptococcus mutans* and *Streptococcus sanguis* in act of dental caries," *Journal of Pharmaceutical Sciences and Research* 7 (2015): 764–66.

Murdock, D. J. E. "The 'biomineralization toolkit' and the origin of animal skeletons," *Biological Reviews* 95 (2020): 1372–92.

Nalepa, C. A. "Origin of termite eusociality: trophallaxis integrates the social, nutritional, and microbial environments," *Ecological Entomology* 40 (2015): 323–35.

Nash, T. H. *Lichen Biology* (Cambridge: Cambridge University Press, 2008).

Nauer, P. A., L. B. Hutley, and S. K. Arndt. "Termite mounds mitigate half of termite methane emissions," *Proceedings of the National Academy of Sciences* 115 (2018): 13306–11.

Nebelsick, J. H., and M. Kowalewski. "Drilling predation on recent clypeasteroid echinoids from the Red Sea," *Palaios* 14 (1999): 127–44.

Neenan, J. M., N. Klein, and T. M. Scheyer. "European origin of placodont marine reptiles and the evolution of crushing dentition in Placodontia," *Nature Communications* 4 (2013): 1621.

Neenan, J. M., C. Li, O. Rieppel, F. Bernardini, C. Tuniz, G. Muscio, et al. "Unique method of tooth replacement in durophagous placodont marine reptiles, with new data on the dentition of Chinese taxa," *Journal of Anatomy* 224 (2014): 603–13.

Nelson, D. L. "The ravages of *Teredo*: the rise and fall of shipworm in US history, 1860–1940," *Environmental History* 21 (2016): 100–124.

Neto, V. D. P., A. H. Parkinson, F. A. Pretto, M. B. Soares, C. Schwanke, C. L. Schultz, et al. "Oldest evidence of osteophagic behavior by insects from the Triassic of Brazil," *Palaeogeography, Palaeoclimatology, Palaeoecology* 453 (2016): 30–41.

Neumann, A. C. "Observations on coastal erosion in Bermuda and measurements of the sponge, *Cliona lampa*," *Limnology and Oceanography* 11 (1966): 92–108.

Nichols, G. *Sedimentology and Stratigraphy* (2nd ed.) (Oxford: Wiley-Blackwell, 2013).

Nielsen, J., R. B. Hedeholm, J. Heinemeier, P. G. Bushnell, J. S. Christiansen, J. Olsen, et al. "Eye lens radiocarbon reveals centuries of longevity in the Greenland shark (*Somniosus microcephalus*)," *Science* 353 (2016): 702–4.

Nixon, M., and J. Z. Young. *The Brains and Lives of Cephalopods* (Oxford: Oxford University Press, 2003).

Nützel, A. "Gastropods as parasites and carnivorous grazers: a major guild in marine ecosystems," in *The Evolution and Fossil Record of Parasitism*, ed. K. De Baets and J. W. Huntley, *Topics in Geobiology* 49 (Cham: Springer, 2021), 209–29.

O'Connor, J. K., Y. Zhang, L. M. Chiappe, Q. Meng, L. Quanguo, and L. Di. "A new enantiornithine from the Yixian formation with the first recognized avian enamel specialization," *Journal of Vertebrate Paleontology* 33 (2013): 1–12.

Odes, E. J., A. H. Parkinson, P. S. Randolph-Quinney, B. Zipfel, K. Jakata, H. Bonney, et al. "Osteopathology and insect traces in the *Australopithecus africanus* skeleton StW 431," *South African Journal of Science* 113 (2017): 1–7.

O'Shea, O. R., M. Thums, M. van Keulen, and M. G. Meekan. "Bioturbation by stingrays at Ningaloo Reef, Western Australia," *Marine and Freshwater Research* 63 (2011): 189–97.

de Oliveira, T. B., E. Gomes, and A. Rodrigues. "Thermophilic fungi in the new age of fungal taxonomy," *Extremophiles* 19 (2015): 31–37.

Oren, A., and S. Ventura. "The current status of cyanobacterial nomenclature under the 'prokaryotic' and the 'botanical' code,'" *Antonie Van Leeuwenhoek* 110 (2017): 1257–69.

Osinga, R., M. Schutter, B. Griffioen, R. H. Wijffels, J. A. J. Verreth, S. Shafir, et al. "The biology and economics of coral growth," *Marine Biotechnology* 13 (2011): 658–71.

Ozeki, C. S., D. M. Martill, R. E. Smith, and N. Ibrahim. "Biological modification of bones in the Cretaceous of North Africa," *Cretaceous Research* 114 (2020): 104529.

Pahari, A., S. Mondal, S. Bardhan, D. Sarkar, S. Saha, and D. Buragohain. "Subaerial naticid gastropod drilling predation by *Natica tigrina* on the intertidal molluscan community of Chandipur, eastern coast of India," *Palaeogeography, Palaeoclimatology, Palaeoecology* 451 (2016): 110–23.

Pallardy, S. G. *The Woody Plant Body* (3rd ed.) (Cambridge, MA: Academic Press, 2008).

Palma, P., and L. N. Samthakumaran. *Shipwrecks and Global "Worming"* (Oxford: Archaeopress, 2014).

Pantazidou, A., I. Louvrou, and A. Economou-Amilli. "Euendolithic shell-boring cyanobacteria and chlorophytes from the saline lagoon Ahivadolimni on Milos Island, Greece," *European Journal of Phycology* 41 (2006): 189–200.

Park, Y., and J. Choe. "Territorial behavior of the Korean wood-feeding cockroach, *Cryptocercus kyebangensis*," *Journal of Ethology* 21 (2003): 79–85.

Park, Y., P. Grandcolas, and J. C. Choe. "Colony composition, social behavior and some ecological characteristics of the Korean wood-feeding cockroach (*Cryptocercus kyebangensis*)," *Zoological Science* 19 (2002): 1133–39.

Parker, W. G., S. R. Ash, and R. B. Irmis (eds.). *A Century of Research at Petrified Forest National Park* (Flagstaff: Museum of Northern Arizona Bulletin No. 62, 2006).

Parkman, E. B. "Rancholabrean rubbing rocks on California's north coast," *California State Parks, Science Notes Number 72* (2007): 1–32.

Parkman, E. B., T. McKernan, S. Norwick, and R. Erickson. "Extremely high polish on the rocks of uplifted sea stacks along the north coast of Sonoma County, California, USA," *Mammoth Rocks and the Geology of the Sonoma Coast, Northern California Geological Society Guidebook* (2010).

Pechenik, J. A., J. Hsieh, S. Owara, P. Wong, D. Marshall, S. Untersee, et al. "Factors selecting for avoidance of drilled shells by the hermit crab *Pagurus longicarpus*," *Journal of Experimental Marine Biology and Ecology* 262 (2001): 75–89.

Pechenik, J. A., and S. Lewis. "Avoidance of drilled gastropod shells by the hermit crab *Pagurus longicarpus* at Nahant, Massachusetts," *Journal of Experimental Marine Biology and Ecology* 253 (2000): 17–32.

Pech-Puch, D., H. Cruz-López, C. Canche-Ek, G. Campos-Espinosa, E. García, M. Mascaro, et al. "Chemical tools of *Octopus maya* during crab predation are also active on conspecifics," *PLOS ONE* 11 (2016): e0148922.

Peris, D., X. Delclòs, and B. Jordal. "Origin and evolution of fungus farming in wood-boring Coleoptera—a palaeontological perspective," *Biological Reviews* 96 (2021): 2476–88.

Perry, C. T., P. S. Kench, M. J. O'Leary, K. M. Morgan, and F. Januchowski-Hartley. "Linking reef ecology to island building: parrotfish identified as major producers of island-building sediment in the Maldives," *Geology* 43 (2015): 503–6.

Pether, J. "*Belichnus* new ichnogenus, a ballistic trace on mollusc shells from the Holocene of the Benguela region, South Africa," *Journal of Paleontology* 69 (1995): 171–81.

Pier, J. "The Devonian monster of the deep," *Palaeontologia Electronica* blog post, related to Z. Johanson et al., "Fusion in the vertebral column of the pachyosteomorph arthrodire *Dunkleosteus terrelli* ('Placodermi')," *Palaeontologia Electronica* 22.2.20 (2019), https://palaeo-electronica.org/content/2011-11-30-22-01-23/2528-the-devonian-monster-of-the-deep.

Pilson, M. E. Q., and P. B. Taylor. "Hole drilling by *Octopus*," *Science* 134 (1961): 1366–68.

Platt, S. G., J. B. Thorbjarnarson, T. R. Rainwater, and D. R. Martin. "Diet of the American crocodile (*Crocodylus acutus*) in marine environments of coastal Belize," *Journal of Herpetology* 47 (2013): 1–10.

Pobiner, B. "Paleoecological information in predator tooth marks," *Journal of Taphonomy* 6 (2008): 373–97.

Pokines, J. T., S. A. Santana, J. D. Hellar, P. Bian, A. Downs, N. Wells, et al. "The taphonomic effects of eastern gray squirrels (*Sciurus carolinensis*) gnawing on bone," *Journal of Forensic Identification* 66 (2016): 349–75.

Polcyn, M. J., L. L. Jacobs, R. Araújoa, A. S. Schulp, and O. Mateus. "Physical drivers of mosasaur evolution," *Palaeogeography, Palaeoclimatology, Palaeoecology* 400 (2014): 17–27.

Pomponi, S. A. "Cytological mechanisms of calcium carbonate excavation by boring sponges," *International Review of Cytology* 65 (1980): 301–19.

Power, A. M., W. Klepal, V. Zheden, J. Jonker, P. McEvilly, and J. von Byern. "Mechanisms of adhesion in adult barnacles," in *Biological Adhesive Systems*, ed. J. von Byern and I. Grunwald (Vienna: Springer, 2010), 153–68.

Pruss, S. B., C. L. Blättler, F. A. Macdonald, and J. A. Higgins. "Calcium isotope evidence that the earliest metazoan biomineralizers formed aragonite shells," *Geology* 46 (2018): 763–66.

Pryor, K. J., and A. M. Milton. "Tool use by the graphic tuskfish *Choerodon graphicus*," *Journal of Fish Biology* 95 (2019): 663–67.

Pu, J. P., S. A. Bowring, J. Ramezani, P. Myrow, T. D. Raub, E. Landing, et al. "Dodging snowballs: geochronology of the Gaskiers glaciation and the first appearance of the Ediacaran biota," *Geology* 44 (2016): 955–58.

Püntener, C., J.-P. Billon-Bruyat, D. Marty, and G. Paratte. "Under the feet of sauropods: a trampled coastal marine turtle from the Late Jurassic of Switzerland?," *Swiss Journal of Geosciences* 112 (2019): 507–15.

Pyenson, N. D. "The ecological rise of whales chronicled by the fossil record," *Current Biology* 27 (2017): R558–64.

Rasmussen, H. W. "Function and attachment of the stem in Isocrinidae and Pentacrinitidae: review and interpretation," *Lethaia* 10 (1977): 51–57.

Rayes, C. A., J. Beattie, and I.C. Duggan. "Boring through history: an environmental history of the extent, impact and management of marine woodborers in a global and local context, 500 BCE to 1930s CE," *Environment and History* 21 (2015): 477–512.

Reich, M., and A. B. Smith. "Origins and biomechanical evolution of teeth in echinoids and their relatives," *Palaeontology* 52 (2009): 1149–68.

Reynolds, C., N. A. F. Miranda, and G. S. Cumming. "The role of waterbirds in the dispersal of aquatic alien and invasive species," *Diversity and Distributions* 21 (2015): 744–54.

Reynolds, S., and J. Johnson. *Exploring Geology* (4th ed.) (New York: McGraw-Hill Education, 2016).

Rice, M. M., R. L. Maher, A. M. S. Correa, H. V. Moeller, N. P. Lemoine, A. A. Shantz, D. E. Burkepile, et al. "Macroborer presence on corals increases with nutrient input and promotes parrotfish bioerosion," *Coral Reefs* 39 (2020): 409–18.

Rindali, C., and T. M. Cole III. "Environmental seasonality and incremental growth rates of beaver (*Castor canadensis*) incisors: implications for palaeobiology," *Palaeogeography, Palaeoclimatology, Palaeoecology* 206 (2004): 289–301.

Ritter, H. Jr. "Defense of mate and mating chamber in a wood roach," *Science* 143 (1964): 1459–60.

Roberts, E. M., C. N. Todd, D. K. Aanen, T. Nobre, H. L. Hilbert-Wolf, P. M. O'Connor, et al. "Oligocene termite nests with in situ fungus gardens from the Rukwa Rift Basin, Tanzania, support a Paleogene African origin for insect agriculture," *PLOS ONE* 11 (2016): e0156847.

Röhl, H.-J., A. Schmid-Röhl, W. Oschmann, A. Frimmel, and L. Schwark. "The Posidonia Shale (Lower Toarcian) of SW-Germany: an oxygen-depleted ecosystem controlled by sea level and palaeoclimate," *Palaeogeography, Palaeoclimatology, Palaeoecology* 165 (2001): 27–52.

Rohr, D. M., A. J. Boucot, J. Miller, and M. Abbott. "Oldest termite nest from the Upper Cretaceous of west Texas," *Geology* 14 (1986): 87–88.

Rouse, G. W., S. K. Goffredi, S. B. Johnson, and R. C. Vrijenhoek. "Not whale-fall specialists, *Osedax* worms also consume fishbones," *Biology Letters* 7 (2011): 736–39.

Rouse, G. W., N. G. Wilson, K. Worsaae, and R. C. Vrijenhoek. "A dwarf male reversal in bone-eating worms," *Current Biology* 25 (2015): 236–41.

Rouse, R. W., and F. Pleijel. *Polychaetes* (Oxford: Oxford University Press, 2001).

Rudkin, D. M., G. A. Young, and G. S. Nowlan. "The oldest horseshoe crab: a new xiphosurid from Late Ordovician konservat-lagerstätten deposits, Manitoba, Canada," *Palaeontology* 51 (2008): 1–9.

Rudolph, D. C., and R. N. Conner. "Cavity tree selection by red-cockaded woodpeckers in relation to tree age," *Wilson Bulletin* 103 (1991): 458–67.

Rutledge, K. M., A. P. Summers, and M. A. Kolmann. "Killing them softly: ontogeny of jaw mechanics and stiffness in mollusk-feeding freshwater stingrays," *Journal of Morphology* 280 (2018): 796–808.

Rützler, K. "The role of burrowing sponges in bioerosion," *Oecologia* 19 (1975): 203–16.

Sagan, L. "On the origin of mitosing cells," *Journal of Theoretical Biology* 14 (1967): 225–74.

Saha, R., S. Paul, S. Mondal, S. Bardhan, S. S. Das, S. Saha, et al. "Gastropod drilling predation in the Upper Jurassic of Kutch, India," *Palaios* 36 (2021): 301–12.

Salamon, M. A., P. Gerrienne, P. Steemans, P. Gorzelak, P. Filipiak, A. Le Hérissé, et al. "Putative Late Ordovician land plants," *New Phytologist* 218 (2018): 1305–9.

Sander, P. M., X. Chen, L. Cheng, and X. Wang. "Short-snouted toothless ichthyosaur from China suggests Late Triassic diversification of suction feeding ichthyosaurs," *PLOS ONE* 6 (2011): e19480.

Sano, K., Y. Beyene, S. Katoh, D. Koyabu, H. Endo, T. Sasaki, et al. "A 1.4-million-year-old bone handaxe from Konso, Ethiopia, shows advanced tool technology in the early Acheulean," *Proceedings of the National Academy of Science* 117 (2020): 18393–400.

Sansom, I. J., N. S. Davies, M. I. Coates, R. Nicoll, and A. Ritchie. "Chondrichthyan-like scales from the Middle Ordovician of Australia," *Palaeontology* 55 (2012): 243–47.

Santos, A., E. Mayoral, C. P. Dumont, C. M. da Silva, S. P. Ávila, B. G. Baarli, et al. "Role of environmental change in rock-boring echinoid trace fossils," *Palaeogeography, Palaeoclimatology, Palaeoecology* 432 (2015): 1–14.

Santos, A., E. Mayoral, C. M. da Silva, and M. Cachão. "Two remarkable examples of Portuguese Neogene bioeroded rocky shores: new data and synthesis," *Comunicações Geológicas* 103 (2016): 121–30.

Sato, K., and R. G. Jenkins. "Mobile home for pholadoid boring bivalves: first example from a Late Cretaceous sea turtle in Hokkaido Japan," *Palaios* 35 (2020): 228–36.

Saunders, W. B., R. L. Knight, and P. N. Bond. "*Octopus* predation on *Nautilus*: evidence from Papua New Guinea," *Bulletin of Marine Science* 49 (1991): 280–87.

Savoca, M. S., M. F. Czapanskiy, S. R. Kahane-Rapport, W. T. Gough, J. A. Fahlbusch, K. C. Bierlich, et al. "Baleen whale prey consumption based on high-resolution foraging measurements," *Nature* 599 (2021): 85–90.

Savrda, C. E., J. Counts, O. McCormick, R. Urash, and J. Williams. "Loggrounds and *Teredolites* in transgressive deposits, Eocene Tallahatta Formation (southern Alabama, USA)," *Ichnos* 12 (2005): 47–57.

Schirrmeister, B. E. "Cyanobacteria and the Great Oxidation Event: evidence from genes and fossils," *Palaeontology* 58 (2015): 769–85.

Schoenemann, B., E. N. K. Clarkson, and M. Høyberget. "Traces of an ancient immune system—how an injured arthropod survived 465 million years ago," *Scientific Reports* 7 (2017): 40330.

Schönberg, C. H. L. "A history of sponge erosion: from past myths and hypotheses to recent approaches," in *Current Developments in Bioerosion*, ed. M. Wisshak and L. Tapanila (Berlin: Springer, 2006), 165–202.

Schönberg, C. H. L., J. K. H. Fang, M. Carreiro-Silva, A. Tribollet, and M. Wisshak. "Bioerosion: the other ocean acidification problem," *ICES Journal of Marine Science* 74 (2017): 895–925.

Schönberg, C. H. L., and J.-C. Ortiz. "Is sponge bioerosion increasing?," *Proceedings of the 11th International Coral Reef Symposium* (2008): 520–23.

Schweigert, G., R. Fraaije, P. Havlik, and A. Nützel. 2013. "New Early Jurassic hermit crabs from Germany and France," *Journal of Crustacean Biology* 33 (2013): 802–17.

Schulp, A. S., H. B. Vonhof, J. H. J. L. van der Lubbe, R. Janssen, and R. R. van Baal. "On diving and diet: resource partitioning in type-Maastrichtian mosasaurs," *Netherlands Journal of Geosciences—Geologie En Mijnbouw* 92 (2014): 165–70.

Schweitzer, C. E., and R. M. Feldmann. "The Decapoda (Crustacea) as predators on Mollusca through geologic time," *Palaios* 25 (2010): 167–82.

Schweitzer, C. E., and R. M. Feldmann. "The oldest Brachyura (Decapoda: Homolodromioidea: Glaessneropsoidea) known to date (Jurassic)," *Journal of Crustacean Biology* 30 (2010): 251–56.

Schwimmer, D. R. *King of the Crocodylians: The Paleobiology of Deinosuchus* (Bloomington: Indiana University Press, 2002).

Scoon, R. N. "Mount Elgon National Park(s)," in *Geology of National Parks of*

Central/Southern Kenya and Northern Tanzania, ed. R. N. Scoon (Cham: Springer, 2018), 81–90.

Scott, A. C. "Trace fossils of plant–arthropod interactions," in *Trace Fossils: Their Paleobiological Aspects*, ed. C. G. Maples and R. R. West, *Paleontological Society Short Course* 5 (1992): 197–223.

Sealey, N. *Bahamian Landscape: Introduction to the Geology and Physical Geography of the Bahamas* (3rd ed.) (New York: Macmillan Publishing, 2006).

Seilacher, A. "Developmental transformations in Jurassic driftwood crinoids," *Swiss Journal of Palaeontology* 130 (2011): 129–41.

Selvaggio, M. M. "Carnivore tooth marks and stone tool butchery marks on scavenged bones: archaeological implications," *Journal of Human Evolution* 27 (1994): 215–28.

Serrano-Brañas, C. I., B. Espinosa-Chávez, and S. A. Maccracken. "*Gastrochaenolites* Leymerie in dinosaur bones from the Upper Cretaceous of Coahuila, north-central Mexico: taphonomic implications for isolated bone fragments," *Cretaceous Research* 92 (2018): 18–25.

Serrano-Brañas, C. I., B. Espinosa-Chávez, and S. A. Maccracken. "*Teredolites* trace fossils in log-grounds from the Cerro del Pueblo Formation (Upper Cretaceous) of the state of Coahuila, Mexico," *Journal of South American Earth Sciences* 95 (2019): 102316.

Servais, T., and D. A. T. Harper. "The Great Ordovician Biodiversification Event (GOBE): definition, concept and duration," *Lethaia* 51 (2018): 151–64.

Sheffield, G., and J. M. Grebmeier. "Pacific walrus (*Odobenus rosmarus divergens*): differential prey digestion and diet," *Marine Mammal Science* 25 (2009): 761–77.

Shipway, J. R., M. A. Altamia, G. Rosenberg, G. P. Concepcion, M. G. Haygood, and D. L. Distel. "A rock-boring and rock-ingesting freshwater bivalve (shipworm) from the Philippines," *Proceedings of the Royal Society, B* 286 (2019): 20190434.

Shorrocks, B. *The Giraffe: Biology, Ecology, Evolution and Behaviour* (West Sussex: John Wiley & Sons, 2016).

Shunk, S. A. *Peterson Reference Guide to Woodpeckers of North America* (New York: Houghton Mifflin, 2016).

Sibley, D. A. *The Sibley Guide to Birds* (New York: Alfred A. Knopf, 2014).

Sigren, J. M., J. Figlus, and A. Armitage. "Coastal sand dunes and dune vegetation: restoration, erosion, and storm protection," *Shore & Beach* 82 (2014): 5–12.

Singh, S., P. Singh, S. Rangabhashiyam, and K. K. Srivastava (eds.). *Global Climate Change* (Amsterdam: Elsevier, 2021).

Siqueira, A. C., D. R. Bellwood, and P. F. Cowman. "The evolution of traits and functions in herbivorous coral reef fishes through space and time," *Proceedings of the Royal Society, B* 286 (2019): 20182672.

Slieker, F. J. A. *Chitons of the World: An Illustrated Synopsis of Recent Polyplacophora* (Ancona: L'Informatore Piceno, 2000).

Smith, C. R., A. G. Glover, T. Treude, N. D. Higgs, and D. J. Amon. "Whale-fall ecosystems: recent insights into ecology, paleoecology, and evolution," *Annual Review Marine Sciences* 7 (2015): 571–96.

Smith, C. R., H. Kukert, R. A. Wheatcroft, P. A. Jumars, and J. W. Deming. "Vent fauna on whale remains," *Nature* 341 (1989): 27–28.

Smith, D. R., J. T. Tanacredi, and M. L. Botton (eds.). *Biology and Conservation of Horseshoe Crabs* (New York: Springer, 2009).

Snively, E., J. M. Fahlke, and R. C. Welsh. "Bone-breaking bite force of *Basilosaurus isis* (Mammalia, Cetacea) from the Late Eocene of Egypt estimated by finite element analysis," *PLOS ONE* 10 (2015): e0118380.

Sokolow, J. A. *The Great Encounter: Native Peoples and European Settlers in the Americas, 1492–1800* (Abingdon: Taylor & Francis, 2016).

Sørensen, A. M., and F. Surlyk. "Taphonomy and palaeoecology of the gastropod fauna from a Late Cretaceous rocky shore, Sweden," *Cretaceous Research* 32 (2011): 472–79.

Southward, A. J. (ed.). *Barnacle Biology* (Boca Raton, FL: CRC Press, 2018).

Spencer, L. H., J. C. Martinelli, T. L. King, R. Crim, B. Blake, H. M. Lopes, and C. L. Wood. "The risks of shell-boring polychaetes to shellfish aquaculture in Washington, USA: A mini-review to inform mitigation actions," *Aquaculture Research* 52 (2020): 438–55.

Stafford, E. S., G. P. Dietl, M. K. Gingras, and L. R. Leighton. "*Caedichnus*, a new ichnogenus representing predatory attack on the gastropod shell aperture," *Ichnos* 22 (2015): 87–102.

Stafford, E. S., C. L. Tyler, and L. R. Leighton. "Gastropod shell repair tracks predator abundance," *Marine Ecology* 36 (2015): 1176–84.

Stark, R. D., D. J. Dodenhoi, and E. V. Jonhso. "A quantitative analysis of woodpecker drumming," *The Condor* 100 (1998): 350–56.

Steemans, P., A. Le Hérissé, J. Melvin, M. A. Miller, F. Paris, J. Verniers, et al. "Origin and radiation of the earliest vascular land plants," *Science* 324 (2009): 353.

Steinmayer, A. G., and J. M. Turfa. "Effects of shipworm on the performance of ancient Mediterranean warships," *International Journal of Nautical Archaeology* 25 (1996): 104–21.

Stock, C. *Rancho La Brea: A Record of Pleistocene Life in California Science Series* 37 (7th ed., rev. by J. M. Harris) (Los Angeles: Natural History Museum of Los Angeles County, 1992).

Stock, S. R. "Sea urchins have teeth? A review of their microstructure, biomineralization, development and mechanical properties," *Connective Tissue Research* 55 (2014): 41–51.

Stolarski, J., M. V. Kitahara, D. J. Miller, S. D. Cairns, M. Mazur, and A. Meibom. "The ancient evolutionary origins of Scleractinia revealed by azooxanthellate corals," *BMC Evolutionary Biology* 11 (2011): 316.

Stout, R. *Darwin and the Barnacle* (New York: W. W. Norton, 2004).

Streelman, J. T., M. Alfaro, M. W. Westneat, D. R. Bellwood, and S. A. Karl.

"Evolutionary history of the parrotfishes: biogeography, ecomorphology, and comparative diversity," *Evolution* 56 (2002): 961–71.

Styrsky, J. D., and M. D. Eubanks. "Ecological consequences of interactions between ants and honeydew-producing insects," *Proceedings of the Royal Society, B* 274 (2007): 151–64.

Summers, A. P. "Stiffening the stingray skeleton: an investigation of durophagy in myliobatid Stingrays (Chondrichthyes, Batoidea, Myliobatidae)," *Journal of Morphology* 243 (2000): 113–26.

Sundberg, A. "Molluscan explosion: the Dutch shipworm epidemic of the 1730s," *Environment & Society Portal, Arcadia* 14 (2015): 1–6.

Surovell, T. A., S. R. Pelton, R. Anderson-Sprecher, and A. D. Myers. "Test of Martin's overkill hypothesis using radiocarbon dates on extinct megafauna," *Proceedings of the National Academy of Sciences* 113 (2016): 886–91.

Sutherland, J. I. "Miocene petrified wood and associated borings and termite faecal pellets from Hukatere Peninsula, Kaipara Harbour, North Auckland, New Zealand," *Journal of the Royal Society of New Zealand* 33 (2003): 395–414.

Syed, R., and S. Sengupta. "First record of parrotfish bite mark on larger foraminifera from the Middle Eocene of Kutch, Gujarat, India," *Current Science* 116 (2019): 363–65.

Talevi, M., and S. Brezina. "Bioerosion structures in a Late Cretaceous mosasaur from Antarctica," *Facies* 65 (2019): 1–5.

Tang, L. M. "Evolutionary history of true crabs (Crustacea: Decapoda: Brachyura) and the origin of freshwater crabs," *Molecular Biology and Evolution* 31 (2014): 1173–87.

Tanke, D. H., and P. J. Currie. "Head-biting behavior in theropod dinosaurs: paleopathological evidence," *Gaia* 15 (1998): 167–84.

Tapanila, L., and E. M. Roberts. "The earliest evidence of holometabolan insect pupation in conifer wood," *PLOS ONE* 7 (2012): e31668.

Tapanila, L., E. M. Roberts, M. L. Bouaré, F. Sissoko, and M. A. O'Leary. "Bivalve borings in phosphatic coprolites and bone, Cretaceous-Paleogene, northeastern Mali," *Palaios* 19 (2004): 565–73.

Tappen, M. "Bone weathering in the tropical rain forest," *Journal of Archaeological Science* 21 (1994): 667–73.

Taylor, B. M. "Drivers of protogynous sex change differ across spatial scales," *Proceedings of the Royal Society, B* 281 (2014): 0132423.

Taylor, P. D., and M. A. Wilson. "Palaeoecology and evolution of marine hard substrate communities," *Earth-Science Reviews* 62 (2003): 1–103.

Taylor, P. D., M. A. Wilson, and R. G. Bromley. "*Finichnus*, a new name for the ichnogenus *Leptichnus* Taylor, Wilson and Bromley 1999, preoccupied by *Leptichnus* Simroth, 1896 (Mollusca, Gastropoda)," *Palaeontology* 56 (2013): 456.

Terry, J. P., and J. Goff. "One hundred and thirty years since Darwin: 'reshaping' the theory of atoll formation," *The Holocene* 23 (2013): 615–19.

Thewissen, J. G. M. *The Emergence of Whales: Evolutionary Patterns in the Origin of Cetacea* (New York: Springer, 2013).

Thoen, H. H., M. J. How, T. -H. Chiou, and J. Marshall. "A different form of color vision in mantis shrimp," *Science* 343 (2015): 411–13.

Thompson, K. M., D. P. W. Huber, and B. W. Murray. "Autumn shifts in cold tolerance metabolites in overwintering adult mountain pine beetles," *PLOS ONE* 15 (2020): e0227203.

Thorington, R. W. Jr., and K. E. Ferrell. *Squirrels: The Animal Answer Guide* (Baltimore: Johns Hopkins University Press, 2006).

Tresguerres, M., S. Katz, and G. W. Rouse. "How to get into bones: proton pump and carbonic anhydrase in *Osedax* boneworms," *Proceedings of the Royal Society, B* 280 (2013): 20130625.

Trouet, V. *Tree Story: The History of the World Written in Rings* (Baltimore: Johns Hopkins University Press, 2020).

Turman, V. Q. P., B. C. P. M. Peixoto, T. da S. Marinho, and M. A. Fernandes. "A new trace fossil produced by insects in fossil wood of Late Jurassic–Early Cretaceous Missão Velha Formation, Araripe Basin, Brazil," *Journal of South American Earth Sciences* 109 (2021): 103266.

Turner, E. C. "Possible poriferan body fossils in early Neoproterozoic microbial reefs," *Nature* 596 (2021): 87–91.

Turner, S. K. "Constraints on the onset duration of the Paleocene-Eocene Thermal Maximum," *Philosophical Transactions of the Royal Society, A* 376 (2018): 20170082.

VanBlaricom, G. R. *Sea Otters* (Stillwater, MN: Voyageur Press, 2001).

Van der Wal, C., S. T. Ahyong, S. Y. W. Ho, and N. Lo. "The evolutionary history of Stomatopoda (Crustacea: Malacostraca) inferred from molecular data," *PeerJ* 5 (2017): e3844.

Van Der Wal, C., and S. Y. W. Ho. "Molecular clock," *Encyclopedia of Bioinformatics and Computational Biology* 2 (2019): 719–26.

van Geel, B., J. F. N. van Leeuwen, K. Nooren, D. Mo, N. den Ouden, P. W. O. van der Knaap, et al. "Diet and environment of *Mylodon darwinii* based on pollen of a Late-Glacial coprolite from the Mylodon Cave in southern Chile," *Review of Palaeobotany and Palynology* 296 (2021): 104549.

van Loo, A. J. "Ichthyosaur embryos outside the mother body: not due to carcass explosion but to carcass implosion," *Palaeobiology and Palaeoenvironments* 93 (2013): 103–9.

Varley, P. M., and C. D. Warren. "History of the geological investigations for the Channel Tunnel," in *Engineering Geology of the Channel Tunnel*, ed. C. S. Harris et al. (London: Thomas Telford, 1996), 5–18.

Vega, F., and R. Hofstetter (eds.). *Bark Beetles: Biology and Ecology of Native and Invasive Species* (Cambridge, MA: Academic Press, 2014).

Velásquez, M., and R. Shipway. "A new genus and species of deep-sea wood-boring shipworm (Bivalvia: Teredinidae) *Nivanteredo coronata* n. sp. from the Southwest Pacific," *Marine Biology Research* 14 (2018): 808–15.

Vermeij, G. J. "Gastropod skeletal defences: land, freshwater, and sea compared," *Vita Malacologica* 13 (2015): 1–25.

Vermeij, G. J., and E. Zipser. "The diet of *Diodon hystrix* (Teleostei: Tetraodontiformes): shell-crushing on Guam's reefs," *Bishop Museum Bulletin in Zoology* 9 (2015): 169–75.

Vetter, T. *30,000 Leagues Undersea: True Tales of a Submariner and Deep Submergence Pilot* (self-published, Tom Vetter Books).

Villa, P., G. Boschian, L. Pollarolo, D. Saccà, F. Marra, S. Nomade, et al. "Elephant bones for the Middle Pleistocene toolmaker," *PLOS ONE* 16 (2021): e0256090.

Ville, S. P., and J. Kearney. *Transport and the Development of the European Economy, 1750–1918* (London: Palgrave Macmillan, 1990).

Villegas-Martín, J., D. Ceolin, G. Fauth, and A. A. Klompmaker. "A small yet occasional meal: predatory drill holes in Paleocene ostracods from Argentina and methods to infer predation intensity," *Palaeontology* 62 (2019): 731–56.

Villegas-Martín, J., J. M. de Gibert, R. Rojas-Consuegra, and Z. Belaústegui. "Jurassic *Teredolites* from Cuba: new trace fossil evidence of early wood-boring behavior in bivalves," *Journal of South American Earth Sciences* 38 (2012): 123–28.

Vishnudas, C. K. "*Crematogaster* ants in shaded coffee plantations: a critical food source for rufous woodpecker *Micropternus brachyurus* and other forest birds," *Indian Birds* 4 (2008): 9–11.

Vogel, K., M. Gektidis, S. Golubic, W. E. Kiene, and G. Radtke. "Experimental studies on microbial bioerosion at Lee Stocking Island (Bahamas) and One Tree Island (Great Barrier Reef, Australia): implications for paleobathymetric reconstructions," *Lethaia* 33 (2000): 190–204.

Voight, J. R. "Xylotrophic bivalves: aspects of their biology and the impacts of humans," *Journal of Molluscan Studies* 81 (2015): 175–86.

Voultsiadou, E., and C. Chintiroglou. "Aristotle's lantern in echinoderms: an ancient riddle," *Cahiers de Biologie Marine* 49 (2008): 299–302.

Vršanský, P., I. Koubová, L. Vršanská, J. Hinkelman, M. Kúdela, T. Kúdelová, et al. "Early wood-boring 'mole roach' reveals eusociality 'missing ring,'" *AMBA Projekty* 9 (2019): 1–28.

Wahl, A., A. J. Martin, and S. T. Hasiotis. "Vertebrate coprolites and coprophagy traces, Chinle Formation (Late Triassic), Petrified Forest National Park, Arizona," *National Park Service Paleontological Research* 3 (1998): 144–48.

Walker, M. V. "Evidence of Triassic insects in the Petrified Forest National Monument, Arizona," *Proceedings of the United States National Museum* 85 (1938): 137–41.

Walker, S. E., and C. E. Brett. "Post-Paleozoic patterns in marine predation: was there a Mesozoic and Cenozoic marine predatory revolution?," in *The Fossil Record of Predation*, ed. M. Kowalewski and P. H. Kelley, *Paleontological Society Papers* 8 (2002): 119–93.

Wang, L., J. T.-M. Cheung, F. Pu, D. Li, M. Zhang, and Y. Fan. "Why do wood-peckers resist head impact injury: a biomechanical investigation," *PLOS ONE* 6 (2011): e26490.

Wang, Y.-H., M. S. Engel, J. A. Rafael, H.-Y. Wu, D. Rédei, Q. Xie, et al. "Fossil record of stem groups employed in evaluating the chronogram of insects (Arthropoda: Hexapoda)," *Scientific Reports* 6 (2016): 38939.

Warner, R. R. "Mating behavior and hermaphroditism in coral reef fishes," *American Scientist* 72 (1984): 128–36.

Warren, C. D., P. M. Varley, and R. Parkin. "UK tunnels: geotechnical monitoring and encountered conditions," in *Engineering Geology of the Channel Tunnel*, ed. C. S. Harris et al. (London: Thomas Telford, 1996), 219–43.

Wassenbergh, S. V., E. J. Ortlieb, M. Mielke, C. Böhmer, R. E. Shadwick, and A. Abourachid. "Woodpeckers minimize cranial absorption of shocks," *Current Biology* 32 (2022): doi: https://doi.org/10.1016/j.cub.2022.05.052.

Weis, J. S. *Walking Sideways: The Remarkable World of Crabs* (Ithaca, NY: Cornell University Press, 2012).

Wilson, M. A. "Macroborings and the evolution of bioerosion," in *Trace Fossils: Concepts, Problems, Prospects*, ed. W. M. Miller III (Amsterdam: Elsevier), 356–67.

Wilson, M. A., and T. J. Palmer. "Domiciles, not predatory borings: a simpler explanation of the holes in Ordovician shells analyzed by Kaplan and Baumiller, 2000," *Palaios* 16 (2001): 524–25.

Wilson, M. A., and T. J. Palmer. "Patterns and processes in the Ordovician Bioerosion Revolution," *Ichnos* 13 (2006): 109–12.

Wisshak, M. *High-Latitude Bioerosion: The Kosterfjord Experiment* (Berlin: Springer, 2006).

Wisshak, M., C. H. L. Schönberg, A. Form, and A. Freiwald. "Ocean acidification accelerates reef bioerosion," *PLOS ONE* 7 (2012): e45124.

Witherington, B., and D. Witherington. *Living Beaches of Georgia and the Carolinas: A Beachcombers Guide* (Sarasota, FL: Pineapple Press, 2011).

Wodinsky, J. "Penetration of the shells and feeding on gastropods by *Octopus*," *American Zoologist* 9 (1969): 997–1010.

Wolf, J. M., J. Luque, and H. D. Bracken-Grissom. "How to become a crab: phenotypic constraints on a recurring body plan," *BioEssays* 43 (2021): 2100020.

Wopenka, B., and J. D. Pasteris. "A mineralogical perspective on the apatite in bone," *Materials Science and Engineering, C* 25 (2005): 131–43.

Wroe, S., D. R. Huber, M. Lowry, C. McHenry, K. Moreno, P. Clausen, et al. "Three-dimensional computer analysis of white shark jaw mechanics: how hard can a great white bite?," *Journal of Zoology* 276 (2008): 336–42.

Xing, L., J. K. O'Connor, L. M. Chiappe, R. C. McKellar, N. Carroll, H. Hu, et al. "A new enantiornithine bird with unusual pedal proportions found in amber," *Current Biology* 29 (2019): 2396–401.

Yarlett, R. T., C. T. Perry, R. W. Wilson, and K. E. Philpot. "Constraining species–size class variability in rates of parrotfish bioerosion on Maldivian coral reefs: implications for regional-scale bioerosion estimates," *Marine Ecology Progress Series* 590 (2018): 155–69.

Zacaï, A., J. Vannier, and R. Lerosey-Aubril. "Reconstructing the diet of a 505-million-year-old arthropod: *Sidneyia inexpectans* from the Burgess Shale fauna," *Arthropod Structure and Development* 45 (2016): 200–220.

Zhang, S.-Q., L.-H. Che, Y. Li, D. Liang, H. Pang, A. Ślipiński, et al. "Evolutionary history of Coleoptera revealed by extensive sampling of genes and species," *Nature Communications* 9 (2018): 205.

Zhang, Z., Z. Zhang, J. Ma, P. D. Taylor, L. C. Strotz, S. M. Jacquet, et al. "Fossil evidence unveils an early Cambrian origin for Bryozoa," *Nature* 599 (2021): 251–55.

Zhao, Z., X. Yin, C. Shih, T. Gao, and D. Ren. "Termite colonies from mid-Cretaceous Myanmar demonstrate their early eusocial lifestyle in damp wood," *National Science Review* 7 (2020): 381–90.

Zhao, Z., X. Yin, C. Shih, T. Gao, and D. Ren. "Termite communities and their early evolution and ecology trapped in Cretaceous amber," *Cretaceous Research* 117 (2021): 104612.

Zipser, E., and G. J. Vermeij. "Crushing behavior of tropical and temperate crabs," *Journal of Experimental Marine Biology and Ecology* 31 (1978): 155–72.

Zonnenveld, J. P., and M. K. Gingras. "*Sedilichnus, Oichnus, Fossichnus*, and *Tremichnus*: 'small round holes in shells' revisited," *Journal of Paleontology* 88 (2015): 895–905.

Online Sources

"Channel tunnel," The Geological Society, accessed November 10, 2021, https://www.geolsoc.org.uk/GeositesChannelTunnel.

"*Cryptochiton stelleri*: giant Pacific chiton," *Animal Diversity Web* (Online), accessed October 28, 2021, https://animaldiversity.org/accounts/Cryptochiton_stelleri/.

"Dermestarium," by Stephen H. Hinshaw of the University of Michigan Museum of Zoology, explains how dermestid beetles are used to prepare skeletons for the museum: https://webapps.lsa.umich.edu/ummz/mammals/dermestarium/default.asp.

"*Lithophaga* Röding, 1798," World Register of Marine Species (WoRMS), MolluscaBase editors (2021), MolluscaBase, *Lithophaga*, accessed October 29, 2021, World Register of Marine Species, https://www.marinespecies.org/aphia.php?p=taxdetails&id=138220.

Mammoth Rocks, California Department of Parks and Recreation, https://www.parks.ca.gov/?page_id=23566.

Murex (Murex) pecten [Lightfoot], 1786, *MolluscaBase*, accessed October 30, 2021, World Register of Marine Species, http://www.marinespecies.org/aphia.php?p=taxdetails&id=215663.

Muricidae: MolluscaBase, "Muricidae Rafinesque, 1815," accessed October 30, 2021, World Register of Marine Species (WoRMS), http://www.marine species.org/aphia.php?p=taxdetails&id=148.

Naticidae: MolluscaBase, 2021, "Naticidae Guilding, 1834," accessed October 30, 2021, World Register of Marine Species (WoRMS), http://www .marinespecies.org/aphia.php?p=taxdetails&id=145, Muricidae: MolluscaBase, 2021.

Teredo, Linnaeus, 1758, MolluscaBase, accessed on November 7, 2021, World Register of Marine Species (WoRMS), http://www.marinespecies.org /aphia.php?p=taxdetails&id=138539.

"Wolf spirit returns to Idaho," M. Cheater, *National Wildlife Federation*, August 1, 1998, https://www.nwf.org/Magazines/National-Wildlife/1998 /Wolf-Spirit-Returns-to-Idaho.

Index

Page numbers in italics refer to figures.

crustaceans: and beaches
(continued)
consumption by, 279n57; on
driftwood, 158; and fossils,
267n23; and insects, 124–25;
and molluscans, 229; and rocks,
45–46; and shells, 97–98, 101–6,
113; and wood, 124–25, 135, 139,
154, 158, 167–68; and wood
bioerosion, 154, 167–68. *See also*
arthropods; barnacles; crabs;
decapods; shrimp, mantis
Cryptocercus. See roaches, wood
Cryptochiton stelleri, giant gumboot
Pacific chiton, 29, 257n3, 339
cryptoendoliths, 22–23
cuttlefishes, 84
cyanobacteria, 16; advent in Archean eon, 26–27; aerobic, 26;
and algae, 14–15, 19–22, 26–27,
38, 171, 247–48; anaerobic,
25–26; and bathymetry, 255n18;
and bioerosion, 13–18, 25,
247–48, 255n18; and borings,
21, 255n18; and climate, 26–27;
endolithic, 15–17, 255n15; as extremely successful single-celled
survivors s, 17–18; and fungi,
25–26, 38, 247; and microborings, 16; and microorganisms,
26; nomenclature, 255n10; and
photosynthesis, 13–15, 17–18,
20–22, 33; and plants, 14–15;
rock-destroying, 18, 26–27, 247;
role in atmospheric cycles,
17–18
Cypraeidae, 268n43
Cyprus, Mediterranean island, 231–
32, 295nn7–8

Danise, Silvia, 187–89
Darwin, Charles, 46, 65–66, 231,
260n59
decapods, 99, 167, 177, 267n23,
271n12. *See also* crabs; cray-

fishes; crustaceans; lobsters;
shrimp
deforestation, 132–33, 143, 233, 248
Deinosuchus crocodilians, 186–87,
222–23, 287n36, 293n91
Demospongiae, 259n31
Demospongia sponges, 40
Denver Museum of Nature & Science, 225
Dermestarium, 289n9, 339
Dermistidae, 198
deuterostomes, 50
Devonian-Carboniferous transition,
283n18
Devonian geologic period: as age
of fishes, 214, 218; barnacles,
46; borings, 44; crinoids, 215,
292n63; and first forests, 154–
55; fishes, 185, 214–19; fossils,
19, 46, 124–25, 218–19; fungi,
mistook for trees, 19; insects,
125; liverwort herbivory and
antiherbivory defense, 276n15;
oceans, 154; placoderms, 216–
18, 292n67; rocks, 124; sponge-
caused bioerosion, 42, 42; trees,
154; vertebrates, 214, 292n67
Dinocardium robustum (Atlantic
cockles), 95, 116
Dinodontosaurus synapsids, 200–201
Dinosaur National Monument
(Colorado and Utah), 201–2,
206, 290n24
dinosaurs, 1–2, 13; birds as, 138–39;
bones of, 198, 201–6, 205, 206,
209, 222–25, 228, 243, 261n70,
289n10, 289n12, 290n28,
291n40, 293n93; Cretaceous,
1, 133–34, 139, 205, 222–23,
225, 228, 279n54, 279n56,
293nn92–93; and crocodilians,
222–23; and crocodylomorphs,
227–28; and crustaceans, consumption of, 279n57; dietary
flexibility of, 228, 279n57; dung,